Smart Things and Femtocells

Smart Things and Femtocells

From Hype to Reality

Fadi Al-Turjman

CRC Press
Taylor & Francis Group
Boca Raton London New York

CRC Press is an imprint of the
Taylor & Francis Group, an **informa** business

Published in 2019 by CRC Press
Taylor & Francis Group
6000 Broken Sound Parkway NW, Suite 300
Boca Raton, FL 33487-2742

First issued in paperback 2020

ISBN 13: 978-0-367-57135-1 (pbk)
ISBN 13: 978-1-138-59389-3 (hbk)

Library of Congress Cataloging-in-Publication Data
Catalog record is available from the Library of Congress

Visit the Taylor & Francis Web site at
http://www.taylorandfrancis.com

and the CRC Press Web site at
http://www.crcpress.com

To whom I owe a lot,

To my mother, Lama and Naciye, Anne ☺ to my father, Ferzat and Tevfik,

To my brothers, Chadi and Osman. To my sisters, Azza and Amal, and their beautiful lovely stars,

To my first lady, Sinem, and best friend, Ferzat ☺

To my wonderful family,

Yours,

Fadi Al-Turjman

Contents

Preface

Because of the rapid increase of mobile connected devices such as smartphones and tablets, the demand for cognitive data traffic especially for mobile multimedia is exponentially increasing. In order to satisfy the gigantic number of mobile users' requests and meet the requirements for high data traffic in next generation communication systems, mobile operators have to increase their network capacities dramatically. One of the promising solutions for the network operators to improve coverage and capacity, and provide high data rate services in a less costly manner is the deployment of femtocells-related technologies. This book provides a comprehensive overview for the use of femtocells in smart Internet of Things (IoT) environments. Femtocells help mobile operators provide a basis for the next generation of services, which are a combination of voice, video, and data services, to mobile users. In this book, we overview femtocells' traffic modeling and deployment strategies and provide a review for their applications in smart environments. Moreover, we highlight efficient real-time medium access, data delivery, caching, and security techniques that can realize smart spaces around us. Finally, we conclude by presenting open research issues associated with smart IoT-femtocell based applications and future research directions.

Author

Prof. Dr. Fadi Al-Turjman is a professor at Antalya Bilim University, Turkey. He received his Ph.D. degree in computer science from Queen's University, Canada, in 2011. He is a leading authority in the areas of smart/cognitive, wireless, and mobile networks' architectures, protocols, deployments, and performance evaluation. His record spans more than 170 publications in journals, conferences, patents, books, and book chapters, in addition to numerous keynotes and plenary talks at flagship venues. He has received several recognitions and best papers' awards at top international conferences, and led a number of international symposia and workshops in flagship ComSoc conferences. He is serving as the lead guest editor in several journals including *IET Wireless Sensor Systems* (WSS), *Sensors* (Multidisciplinary Digital Publishing Institute), and *Wireless Communications and Mobile Computing* (Wiley). He is also the publication chair for the IEEE International Conf. on Local Computer Networks (LCN'18). He is the sole author for three recently published books about cognition and wireless sensor networks' deployments in smart environments with Taylor & Francis/CRC Press.

Chapter 1

Introduction

Fadi Al-Turjman
Antalya Bilim University

Content

Throughout the last decade, cellular networks have been experiencing a dramatic increase in mobile data traffic and the number of mobile-connected devices. According to Cisco Visual Networking Index (VNI), mobile data traffic has increased 18 times over the past 5 years, and by 2021, it will increase to nearly seven times as much as it was in 2016. In addition, it is expected that the number of mobile-connected devices will be 11.6 billion by 2021, which will be more than the world's estimated population at that time. Similar to the trends in this explosive growth in mobile broadband, cellular networks are also becoming more important as a significant part of the Internet of Things (IoT). The IoT is envisaged as a network in which everyone and all objects are connected anytime and anywhere. To cope with the rapid growth in the number of communication devices as well as the dramatic increase in mobile data traffic in the IoT era, Mobile Network Operators (MNOs) have been trying to find solutions to increase capacity and coverage of the networks, and provide services at higher quality levels. Femtocells proved to be a promising solution in this regard. A femtocell is a cell that provides cellular coverage and is served using a Femto Base Station (FBS). At first, FBSs were generally installed by the end users in indoor environments in order to have better mobile coverage and capacity. Nowadays, MNOs are utilizing FBSs in rural and highly populated outdoor environments as well. From the operators' point of view, deployment of femtocells can reduce the need for adding costly macro base station towers, which is one of the most important advantages of femtocells' deployment.

In this book, we overview further advantages for femtocells in improving the MNOs experience and spotlight key challenges in modelling their data traffic characteristics, deployment and caching strategies, data delivery, and provided smart spaces. Accordingly, our main contributions in this work can be summarized as follows:

- We start by a comprehensive overview about the femtocells and their traffic modeling techniques. We further classify various deployment strategies that can be applied in IoT applications.
- Various techniques and tools available for performance evaluation of femtocells in IoT environments are also presented in order to realize more energy-efficient solutions. These solutions and energy-related challenges are discussed in detail.
- We describe prominent performance metrics in order to understand how energy efficiency is evaluated. Then, we elucidate how energy can be modeled in terms of femtocell and provide some models from the literature.
- Then, we present energy-based analysis of femtocells in Ultra-Large Scale (ULS) applications such as the smart grid. The potential reduction of the consumed energy and service capacity due to mobility effects are considered as well as various performance metrics such as throughput, mean queue length, and response time.
- A framework employing the standardized Long-Term Evolution (LTE) while improving the Quality of Service (QoS) in modern mobile applications is also proposed. It minimizes the communication delay in sensitive real-time applications while maintaining the highest network throughput via an Enriched–real-time Polling System (E-rtPS).
- Next, we present the main motivations in carrying smart devices and the correlation between the user surrounding context and the application usage. We focus on context-awareness in smart systems and space discovery paradigms; online versus offline, the femtocell usage, and energy aspects to be considered in social IoT applications.
- We also propose a bio-inspired Particle Multi-Swarm Optimization (PMSO) routing algorithm to construct, recover, and select k-disjoint paths that tolerates the failure while satisfying QoS parameters. Multi-swarm strategy enables determining the optimal directions in selecting the multipath routing while exchanging messages from all positions in the network.
- Next, we propose a hybrid fitness function for a genetic-based Data Collectors (DCs) selection approach. Our function considers resource limitations in terms of count, storage capacity, and energy consumption as well as the targeted application characteristics.
- In addition, we propose a cognitive caching approach for the future femtocell networks (e.g., Fog networks) that takes into consideration the value of the exchanged data. Our approach depends on four functional parameters. These

four main parameters are as follows: age of the data, popularity of on-demand requests, delay to receive the requested information, and data fidelity.
- Finally, we conclude this book by highlighting open research issues in green femtocells and IoT-related applications.

1.1 Book Outline

The rest of this book is organized as follows. In Chapter 2, we delve into an overview for the femtocells in the IoT era. In Chapter 3, we describe prominent performance metrics in measuring the femtocell energy efficiency in IoT setups. Chapters 4 and 5 provide energy-based analysis of static and mobile femtocells in ULS applications such as the smart grid. Chapter 6 proposes a framework employing the standardized LTE while improving the QoS in modern mobile applications. Chapter 7 emphasizes the context-awareness concept in smart spaces using online versus offline mobile applications. In Chapter 8, we propose a bio-inspired PMSO data delivery approach to construct, recover, and select k-disjoint paths that tolerates the failure while satisfying QoS parameters. In Chapter 9, a hybrid fitness function for a genetic-based DCs selection approach is discussed. Chapter 10 debates a cognitive caching approach for the future small cell networks (e.g., Fog networks) that take into consideration the value of the exchanged data in the IoT paradigm. In Chapter 11, we further investigate the green femtocell topic and highlight open research issues in IoT-related smart applications. Finally, we conclude this book in Chapter 12 with potential viewpoints on the proposed work and future directions.

Chapter 2

Small Cells in the Forthcoming 5G/IoT: Traffic Modeling and Deployment Overview

Fadi Al-Turjman, Enver Ever, and Hadi Zahmatkesh
Middle East Technical University

Contents

2.1 Introduction

Cellular networks have been experiencing an explosive growth in mobile data traffic and number of mobile-ready devices as well as connections. According to Cisco Visual Networking Index (VNI) [1], mobile data traffic has increased 18 times over the past 5 years, and by 2021, it will increase to nearly seven times as much as it was in 2016. Moreover, it is expected that the number of mobile connected devices will become 11.6 billion by 2021, which will be more than the world's estimated population at that time. It is also predicted that 75% of the world's mobile traffic will be video by 2020 which accounts for three-fourth of total mobile data traffic in that year [1]. With this rapid increase of mobile broadband, wireless cellular technologies are becoming more important as a significant part of the Internet of Things (IoT) since the main part of IoT communications are designed over cellular technologies [2]. Cellular networks are already being used in IoT communications today in various scenarios and use cases, and will be used even more in the future due to its ubiquitous connectivity and mobility [2]. To deal with this explosive growth in mobile data traffic, especially in mobile video and the number of connected devices, mobile operators have been searching for new solutions to boost the capacity and coverage as well as to provide services at higher quality levels.

A recent development in this regard is the deployment of femtocells. A femtocell is a cell that provides cellular coverage and is served using a Femtocell Base Station (FBS). General specifications for the femtocells can be seen in Table 2.1. FBSs are typically installed by the consumers in indoor environment to provide better mobile coverage and capacity [3,4]. In recent years, many mobile operators have started using FBSs in outdoor environments in rural and densely populated areas [5]. In addition, femtocells can also be used in public transportation vehicles such as buses and trains to offload mobile traffic from loaded macrocell networks [6,7] and enhance the user experience for voice and video services while they are on the move. From the operators' point of view, deployment of femtocells will also decrease the need for adding expensive macro Base Station (BS) towers [3].

Future communication is expected to be a network of a large number of devices that will interact together in real time. This is referred to as the "IoT" [11]. An example of such devices includes thermostats and Heating, Ventilation, and Air Conditioning (HVAC) monitoring systems, which enable smart homes [12]. There are also other environments such as health [13,14] and agriculture [15] where IoT plays a significant role to improve the quality of life of people. IoT enables physical devices to talk to each other and share information using technologies such as ubiquitous and pervasive computing, embedded devices, and sensor networks

TABLE 2.1 General Femtocell Specifications

Characteristics	*Values*
Data range capabilities	7.2–14.4 Mbps
Power output	1.9–2.6 GHz
Operating frequency	10,100 mW
Range	20–30 m
Characteristic	*Interference control method*
Uplink interference control methods	Adaptive uplink attenuation [8]
	Autonomous component carrier selection [9]
Down-link interference control methods	Down-link power control [10]
	DL Tx power self-calibration [8]

[12,16]. The number of such devices is expected to increase considerably by 2020 and reach to the order of tens of billions [17]. Therefore, there will be a continually increasing demand for wireless communications, which are consistently good in quality and performance. IoT can be viewed as a dynamic networked system composed of a large number of smart object that can communicate with each other and/or with users through wireless connections. Many IoT applications can be considered in indoor environments. These applications include smart homes, smart museum, smart offices, smart industrial plant, etc. In these cases, femtocells can be used as communication devices for the smart objects in the cities to access the network [18]. By deployment of femtocells, a significant amount of traffic can be off loaded from the macrocell. Therefore, the operator's cost can significantly be reduced. In addition, due to the short distance between receivers and transmitters, the power consumption and battery of user devices can be saved [18]. In terms of security and privacy, the calls and data of mobile users are encrypted when handling through femtocells. Therefore, the security of the user data can be increased. Based on the discussions around 5G, two views are specified regarding the requirements and operational characteristics [19]. The first view considers 5G as a superset of 2G, 3G, 4G, Wireless Fidelity (Wi-Fi), and other wireless communication technologies. The other view predicts higher data speeds and considerable reduction in latency. In any case, people expect to see services from 5G that are considerably better in terms of quality and performance compared to those services they have received so far ranging from voice services in 1G to high data speed services in 4G. 5G is expected to be available by 2020 and it will be able to deal with 1000 times more mobile data traffic than today's cellular networks such as 3G and 4G. 5G is also expected to be the backbone of the IoT by connecting a large number of fixed and mobile devices [17]. Another support expected from 5G-based technologies is for Ultra-Reliable and Low-Latency Communication (URLLC), which aims to support applications that require highly low packet loss rate and extremely reduced end-to-end latency [20,21]. Intelligent transportation systems, industrial real-time controlled automation, and tactile internet are some of the use cases of URLLC [20].

There are several published survey papers that cover various aspects of the femtocell and IoT technology but none of them considers the use of femtocells and their applications in traffic modeling and deployment in the 5G/IoT environment. For example, a detailed description of femtocell networks, the main benefits of femtocells, and technical and business challenges are comprehensively presented in Refs. [3,22]. In addition, an overview of the IoT with emphasis on enabling technologies, protocols, and application issues was provided in Ref. [12]. The main communication enabling technologies, wired and wireless, and the elements of wireless sensor networks were covered in Ref. [23]. IoT architecture and the challenges of developing and deploying applications were addressed in Ref. [24]. The authors in Ref. [25] also presented the current Internet Engineering Task Force (IETF) standards and challenges for the IoT.

The contributions of this paper relative to the recent literature in the field are summarized as follows:

- To the best of our knowledge, this study is the first survey paper that provides a summary of the use of femtocells and their applications in traffic modeling and deployment issues in the 5G/IoT environment.
- We provide a classification for critical design factors while focusing on data in IoT-femtocell based applications for better femtocell-traffic modeling and deployment strategies in IoT environments.
- We present and compare different performance evaluation techniques and tools that enable researchers to have accurate traffic-modeling strategies in order to predict the performance of the femtocells in IoT environments.
- We also present different femtocell deployment strategies and provide a summary of related research works to have better deployment planning approaches.
- We provide tabular summaries about the existing literature for the following:
  - Objectives and constraints of the femtocell design aspects,
  - Femtocell Access Types (FATs),
  - Design factors for the IoT-femtocell based applications, and
  - Tools used for modeling traffic in femtocells.
- Finally, we discuss some open research issues and provide future research directions.

In order to assist the readers, a list of abbreviation along with brief definitions used throughout the paper is provided in Table 2.2. The rest of this article is organized as follows. The importance of femtocells in IoT environments is discussed in Section 2.2. Section 2.3 provides an overview of the market opportunity for femtocell-based systems. Elements of femtocells in the 5G/IoT era are discussed in Section 2.4. Section 2.5 discusses and compares different standards for femtocells and IoT. Section 2.6 describes objectives and constraints of a number of IoT-femtocell based applications in the literature, and a classification of design factors for such applications. Following a comprehensive overview through Sections 2.1–2.6 for the targeted problem in this survey, modeling traffic in femtocells and femtocell deployment strategies are investigated in Sections 2.7 and 2.8 respectively. Section 2.9 gives future research directions and discusses some open research issues. Finally, Section 2.10 concludes this survey article.

2.2 Femtocells in IoT

Femtocell technology is one of the popular infrastructures employed in HetNet environments. As an example, femtocell specifications for Samsung wireless extender provided by Verizon is presented in Table 2.3 [26]. Different

TABLE 2.2 Abbreviations

Abbreviation	*Name*
3GPP	3rd Generation Partnership Project
AP	Access Point
ASN	Ambient Sensor Network
AWGN	Additive White Gaussian Noise
BAN	Body Area Network
BAS	Building Automation System
BLE	Bluetooth Low Energy
BS	Base Station
BSN	Body Sensor Network
CAC	Call Admission Control
CDF	Cumulative Distribution Function
CDMA	Code Division Multiple Access
CR	Cognitive Radio
CSG	Close Subscriber Group
D2D	Device to Device
DSA	Dynamic Spectrum Access
Emtc	Enhanced Machine Type Communication
EC-GSM-IoT	Extended Coverage GSM for IoT
EE	Energy Efficiency
FAP	Femtocell Access Point
FAR	Femtocell Adaptability and Reliability
FAT	Femtocell Access Type
FBS	Femtocell Base Station
FDD	Frequency Division Duplex
FPGA	Field Programmable Gateway Array
FT	Fault Tolerance

(*Continued*)

TABLE 2.2 (*Continued*) Abbreviations

Abbreviation	*Name*
GSM	Global System for Mobile Communications
GSPN	Generalized Stochastic Petri Nets
HAVC	Heating, Ventilation, and Air Conditioning
HBS	Home Base Station
HetNet	Heterogeneous Network
HM	Handover Management
IM	Interference Management
IoT	Internet of Things
ISP	Internet Service Provider
LED	Light Emitting Diode
Li-Fi	Light Fidelity
LoRaWAN	Long Range Wide Area Network
LQE	Link Quality Estimator
LPWA	Low Power Wide Area
LR-WPAN	Low Rate Wireless Personal Area Network
LTE	Long-Term Evolution
LTE-A	Long-Term Evolution—Advanced
M2M	Machine to Machine
MFemtocell	Mobile Femtocell
MGM	Matrix Geometric Method
MOS	Mean Opinion Score
MQL	Mean Queue Length
MTC	Machine Type Communication
NB-IoT	NarrowBand-IoT
NFC	Near Field Communication
NTP	Network Time Protocol

(*Continued*)

TABLE 2.2 (*Continued*) Abbreviations

Abbreviation	*Name*
PA	Power Amplifier
PPP	Poisson Point Process
PSO	Particle Swarm Optimization
QoE	Quality of Experience
QoS	Quality of Service
RAM	Random Access Memory
RF	Radio Frequency
RFID	Radio Frequency Identification
RSS	Received Signal Strength
SE	Spectral Efficiency
SINR	Signal to Interference plus Noise Ratio
SIR	Signal to Interference Ratio
SMS	Short Message Service
SON	Self-Organizing Network
SS	Signal Strength
TD-SCDMA	Time Division Synchronous Code Division Multiple Access
TDD	Time Domain Duplex
UE	User Equipment
UHF	Ultra High Frequency
UMTS	Universal Mobile Telecommunications System
URLLC	Ultra Reliable and Low Latency Communication
UWB	Ultra-Wide Bandwidth
VLC	Visible Light Communication
VNI	Visual Networking Index
WCDMA	Wideband Code Division Multiple Access
Wi-Fi	Wireless Fidelity

(*Continued*)

TABLE 2.2 (*Continued*) Abbreviations

Abbreviation	*Name*
WiMAX	WorldWide Interoperability for Microwave Access
WPAN	Wireless Personal Area Network
WSN	Wireless Sensor Network
WSP	Wireless Service Provider

TABLE 2.3 Samsung Wireless Network Extender Femtocell Specification

Model	*Samsung Wireless Network Extender (SCS-2U01)*
Frequencies	800/1900 MHz
Air interface	CDMA2000 1× Rel 0, CDMA2000 EvDO 0/A
Traffic channel	Up to six simultaneous users (a seventh is reserved for emergency calls)
Transmission	10/100 Base-T ethernet/nework
Standards	IEEE 802.3, IEEE 802.3u for ethernet IEEE 802.11g, IEEE 802.11b for wireless
Power range	10–30 mW

applications in IoT environments can utilize femtocells to reduce the traffic load from the macrocell layer as well as to improve coverage and capacity of the entire network.

IoT is a novel paradigm where various objects such as sensors, actuators, smartphones, and other objects connect together and become part of the Internet. IoT touches every aspect of our daily lives and has potential to cover a broad range of applications such as smart environments, healthcare, transportation and logistics, futuristic, and personal and social life. These applications can significantly improve the quality of life of people in different places. Using femtocells in IoT can significantly improve Energy Efficiency (EE) of the network [27]. With deployment of femtocells, more users can be packed into a given area on the same radio spectrum that allows for a greater area Spectral Efficiency (SE). Moreover, because of the shorter distance between User Equipments (UEs) and the serving BS, these devices can lower transmit power and therefore, significant power savings can be achieved for the UE by deployment of femtocells. For instance, femtocells can be used in smart homes as a communication mechanism in order to manage EE. In addition, femtocells can be used in various IoT applications in order to provide efficient communication. For example, in Ref. [28], the authors proposed a femtocell-based communication mechanism in home area network and discussed security related issues about using femtocells in the smart grid. In addition, using femtocells in home area network as a cost-effective solution was discussed in studies such as Refs. [29,30].

Healthcare applications are another example of using femtocells in IoT environments. The authors in Ref. [31] introduced an IoT-oriented healthcare monitoring system where sensors collect the data from the android application and then send the data with new scheduling approach using Long-Term Evolution (LTE)-based femtocell network. In Ref. [32], the authors discussed a combination of wireless Body Area Networks (BANs), which are largely used in the passive healthcare data collection with the concept of mobile cloud computing that provides flexibility in large storage spaces and computing in healthcare monitoring systems.

The benefits of using femtocells for supporting indoor generated IoT traffic were discussed in Ref. [33]. The authors highlight that supporting the traffic generated from IoT is the main challenge for 5G since a considerable amount of the traffic is generated indoor. Therefore, it can be underlined that femtocells can be used in most of the IoT applications since there is a need for evolutionary telecommunication mechanisms. Moreover, it is expected that femtocells will be a central point in 5G architectures both for human users and for the IoT [27].

2.3 Market Opportunity

Femtocells create an exciting and promising market opportunity for Wireless Service Providers (WSPs). WSPs can benefit from the new femtocell services as well as the use of femtocells for increased macrocell user satisfaction because of traffic offloading. The delivery of services through femtocells have effects on the economics of the services for WSPs in different aspects such as decreasing cost, increasing revenue, reducing energy consumption, and increasing the speed of deployment [34]. Economic growth of IoT-femtocell based services is fairly large for businesses as well. For example, healthcare applications and other IoT-based services such as mobile health (m-health) together with femtocells can be used to monitor a set of medical parameters such as blood pressure, body temperature, and heart rate in elderly people and also enable medical wellness and treatment services to be efficiently delivered using electronic media. These application and services are expected to have annual growth of $1.1–$2.5 trillion by the global economy by 2025 [35]. On the other hand, femtocells play a significant role in smart homes to build a system that efficiently monitors a house's temperature, humidity, and light and also to have a Building Automation System (BAS). According to [36], the BASs market is expected to reach $100.8 billion by 2021 which is a 60% increase compared to 2013.

All these statistics reveal a potential of significant growth for the IoT-femtocell based applications and services in the near future. Internet Service Providers (ISPs) should of course provide networks that have capability to support Machine-to-Machine (M2M) traffic in order to make IoT a reality. In addition, this potential provides a great opportunity for equipment manufacturers to transform into "smart products."

2.4 Elements of Femtocells in IoT

In the following sections, we discuss three main elements of femtocells in order to have a better understanding of femtocell building blocks in the IoT era.

2.4.1 Communication

A significant part of IoT traffic is generated in indoor environments [37] and is designed over cellular technologies [2]. Therefore, communication between various objects in the IoT environments can be done through cellular networks [38]. Data from smart devices (e.g., smart meters for electricity and water), and from home monitoring sensors (e.g., to control temperature, humidity, light, and pollution level inside a building) are a few examples of these kind of communications. As mentioned earlier, IoT traffic can also be generated from m-health applications such as systems that are gathering data of elderly people and transferring them into health centers [39]. In this regard, Femtocell Access Points (FAPs) can be utilized to handle these indoor traffics and decrease the load of macro BSs to meet Quality of Service (QoS) requirements of indoor users. The IoT provides the interconnection of different devices to the Internet [40]. There are multiple IoT devices available for indoor environments. These devices include electricity-, water-, and gas- smart meters, home monitoring sensors, and BANs for m-health applications. Examples of communication technologies that these IoT devices use in order to communicate with the network are Bluetooth, Wi-Fi, ZigBee, Ultra-Wide Bandwidth (UWB), LTE—Advanced (LTE-A), and Light Fidelity (Li-Fi). With the development and existence of fifth generation of mobile network (5G) and the expected increase in the number of IoT devices, these devices, using cellular technology, can communicate with an FAP. Figure 2.1 shows a typical scenario where several indoor IoT devices communicate by using the infrastructure provided by the FAP.

For example, in the case of BAN, the sensors use the technologies such as Bluetooth, Wi-Fi, or ZigBee to communicate with the smartphone of the patient and then the smartphone communicates with the FAP. With the LTE-A, this can also happen using Device-to-Device (D2D) communications. Other devices in Figure 2.1 such as laptops and mobile phones can directly communicate with the

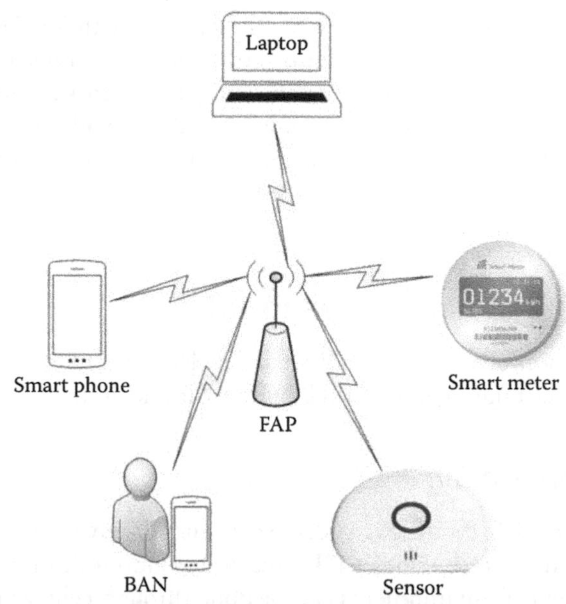

FIGURE 2.1 Communication of multiple IoT devices with a Femtocell Access Point (FAP).

FAP. This allows these IoT devices to make profits from 5G features, which guaranty high levels of QoS and provides wireless connectivity in indoor environment without any extra costs. This is because the communication between an FAP and IoT devices can be free of charge [37].

Wi-Fi is another alternative for communication of IoT devices within a range of up to 100 m [41]. It is also possible to employ popular Wi-Fi technologies in ad-hoc configuration for the cases where a router is not available to support the infrastructure. Bluetooth which uses short wavelength radio can also be employed to exchange information between devices in order to reduce power consumption [42] as long as the distances in between are relatively shorter. LTE, on the other hand, is a communication technology that is originally used for high-speed transfer between mobile devices based on GSM/UMTS [43]. Enhanced version of LTE is called LTE-A, which supports higher bandwidth up to 100 MHz and provides enhanced coverage, higher throughput, and lower latencies. The ZigBee is a communication technology designed and created for wireless controls and sensors, and is based on IEEE 802.15.4 standard. ZigBee allows smart devices to communicate within a range of typically 50 m and is designed to provide low data rate and low power consumption communication [44]. The UWB is another communication technology that supports communication between devices within a low range coverage area using high bandwidth and low energy [45]. Li-Fi is a cost-effective and alternative communication technology that was introduced to improve the limitations of the Wi-Fi technology. Li-Fi uses Visible Light Communication (VLC)

TABLE 2.4 Summary of the Communication Technologies for the IoT Devices

IoT Devices	*Communication Technology*	*Application*
Laptop	Wi-Fi	Indoor environment
Smartphone	LTE, LTE-A, Wi-Fi	Indoor environments, smart-cities
Smart meter	LTE, LTE-A	Smart homes
Sensor	ZigBee, UWB	Smart homes, e-health
BAN	Bluetooth, Wi-Fi, ZigBee	e-health

and LED concept for data communications [46]. The concept of small cells such as femtocells can easily be extended to VLC in order to successfully mitigate the high interference of Radio Frequency (RF) spectrum in Heterogeneous Network (HetNets) [47]. Details of the mechanism used by Li-Fi using femtocells can be found in Ref. [46]. A summary of the communication technologies for the IoT devices is presented in Table 2.4.

2.4.2 Computation

Hardware units and software applications are important parts of FAPs. For example, a good understanding of the femtocell hardware is needed to design algorithms to optimize the sleep cycle that can be used to switch off different hardware components in low traffic periods. Figure 2.2 shows a typical hardware model for an FBS [48]. This model contains a microprocessor that has the responsibility of implementation and management of radio protocol stack and baseband processing. It is also responsible for administration of the backhaul connection to the core network. In addition to on-chip memory, some Random Access Memory (RAM) components are also connected to the microprocessor. RAM components are needed to handle different data and system boot-up.

This model also includes a Field Programmable Gate Array (FPGA) and a number of integrated circuitry for the implementation of different features such as hardware authentication, data encryption, and Network Time Protocol (NTP). In the FPGA, the radio component works as an interface between the microprocessor and RF transmitter/receiver. Each of these RF components (RF transmitter and RF receiver) consumes a certain amount of power. Therefore, an RF Power Amplifier (PA) is utilized to pass high power signal to the antenna. As the current femtocell hardware components are cooled through natural convection, a cooling component is not included in this model. The minimum power consumption profile of the femtocell hardware components presented in Figure 2.2 can be found in Ref. [48].

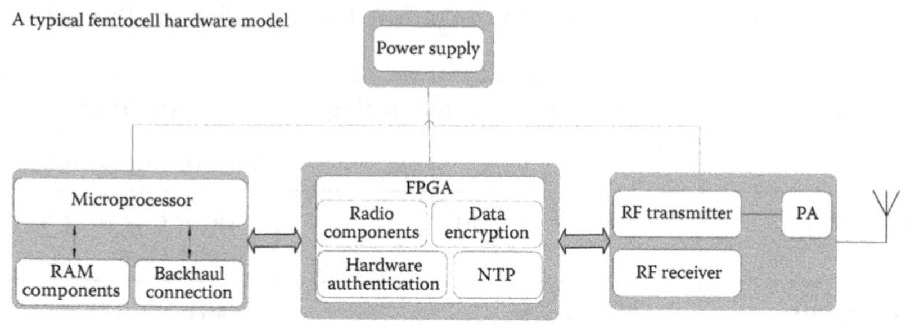

FIGURE 2.2 A typical hardware model for an Femtocell Base Station (FBS).

Considering EE of femtocell users, in Ref. [49], the authors investigated energy efficient uplink power control and subchannel allocation in two-tier femtocell networks. Taking transmit power and circuit power into account, they model the power control and subchannel allocation problem as a supermodular game in order to maximize the EE of femtocell users. To reduce the co-channel interference from femtocell users to neighboring femtocells and macrocells, they introduce a convex pricing scheme to curb their selfish behavior.

Femtocell applications can be implemented either in the FAP itself for most of the applications or in the mobile device that communicates with FAP [34]. For the ones implemented in the FAP, the development of the applications should be performed by the manufacturers. Therefore, a good option is an Application Programming Interface (API), which is provided by the operators to allow external companies to develop their software easily [50]. iPhones from Apple are examples where a successful API is utilized by many third-party companies. The Android operating system for smartphones is another example, which was released by Google in 2008. Femtocell manufacturers can provide the API along with a software development kit and documentation. Linux [51] is yet another operating system that can be used in communication with FAP since all of the femtocells analyzed so far run some lightweight sort of the Linux operating system [52].

2.4.3 Services

Femtocell-based infrastructures are becoming one of the key components in the HetNet deployments [53]. Many applications can be enabled by deployment of femtocells to provide better coverage and capacity as well as to reduce the traffic loads from the macrocell layer. In this section, we discuss some of the configurations where femtocells are commonly employed.

1. *Indoor environments:* Femtocells are commonly installed in indoor environments to improve coverage and provide higher data rates especially for mobile applications such as video conference that require high data rates as well as

a delay sensitive nature. Indoor femtocells bring receivers and transmitters closer to each other, which significantly decrease penetration loss as well as packet loss. Therefore, energy consumption can be reduced significantly [54]. A comparison between different indoor coverage techniques is presented in Table 2.5. Femtocells can also be deployed in public places such as shopping malls or airports to improve the users' internet experience.

Nowadays, electronic health (e-health) monitoring systems are becoming more popular since they can make a person aware of his/her physical and psychological fitness while being far from the health center [56,57]. Femtocell technology together with other relevant technologies can be used in order to utilize e-health monitoring systems [31,58–62]. Details of the e-health monitoring systems utilizing femtocell technology will be discussed in the next section.

2. *Outdoor environments:* Femtocells can be deployed in public transportation vehicles such as busses and trains to provide better coverage and internet experience in terms of voice and video services for the users while on the move [6,7,53,62–66]. Using femtocell technology inside a vehicle such as a bus or a train is called mobile femtocell (MFemtocell) [6]. According to [6], SE of the whole network can be improved by the implementation of the MFemtocells. In addition, the battery life of the mobile users can be prolonged significantly because of the shorter-range communication with the serving MFemtocell. It should also be noted that MFemtocells are placed inside vehicles and the antennas are placed outside of the vehicles. This provides better signal quality inside the vehicles.

TABLE 2.5 A Comparison between Different Indoor Coverage Techniques [55]

Macro/Microcell	*Expensive Price*	*Expensive Installation*	*High Power*	*Bad Indoor Coverage (High Risk of Path Loss)*
Repeater	Convenient price	Difficult installation	Low power	Acceptable coverage
Distributed Antenna System (DAS)	Convenient price	Easy installation	Low power	Good coverage
Radiating cable	Convenient price	Difficult installation	Low power	Good coverage
Femtocell	Cheap	Very easy installation	Very low power	Good coverage

In smart cities, cellular networks play an important role to support connectivity anytime and anywhere. Different technologies are available in smart cities in order to provide real-time services. These technologies include Wi-Fi, GSM, LTE-A, and heterogeneous cellular network which is defined as the combination of macrocells and small cells such as femtocells. In addition, in smart cities, a broad range of services will be available to users. These services include e-commerce, e-health, e-banking, e-government, and intelligent transportation systems. Smart cities are extremely dependent on the network infrastructure [67]. Mobile networks have to support increased coverage and excellent QoS. In this regard, femtocells play a significant role in smart-city projects where they can be deployed in streets, shopping malls, bus stations, and airport to provide better coverage and capacity to mobile users. For example, one of the major applications of the IoT for a smart city is related to "smart parking lots" where arrival and departure times of various vehicles are traced all over the city. Therefore, these parking lots have to be planned in a way that to take the number of cars in every region into account. This service is applicable based on sensors deployed on the roads of the city and intelligent displays which inform drivers about the best place for parking in the nearby. In this way, drivers can find a place for parking faster, which means fewer CO emissions from the car and less traffic congestion in the city that make the life of citizens happier. In addition, different IoT technologies for smart cities such as Radio Frequency Identification (RFID) and Near-Field Communication (NFC) can be utilized to realize an electronic verification of parking permits, which in turn allows for providing better services to citizens [68]. Moreover, wireless-networking requirements of smart cities cannot be fulfilled with traditional macro-only networks because of the spectrum efficiency and regularity issues to indoor coverage [67].

In general, Femtocell services can be classified into four categories [34] as illustrated in Figure 2.3: Carrier services, Internet services, Femtozone services, and connected home services. Carrier services are the common and basic services offered by mobile operators. Examples of these services are voice calls, Short Message Service (SMS), and push-to-talk. Internet services such as e-mail, web browsing, ftp, and mobile TV are accessible by any Internet access method such as mobile broadband, and are based on third-party platforms.

Femtozone services are services where the femtocell provides or helps in the delivery of the service. This is done through service information from the femtocell or service execution at the femtocell. Virtual home phone, and home presence are few examples for this type of services. Connected home services are delivered to a user's mobile device if he/she is in the home environment. Examples of this type of services are automated music synchronization, use of the mobile device as a media player to listen/watch home music/videos collection, and use of the mobile device as a home automation remote control. A few examples of femtozone services and connected home services are illustrated in this section [34]. For example, Figure 2.4 shows that when a mobile user enters the femto-zone, the presence of the user is detected by the femtocell. The femtocell then updates a Web 2.0 application such as

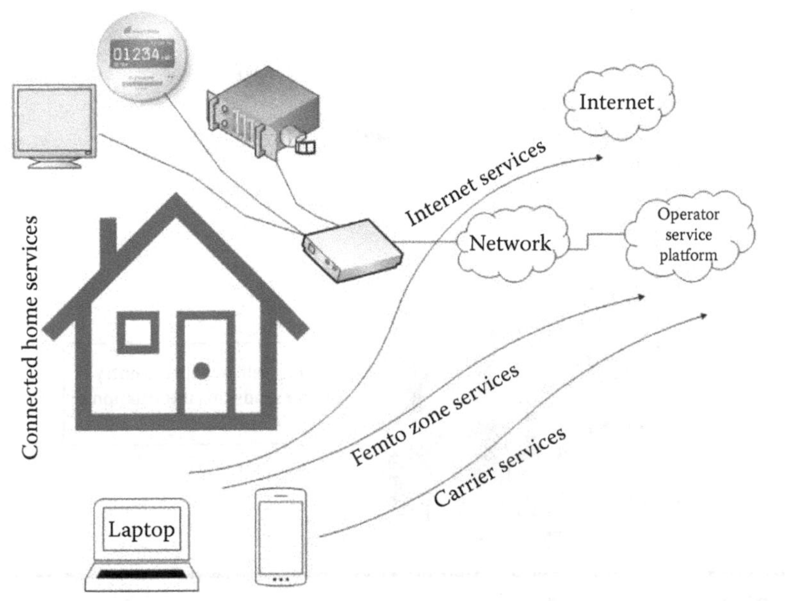

FIGURE 2.3 Four types of femtocell services.

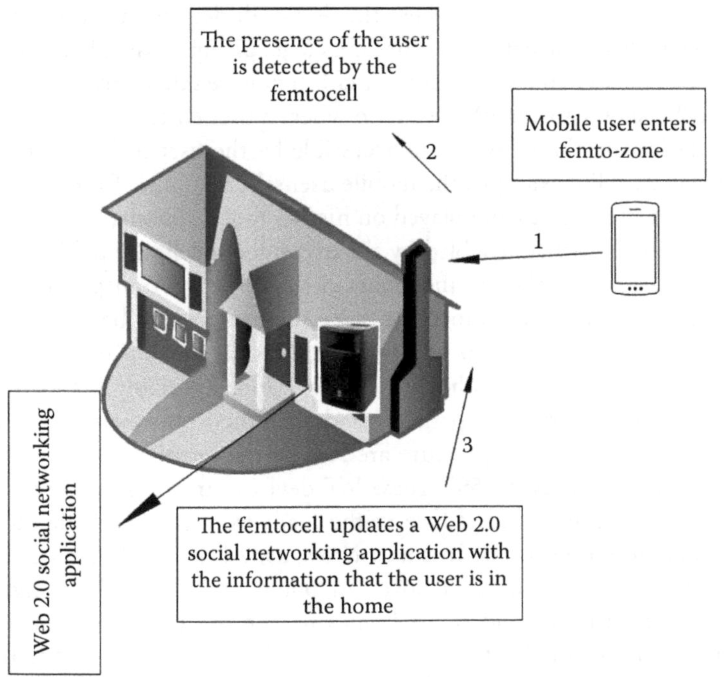

FIGURE 2.4 Presence and Web 2.0 applications.

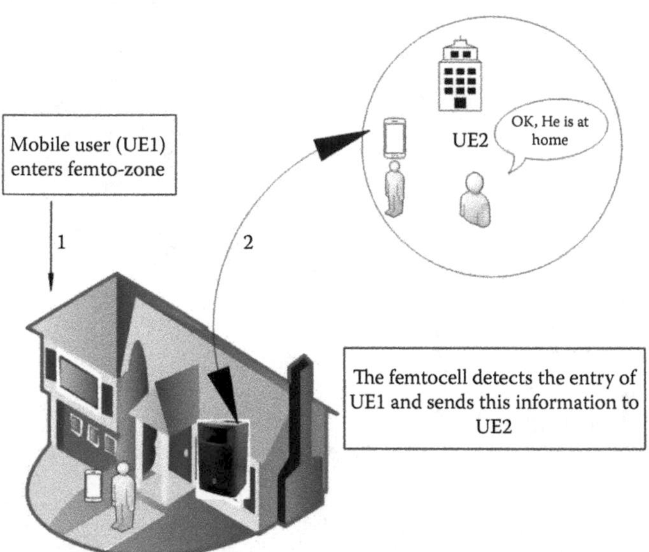

FIGURE 2.5 "I'm home" service.

a social networking application with the information that the user is in the home. This information can be shared with trusted and close friends. Another service example is shown in Figure 2.5 where the femtocell detects the entry of a mobile user and sends this information to other UEs (e.g., parents' UEs). For example, this can allow parents to make sure that their children have safely arrived home.

Figure 2.6 shows an example scenario where a user enters the femto-zone, and different devices on the network are accessible by the user to be controlled from the mobile device. For example, the mobile user selects a music from a home media server and it is streamed to and played on his/her mobile phone.

Other example services enabled by the use of femtocells are public transportation systems, smart healthcare, and smart cities. In public transportation vehicles such as busses and trains, femtocells can be used to provide better coverage and Internet experience for the users while they are on the move [7,63]. Smart healthcare plays an important role in healthcare applications utilizing IoT environments where sensors are embedded in patients' body and their information such as blood pressure, heart rate, and temperature are sent to the health centers for analyzing and monitoring purposes [58,59]. These IoT devices can communicate with FAPs to handle indoor traffics and reduce traffic loads from macrocell BSs to provide better services to indoor users. In smart cities, as well [69–73], femtocells will be available on roads and can be deployed in public and private buildings such as shopping malls, airports, and smart homes to improve the quality of life in the city and provide required services such as health and transportation. In such a HetNet environment, an energy-efficient policy is required [74,75] to motivate the use of femtocells in providing services to incoming requests in a green framework.

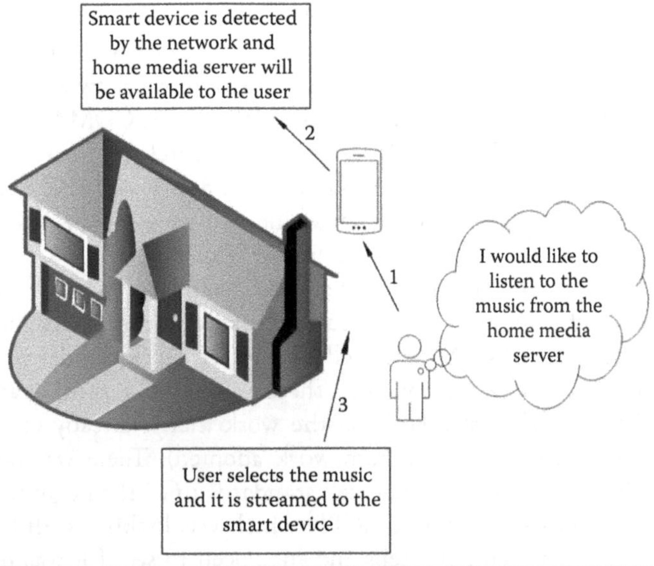
Smart device is detected by the network and home media server will be available to the user
2
1
I would like to listen to the music from the home media server
3
User selects the music and it is streamed to the smart device

FIGURE 2.6 Network music server.

TABLE 2.6 Building Blocks and Technologies of Femtocells in IoT

Element of Femtocells in IoT	*Samples*
Communication	Bluetooth, Wi-Fi, ZigBee, UWB, LTE-A, Li-Fi
Computation (hardware components)	Power supply, microprocessor FPGA, RF transceiver, memory elements, backhaul circuitry
Computation (operating system)	Apple IOS, android, linux
Service	Carrier services, Internet services, femto-zone services, connected-home services

Table 2.6 provides a summary of the femtocell's elements and building blocks in the IoT era and list a few examples for each element.

2.5 Femtocell/IoT Standards

Globally, standards are a significant factor for cost effective and efficient deployment of femtocells in IoT. The ability of femtocells to provide services to a very large number of mobile devices without any changes requires full support of standards to enable quick introduction of huge number of customers. The standards also provide support for various femtocell deployment scenarios such as residential, enterprise, and outdoor environments. In this section, we show and compare different standards for femtocells including GSM, WCDMA, CDMA, TD-SCDMA, Mobile WiMAX, and LTE [34]. In addition, current IoT technologies such as RFID, Bluetooth Low Energy (BLE), NFC, IEEE 802.15.4, IEEE 802.11ah, and Low Power Wide Area (LPWA) technologies including the LoRaWAN (Low Power WAN protocol for the IoT) are briefly discussed [76].

Operators are interested in GSM femtocell standards because GSM is the most common technology being used currently around the world; but, there are no specific standards for GSM femtocells. However, since the demand increases for applications in developing markets, there would be a considerable number of opportunities to define standards on the work that is already carried out for WCDMA (e.g., the management framework adopted). There was no dedicated intention in 3GPP to build any femtocell standards until the beginning of 2008. However, there was a study item in 3GPP, which was looking at different issues, but mainly at RF between femtocells and macrocell to see if it was manageable. The study item called "essentially a feasibility study in 3GPP parlance" finally came up with a technical report called "3G Home NodeB Study Item Technical Report" TR 25.820 [77]. It should be noted that "Home NodeB" and "Home eNodeB" are referred to a femtocell in 3GPP for WCDMA and LTE, respectively.

Another standard for femtocell is CDMA that has a strong support among operators. Actually, a CDMA-based femtocell was the first commercial femtocell deployed by Sprint [34], which is an ISP and is an American telecommunications holding company that provides wireless services. 3GPP2 has released an overview document of femtocell systems for cdma2000 communication systems [78]. This document describes the main features of the 3GPP2 femtocells that have been specified in different 3GPP2 specifications. The 3GPP2 TSG-S group published a document in 2008 [79] about the system requirements, which proposes the development of standards in two phases. Phase 1 is for the support of residential use for legacy mobiles and femto-macro mobility. In this phase, it is assumed that the same radio interface is used for both femto and macro layers. Phase 2 is the enhancement phase, which allows femto-femto mobility, mobility between different radio interfaces, and femtocells denser deployment. The document contains requirements in different categories such as system requirements, radio requirements, mobility requirements, security requirements, and operation, administration, maintenance, and provisioning requirements. 3GPP2 Release 9 contains a Radio Access Network 4 (RAN4) study item to investigate and analyze interference scenarios for TD-SCDMA in the same approach to those investigated for WCDMA. It is then expected to continue with a set of femtocell standards for TD-SCDMA defining closely on the work carried out for WCDMA in Release 8.

Most of the work conducted for standardization of LTE femtocells is included in Release 9 of 3GPP. LTE was directly built on the work carried out for WCDMA femtocells. However, in some aspects LTE has an advantage over WCDMA femtocells (e.g., in having an opportunity consisting of a number of optimized specifications for femtocells, which are already defined within the mobile sets). For instance, Release 8 LTE mobiles consist of Close Subscriber Group (CSG) support that is important to limit a set of users with connectivity access to a femtocell network due to some security reasons or because of the limited source of the backhaul network [55]. For LTE, standards are giving the same importance to both modes of operations called Frequency Division Duplex (FDD) and Time Domain Duplex (TDD). This is envisaged to extend to femtocells, which provides new opportunities for operators. Mobile WiMAX is another femtocell standard that permits sellers to begin developing femtocells and related equipment based on WiMAX 802.16e standard. These features permit WiMAX networks to support many different Access Points (APs) through standard IPSec-based security gateways. It also consists of Self-Organizing Network (SON) capabilities to permit configuration of several femtocells automatically. Moreover, the standard includes support for different femtocell deployment scenarios in residential, enterprise, and outdoor environments.

The current IoT technologies for short-range applications include RFID [80], BLE [81], NFC [82], IEEE 802.15.4 [83], and IEEE 802.11ah [84]. For long-range applications, the LPWA technologies include the LoRaWAN [85] and future cellular IoT.

RFID [80] that is based on receiving and transmitting data remotely using RF waves has attracted significant amount of attention in the last decade both in the research and development sectors. RFID technology provides automated tracking capability and wireless identification, and has revealed a global deployment following frequency allocation in the Ultra High Frequency (UHF) band. BLE [81] is a Wireless Personal Area Network (WPAN) technology that provides reduced cost and power consumption while maintaining a similar communication range. Moreover, BLE provides high-speed and IP connectivity making it appropriate for IoT nodes [12].

NFC [82] is a set of short-range protocols working at the range of about 10 cm and its targets can be simple devices such as stickers, cards, or unpowered tags. In addition, NFC allows peer-to-peer communication as well in which both devices should be powered. The IEEE 802.15.4 standard defines the operation for physical and data link layers for Low-Rate Wireless Personal Area Networks (LR-WPANs) products. It provides low-power and low-cost wireless connectivity within short ranges of up to 20 m which makes it suitable for use in Wireless Sensor Networks (WSNs), M2M and IoT. Another standard of IoT is IEEE 802.11ah [84], which is a competitor of IEEE 802.15.4. It is a standard that targets IoT nodes using larger range and lower power consumption. A comparison between performance of IEEE 802.15.4 and IEEE 802.11ah reveal that IEEE 802.11ah outperforms IEEE 802.15.4 in case of congested network; however, IEEE 802.15.4 performs better in terms of energy consumption [86].

The use of low power, low data requirements, and long-range operation devices is supported by LPWA technologies. These technologies include LoRaWAN [85] and the current and future cellular technologies such as 4G and 5G. The LoRaWAN is a data link layer technology with long range, low power, and low bit rate. It is considered a promising solution for IoT in which end devices use LoRa (Long Range) to communicate with gateways through a single hop. The LoRaWAN is expected to solve the connectivity problem of billions of smart devices in the IoT environment in the next decade [87].

There are also two new technologies that will lead to new standards. These include enhanced Machine-Type Communication (eMTC) and NarrowBand-Internet of Things (NB-IoT) [76]. eMTC was released in LTE-A pro release 12 in 2014 while additional features are included in release 13 with the objective of reducing cost and power consumption [88]. These new features include extended coverage, narrow-band and simplified operation, and low cost [88]. NB-IoT is another technology, which is expected to be the main LTE technology over LTE in the next years [76]. NB-IoT has a reduced bandwidth compared to eMTC that in turn causes reduction in device complexity as well as peak data rate [76]. Moreover, a new standard called Extended Coverage GSM for IoT (EC-GSM-IoT) has emerged that supports older 2G GSM networks. It is designed as a low energy, long range, high capacity, and low complexity technology [76]. A summary of all the discussed standards in this Section is presented in Table 2.7.

2.6 Design Factors of IoT-Femtocell Based Applications

5G is expected to be available very soon. In fact, Japan is gearing up for 2020 Olympics with 5G trials as 4.5 and 28 GHz [89]. Compared to the 4th generation of wireless communication systems, 5G will achieve higher system capacity, SE, EE, data rate, QoS, and system throughput [90]. The objective is to connect the entire world, and have seamless and ubiquitous communication anytime and anywhere between anybody and anything. This means that 5G will support communications for some scenarios which were not supported by the previous generation of wireless communication systems such as 3G and 4G (e.g., for mobile users in high-speed trains). For example, 4G networks can only support communications up to 250 km/h, while modern high-speed trains can reach to the speed of up to 500 km/h easily [90].

In this section, objectives and constraints of a number of IoT-femtocell based applications in the literature are considered. We classify these objectives and constraints into sub-categories as primary and secondary. Primary objectives are the most important goals of the targeted femtocell applications, and secondary objectives are those that are accomplished in addition to primary objectives in order to have more realistic 5G systems. Similarly, primary constraints are the most important limitations of the study in order to achieve accurate results, and secondary constraints are those restrictions, which have less importance compared to primary

TABLE 2.7 Summary of the Femtocell/IoT Standards

Current Standards	*Features*
WCDMA	• supports high data rate transmission, high service flexibility, handover
CDMA	• supports residential use for legacy mobiles and femto-macro mobility, femto-femto mobility and mobility between different radio interfaces
Mobile WiMAX	• permits sellers to develop femtocells and related equipments based on WiMAX 802.16e • supports many different APs through standard IPSec-based security gateways • consists of SON capabilities to permit configuration of several femtocells automatically
LTE	• was built on the work carried out for WCDMA • consists of CSG support to limit a set of users with connectivity access to a femtocell network
RFID	• provides automated tracking capability and wireless identification
BLE	• provides reduced cost and power consumption • provides high-speed and IP connectivity
NFC	• is a set of short range protocols working at the range of about 10 cm • allows peer to peer communication
IEEE 802.15.4	• defines operation for physical and data link layers for LR-WPANS products • provides low-power and low-cost connectivity within short ranges up to 20 m
IEEE 802.11ah	• is a competitor of IEEE 802.15.4, and targets IoT nodes using larger range and lower power consumption
Future/Enabling Standards	*Features*
LoRaWAN	• long range, low power, and low bit rate • is expected to solve connectivity problem of billions of smart devices in the IoT environment in the next decade
eMTC	• provides extended coverage, narrow-band and simplified operation, low cost, and low power consumption
NB-IoT	• has reduced bandwidth, reduced device complexity as well as peak data rate
EC-GSM-IoT	• is designed as a low energy, long range, high capacity, and low complexity technology

constraints but still can significantly affect the performance of the system under study. At the end, design factors of IoT-femtocell based applications are categorized according to these objectives and constraints.

2.6.1 Primary/Secondary Objectives

In recent years, energy consumption of HetNet environments has become one of the main concerns for network deployment strategies. The impacts of deploying femtocells in a cellular network were studied in Ref. [91]; it was shown in the study that deployment of femtocells can definitely improve EE of the network. However, performance of the network may be degraded due to the interference between femtocell and macrocell, especially in densely deployed networks. Therefore, there is a trade-off between EE and system throughput (or SE) in the deployment of femtocells. In Ref. [91], the primary objective of the study was EE, and the secondary objective was Interference Management (IM) and resource allocation, which have significant effects on the overall system performance. Another similar study in Ref. [92] showed that in a HetNet architecture, combining cellular network with appropriate number of femtocells significantly reduce energy consumption of the network while maintaining high QoS performance. In Ref. [92], EE and QoS were considered as the primary factors in designing the system. Also, EE as the main concern in the deployment of femtocells was studied in Ref. [93]. According to [91–93], by deployment of femtocells in existing macrocells, achievable gain in the system's EE is significant as a result of smaller path loss and lower transmission power, and this results in lower energy requirement. In Ref. [94], the authors introduced a novel idle mode energy saving procedure to improve EE as the primary objective of femtocellular BSs. In this work, femtocellular BSs completely switch off their radio transmissions and associated processing when they are not involved in an active call. The average reduction in femtocell's power consumption is described as a percentage of the total power as follows:

$$\Omega = \frac{P_{\text{saved}}}{P_{\text{ACT}}} \times (1 - \overline{\eta}) \times 100, \tag{2.1}$$

where P_{saved} and P_{ACT} are the power saved by switching off the hardware components of the femtocell BS in the IDLE mode and power consumption in ACTIVE mode, respectively, and $\overline{\eta}$ is the average duty cycle of the femtocell.

IM and Handover Management (HM) were the secondary objectives considered in Ref. [94]. A new Wireless over Cable architecture for femtocells (Femto WoC architecture) as a green solution for improving EE in femtocell systems was proposed in Ref. [95]. In this study, EE is a primary objective, and IM is the secondary objective in the system to improve throughput performances of the system.

In Ref. [96], the effects of cell size on EE of the future communication systems were studied. The results show that deployment of small cell BSs such as femtocell can significantly reduce energy consumption of the system. However, it may negatively affect performance of the system due to interference of the small cells. In this study, EE and system capacity are considered as the primary and secondary objectives respectively. In Ref. [97], a framework to analyze the EE and SE of a cellular HetNet composed of macro- and femtocell BSs was proposed. The trade-off between SE and EE was investigated and quantified using a Lebesgue measure. It is shown that the improvement in EE and SE depends on the traffic load and the BS power consumption attributes. EE and SE are the primary objectives of the study. The SE obtained in the two-tier network is defined as the total throughput achieved by all the users in the macro and femto layers per unit area per unit bandwidth.

$$\mathrm{SE} = \frac{\xi_m E\left[b_m \ln(1+\mathrm{SIR}_m)\right] + \xi_f \pi R_f^2 \lambda_f E\left[b_f \ln\left(1+\mathrm{SIR}_f\right)\right]}{B}, \tag{2.2}$$

where $E\left[b_m \ln(1+\mathrm{SIR}_m)\right]$ and $E\left[b_f \ln\left(1+\mathrm{SIR}_f\right)\right]$ are the average throughput achieved in the macro and femto tiers, respectively, ξ_m and ξ_f are the user densities in the macro and femto tiers, and b_m and b_f are the bandwidth allocated per user on average in each cell in the macro and femto tiers.

Moreover, the EE of the network is defined as the total throughput achieved with the available spectrum, normalized by the total power consumed in the network.

$$\mathrm{EE} = \frac{\xi_m E\left[b_m \ln(1+\mathrm{SIR}_m)\right] + \xi_f \pi R_f^2 \ \lambda_f E\left[b_f \ln\left(1+\mathrm{SIR}_f\right)\right]}{\lambda_m (q_m \Pi_{m,v} + (1-q_m)\Pi_{m,z}) + \lambda_f \Pi_f}, \tag{2.3}$$

where q_m is the activation probability of the macro tier whose coverage areas of the BSs contain users, $\Pi_{m,v}$ and $\Pi_{m,z}$ are the total power consumed by the macro tier in active mode and sleep mode respectively, and Π_f is the power consumed in the femto tier.

The study in Ref. [98] presented an energy efficient architecture utilizing femtocells to make the indoor coverage of cellular mobile devices better in the areas where WiMAX coverage is low. It is demonstrated that by deployment of femtocells in low coverage areas of WiMAX, the coverage is significantly enhanced and network capacity can significantly be improved when compared to the macrocell networks. EE as the primary objective and coverage and network capacity as the secondary objective of the study were investigated in Ref. [98].

Mobile femtocells can be used in public transportation vehicles such as busses and trains to improve service quality for the users inside vehicles [65,90]. For instance, in Ref. [7], a regular FBS was installed in a bus to offload traffic from the overloaded macrocell via Wi-Fi access points. An instance of the proposed framework is shown in Figure 2.7. The results presented revealed the efficiency of the

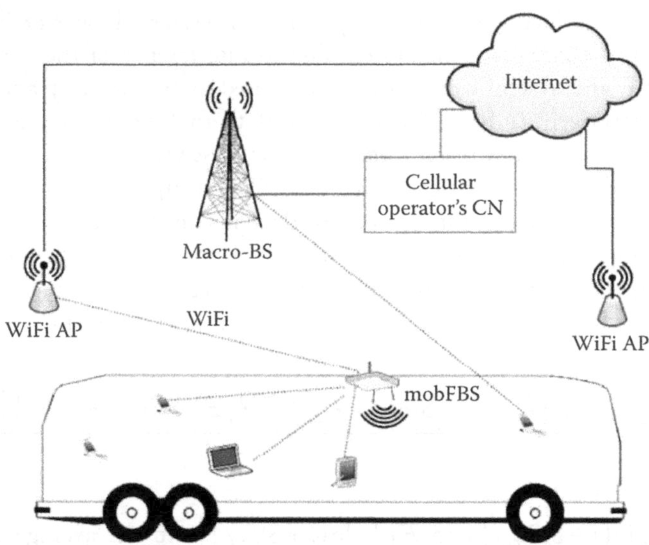

FIGURE 2.7 Mobile femtocell utilizing Wireless Fidelity (Wi-Fi).

proposed framework in terms of data traffic offloading. In this study, QoS is considered as the main objective and capacity enhancement is the secondary objective of the study.

The impacts of deploying MFemtocells in LTE cellular networks were studied in Ref. [99]. The authors investigate the system performance in terms of access delay, capacity, and signaling overhead where the macro users and MFemtocell are served over K resource blocks using multiuser scheduling. The average delivered rate in the past, measured over a fixed window of observation is calculated by the following:

$$\bar{R}_n(t) = \left(1 - \frac{1}{T}\right)\bar{R}_n(t-1) + \frac{1}{T}\sum_{k=1}^{N} R_n(k,t)d_n(k,t), \quad (2.4)$$

where T is the time window constant, $d(k,t)$ is a binary indicator that is set to 1 if the user n is scheduled on resource block k at time t and to 0 otherwise, and $R(k,t)$ is the instantaneous achievable rate on radio block k.

The results presented showed that by deployment of MFemtocells in an LTE cellular network, the signaling overhead can be reduced which causes an increased system performance especially for the uplink. Primary objective of Ref. [99] is to improve SE, and the secondary ones are related to latency and capacity. The MFemtocells can also be used to improve outage performance in high-speed mobility environments [63]. In Ref. [63], primary and secondary objectives were QoS and coverage, respectively. Performance evaluation of LTE networks using fixed/mobile femtocells was studied in Ref. [65]. Fixed FBSs are placed near to the threshold of the macrocell, and mobile FBSs are installed inside public transportation vehicles such as busses. It is clear from the results that the performance of mobile users has been improved after adding fixed/mobile FBSs. The results also indicate that in the case of high load traffic, adding mobile FBSs in the moving vehicles improve throughput of the user equipment. In this study, QoS and SE are the primary objectives and HM is the secondary objective of the study. In Ref. [64], femtocells were used to provide LTE wireless services for the passengers of a high-speed train through FBSs. The results obtained indicate that LTE users in the train with high mobility can have continuous wireless access with high QoS. In this study, QoS is the primary objective and similar to the study in Ref. [65]. HM is the secondary objective of the study. Another similar study to improve mobile services on trains was presented in Ref. [66]. The results of this study show that deployment of femtocells on trains has an advantage to the other architecture options (IP data access points and wideband repeaters) in terms of coverage, cost, revenue, and human exposure. In Ref. [66], QoS and network coverage were considered as the primary and secondary objectives of the study, respectively.

In recent years, use of femtocells in electronic healthcare systems has attracted attention worldwide. In Refs. [56,57], a femtocell-based mobile health monitoring system was proposed by using the concept of mobile cloud computing. Using Body Sensor Networks (BSNs), health information of a mobile user is captured and sent to the mobile device, which is under registration of a femtocell. The information is checked using a database inside the femtocell. If the user's health is abnormal, the data is sent to the cloud for access by the health center. Simulation results indicate that by using femtocells in e-health monitoring system, power consumption for accessing data in cloud will significantly be reduced compared to using macrocell, microcell, or picocell BSs. In Refs. [56,57], monetary cost consumption and EE were considered as the primary objectives of the application under study. Another use of femtocells in health monitoring system was presented in Ref. [60]. This time, femtocell and Machine-Type Communication (MTC) concepts are applied to personal telehealth system in the Continua Health Alliance or simply "Continua" framework [100], which is the most important open industry group that provides guidelines and certification programs for taking action in personal telehealth systems. EE and QoS are considered as the primary objectives of this study, and coverage and network capacity are the secondary ones. In Ref. [58], the authors proposed a smart femtocell-based sensor network for indoor monitoring of elderly people to control a number of parameters such as blood pressure, body temperature, and heart rate. EE is considered as the primary objective of this study. Another similar study that focused on the application of the IoT in the modern healthcare systems using sensors and LTE femtocell networks was presented in Ref. [31]. By developing an android application to act as a communication interface between LTE femtocell networks and the sensors, the proposed approach measures various parameters such as blood pressure and heart rate. EE and SE are the primary objectives of the study, and latency is considered as the secondary objective. In Ref. [59], an intelligent hybrid sensor network based on femtocells was proposed. The femtocell contains two important layers, namely BSN, and Ambient Sensor Network (ASN). In this study, ambient sensors and body sensors are integrated into the network for indoor monitoring. In the system, ambient sensors control the house of a patient by monitoring temperature, humidity, pressure, and light. Besides, body sensors control various parameters such as temperature and heart rate of the patient. The femtocell implementation of the system is shown in Figure 2.8. Similar to the study in Ref. [31], EE is the primary objective of the study.

The authors in Ref. [61] introduced a proposal for the use of available femtocell networks for emergency telemedicine applications in indoor environments. The results presented show significant performance improvements when femtocells are utilized compared to macrocellular networks. QoS is the primary objective of the study. Use of femtocells for medical issues in moving vehicles such as ambulance was studied in Ref. [62]. A femtocellular BS is placed inside the ambulance. Data are transmitted/received to/from the backhaul macrocellular network through a transceiver installed on the roof of the ambulance in which it is connected to FBS by the wired network. The simulation results presented reveal that significant performance advantages in terms of throughput, packet loss, and delay are achieved by the use of femtocell in comparison to the macrocellular-based networks. In Ref. [61] as well, QoS was the primary objective of the study.

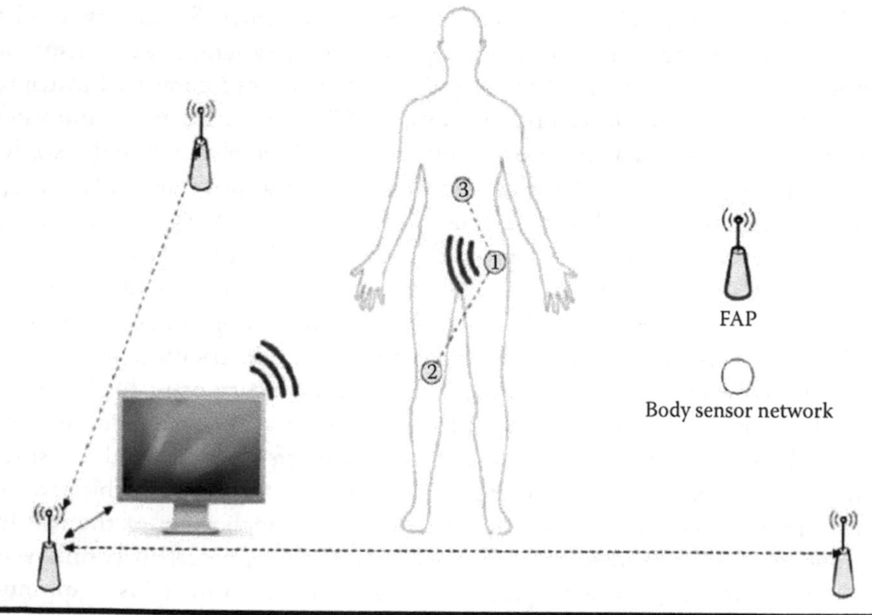

FIGURE 2.8 Femtocell implementation of intelligent hybrid sensor networks.

2.6.2 Primary/Secondary Constraints (PC/SC)

In Ref. [91], fixed transmission power was the primary constraint of the study since there is a trade-off between transmission power and EE. The secondary constraint of the study is the inter-site distance that has a fixed value of 500 m. The primary constraint of the study in Ref. [92] was average usage of the mobile equipment, which was considered as 0.5 hours per 12 hours of daytime. In Ref. [93], the primary constraint was related to the user date rate, which was considered to be 64, 128, and 256 kbps. The effects of daily traffic variations should also be taken into account, which can be the secondary constraint of the study in Ref. [93]. In Ref. [94], the primary constraint of the study was the walking speed of the users, which is considered to be 1 m/s, and the secondary one can be the number of registered users considered for the femtocell average usage and power consumption. Similar to the study in Ref. [91], fixed transmission power is the primary constraint of this study. The secondary constraint is the inter-site distance, which has a fixed value of 1732 m. A detailed description of femtocell networks, the main benefits of femtocells, technical and business challenges, and research opportunities were thoroughly investigated in Ref. [3]. The investigations in this study show that femtocells have the potential to provide network access to indoor users with higher quality at low cost, while reducing the load of the whole system simultaneously. Similar to studies in Ref. [91,95], the primary constraint of the study in Ref. [97] was that each BS transmits with a fixed transmission power. Fixed number of mobile nodes can be considered as the primary constraint of the study in Ref. [98]. The primary constraint in Ref. [7] was the fixed maximum number of user equipment, which is equal to 40 users, and the secondary one is the fixed capacity of mobile FBS, which is set to 5 Mbit/s.

The primary constraint of the study in Ref. [99] was the fixed transmission power for the users and MFemtocell. In Ref. [65], the primary constraint of the study was the fixed speed of the vehicles under study. The primary constraint of the study in Ref. [64] was the number of users in the high-speed train cabins that is considered to be 100. In Ref. [58], data transmission rate from the BSN was the primary constraint of the study. The primary constraint of the study in Ref. [31] was its limitation to have an android application. The primary constraint of the study in Ref. [61] was the fixed data service rate in the study. In Ref. [62], the primary constraint of the study was the fixed transmission power rate, and the secondary one is the number of users per femtocell, which considered to be maximum of 8. The summary of the objectives and constraints in these studies can be found in Table 2.8.

In this study, according to the mentioned objectives and constraints, we classify design factors of IoT-femtocell based applications into categories as primary and secondary. In the primary category, FAT, EE, QoS, SE, and security are the most important factors. In the secondary category IM, coverage and network capacity, HM, cost, and Femtocell Adaptability and Reliability (FAR) have higher

TABLE 2.8 Summary of the Objectives and Constraints of IoT-Femtocell Based Applications

References	*PO*	*SO*	*Primary Constraint*	*Secondary Constraint*
[91]	EE	IM	Fixed transmission power of BS	Inter-site distance
[92]	EE, QoS	–	Average usage of mobile equipment	–
[93]	EE	–	User data rates	Daily traffic variations
[94]	IM, HM	–	Walking speed of the users	No of registered user
[95]	EE	IM	Fixed transmission power of BS	Inter-site distance
[98]	EE	Coverage/ network capacity	Fixed transmission power of BS	–
[7]	QoS	Network capacity	No of UEs	Fixed capacity of mobile FBS
[99]	SE	Latency, network capacity	Fixed transmission power for the user and MFemtocell	–
[63]	QoS, SS	HM	Fixed speed of the vehicles	–
[64]	QoS	HM	No of users	–
[66]	QoS	Network coverage	–	–
[96]	EE	System capacity	–	–

(Continued)

TABLE 2.8 (*Continued*) Summary of the Objectives and Constraints of IoT-Femtocell Based Applications

References	*PO*	*SO*	*Primary Constraint*	*Secondary Constraint*
[97]	EE, SE	–	Fixed transmission power of BS	–
[56–57]	Monetary cost consumption, EE	–	–	–
[60]	EE, QoS	Coverage/ network capacity	–	–
[58]	EE	–	Data transmission rate	–
[31]	EE, SE	–	Android application	–
[59]	EE	–	–	–
[61]	QoS	–	Data service rate	–
[62]	QoS	–	Fixed transmission power	No of users per femtocell

importance while designing the applications. These design factors are shown in Figure 2.9 and discussed in detail in the following subsections.

2.6.3 Primary Design Factors

The primary design factors for IoT-femtocell based applications are those factors having the utmost importance regarding the applications.

1. *Femtocell Access Type (FAT):* In femtocell networks, the number of handovers depends on many different issues and one of them is access type. The choices of access mode depend highly on the cellular user density, with both operator and owner preferences [101]. There are three different access modes in femtocell networks: open, closed, and hybrid. The comparison of these modes is shown in Table 2.9.
   - Open Access: In open access mode, all available resources are shared between users and everyone can connect to the network. It provides

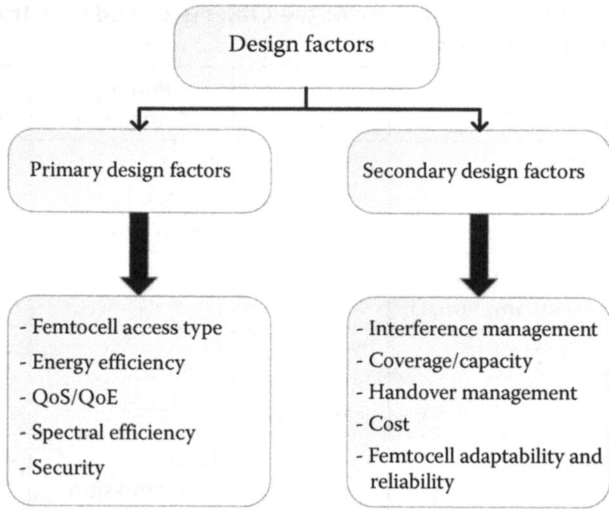

FIGURE 2.9 Design factors of Internet of Things (IoT)-femtocell based applications.

TABLE 2.9 Comparison of Femtocell Access Types

Item	*Open Access*	*Closed Access*	*Hybrid Access*
Deployment	Public access	Residential deployment	Enterprise deployment
No of handovers	High	Small	Medium
Provider cost	Inexpensive	Expensive	Expensive
Owner preference	No	Yes	Yes
High user densities	No	Yes	Yes
QoS	Low	High	High
Femto-to-macro interference	Increase	Decrease	Decrease

better network performance in terms of QoS and throughput [102] but the number of handovers is very high since they are deployed in public places such as airports and shopping malls, and there is no restriction to connect the network.

- Closed Access: In closed access mode, only CSG users can connect to the network but there can be different service levels between users. In this mode, the owner of femtocell does not want to share the femtocell because of some security reasons or due to the limited source of the backhaul. Therefore, any UE that is not the CSG will be rejected by the femtocell [55]. The number of handovers is very low in closed access mode since they are mainly used in individual home deployment.

- Hybrid Access: Hybrid (or semi-open) access mode merges open and closed access so that it allows particular outside users to access a femtocell. However, the conditions to connect the femtocell by a user from outside of the CSG are defined by the operator and new entries to the system are requested by the owner [103]. Non-CSG users can get only limited services depending on the operator management [104]. In hybrid access mode, the numbers of handovers are less than open access but more than closed access. More information about hybrid access can be found in Ref. [105].

In IoT environments, it is important to choose the appropriate FAT since energy consumption would be a challenging issue in HetNets considering dense deployment of femtocells.

2. *Energy Efficiency (EE):* The concept of energy efficiency has recently received much attention due to many different reasons such as ecological and economic ones, especially for network operators. EE is a significant issue in heterogeneous cellular networks where a large number of small cells such as femtocells are deployed within the coverage area of the macrocell in order to provide better services to the users [106,107]. EE can be defined either as the ratio of the efficient output energy to total input energy or as the performance per unit energy consumption [108]. Using femtocells in IoT can improve EE of the network since femtocells can be used in many IoT applications as a communication mechanism in order to manage EE [27]. EE has been investigated as one of the main factors in many femtocell-based applications in studies such as Refs. [31,56–60,91–98].
3. *Quality of Service (QoS)/Quality of Experience (QoE):* Apart from EE, QoS is also a critical issue in modern wireless communication systems [92], especially in the IoT environments. Using femtocells in IoT is an effective solution, which provides longer battery life, larger coverage area, and therefore, better QoS in the IoT environment. In order to measure QoS of a system, there are certain parameters introduced to specify the system's performance such as the system bandwidth [92], outage probability [63], maximization of the service rate [64], and Signal Strength (SS) [65]. QoS variation can largely affect the users' QoE because any possible degradation in QoS, may result in degradation in the users' QoE as well. QoE can be assessed in the form of a metric such as Mean Opinion Score (MOS) that shows average level of user satisfaction [109]. In Refs. [7,61–66,92], QoS was considered and investigated as an important issue in designing IoT-femtocell based applications.
4. *Spectral Efficiency (SE):* SE is defined as the information rate that can be transmitted over a given bandwidth in a specific communication system or as

the sum of the maximum average data rates per unit bandwidth [110]. There was a trade-off between EE and SE in deploying femtocells as shown in Ref. [91]. Therefore, considering SE in designing femtocell-based applications in IoT environments is an important issue. SE was investigated in studies such as Refs. [58,91,99] as the primary design factor of the applications.

5. *Security:* As the IoT becomes the key element for the future Internet, it becomes more important to provide adequate security for the IoT-based infrastructures [111]. Security is one of the most important issues in femtocell deployments since most femtocell customers are concerned about their privacy [55]. For example, if femtocells are used in open access, any mobile user can be connected to femtocells. In this case, femtocells need to be secured in order to avoid unauthorized users to use the femtocell illegally, and also to protect private information of the authorized femtocell subscribers [55]. Security, as one of the most important and primary design factors for IoT-femtocell based applications was considered in studies such as Refs. [52,112–114].

2.6.4 Secondary Design Factors

Other important factors in designing IoT-femtocell based applications are classified as secondary design factors in this article.

1. *Interference Management (IM):* The dense deployment of femtocells in HetNet environment can cause serious interference issues [103,115]. Therefore, IM is crucial in order to reduce the interference between the macrocells and femtocells. A typical HetNet in IoT environment includes several macrocells and femtocells and since many of these femtocells and macrocells are using the same portion of the spectrum, this may result in interruptions in the system [103]. The study in Ref. [95] considered IM as the secondary design factor of a femtocell-based application. Moreover, different techniques and approaches used to mitigate interference in femtocell-based networks were explained in detail in Ref. [103].
2. *Coverage and Network Capacity:* With femtocells, it is expected that the mobile users enjoy better signal qualities because of the closeness of transmitters and receivers. From the operators' perspective, femtocells provide better coverage and enhance network capacity [55,34]. This is because the distance between transmitter and receiver is shorter and therefore, femtocells can significantly reduce transmitting power, increase the battery life of the mobile devices, and achieve a higher level of Signal to Interference plus Noise Ratio (SINR). These actually mean improvement in coverage as well as higher capacity [3], which provides better QoS in the IoT era. Coverage and network capacity were considered as the secondary design factors of the femtocell-based applications in studies such as Refs. [7,63,96,98,99].

3. *Handover Management (HM):* HM is one of the most challenging issues in deployment of femtocells in IoT environments since it is a very important feature for the reduction of power consumption and signaling load [55]. In an IoT environment where a large number of femtocells are deployed within the coverage area of a macrocell, mobile users inside public or private vehicles may perform multiple handovers that can lead to significant increase in signaling load and degrade the network performance [65]. Studies such as Refs. [64,65] investigated HM as the secondary design factor of the IoT-femtocell based applications. Handover decision [116–118] is application specific in any femtocell-based applications and can be performed based on UE speed [119–122], signal strength [123–130], traffic type [121,131], and/or path loss [132,133].
4. *Cost:* From the operator's point of view, one of the main advantages of the integration of femtocells with the current macrocell technology is the ability to offload a huge amount of traffic from the macrocell to femtocell network to reduce the burden from the macrocell network. This reduces the investment capital, the maintenance expenses, and the operational costs and also increase the reliability of the cellular networks [55,134]. In addition, the average cost for each megabyte of traffic operated in femtocell is expected to be cheaper than that of a macrocell [31]. Therefore, sensors and actuators, which are largely deployed in IoT environments can have the benefits of the reduction of costs related to their generated traffic [135]. Cost as one the important factors in any femtocell-integrated applications was investigated in studies such as Refs. [134,136–139].
5. *Femtocell Adaptability and Reliability (FAR):* In this section, we investigate the ability of femtocells to adapt for the following:
   a. Resource Management (RM)
   b. Call admission and service provisioning

Resource Management (RM): In IoT environment, heterogeneous deployment of femtocells which overlays the existing macrocells enhance coverage and capacity for the indoor users in the 5G wireless networks [140]. However, since macrocells and femtocells share the available radio resources, it may cause interference between femtocell and macrocell (cross-layer interference) and between neighboring femtocells (co-tier interference) [140,141]. Therefore, it is crucial that RM is done in a way to mitigate these interferences. The following approaches are proposed in the literature for RM in the integrated macrocell/femtocell networks: resource partitioning based method [142–151], transmit power control based methods [102,152–156], cognition-based methods [157], and self-organized and learning-based methods [158–162]. RM in femtocells is a crucial factor to enhance SE and improve QoS of end users, especially in terms of video streaming as well [163,164]. Therefore, many researchers investigate new RM approaches and techniques to deal with video streaming applications [164].

Call admission and service provisioning: Alongside RM, Call Admission Control (CAC) is a policy that ensures QoS guarantees for new and handover requests [165]. CAC is a method that provides QoS in networks by restricting network resources and their access [166]. If there are enough network resources available to satisfy the call or handover requests in terms of QoS, and without compromising the QoS needs of the applications that are already running, CAC allows the new call or handover requests. CAC plays a significant role in QoS provisioning of wireless communications for different data and multimedia services in terms of signal quality, call-blocking probability, call-dropping probability, packet delay and loss rate, and transmission rate [167]. In the IoT environment, where millions of devices are connected to the network, CAC is a way that offers an effective solution to avoid network congestion and can play a significant role in the provision of guaranteed QoS [168]. The performance of CAC methods has been addressed in various related works [169–171]. Different CAC policies have been considered for femtocell operation in several related works [152,172–178]. In Ref. [178], CAC was used for inter-cell communication and proper RM, which enhances the system capacity and provides better user experience within a cell. In Ref. [176], CAC was used as a backhaul mechanism to control the congestion and limit the unnecessary calls in the femtocells. This mechanism increases the capacity of the cell. In Ref. [177], the authors addressed various CAC policies for femtocell operation and proposed an operation area where FBS has willingness to operate in open access mode. In studies such as Refs. [152,175], a CAC mechanism in hybrid access mode femtocell systems was described on the differences for service between CSG and non-CSG members. In Ref. [173], a CAC strategy in the integrated LTE femtocell/microcell was proposed to minimize the unnecessary handovers. A similar study to reduce unnecessary handovers between WiMAX and femtocells was presented in Ref. [172]. An adaptive CAC policy was proposed in Ref. [174] using Markov decision model for LTE femtocell networks to guarantee the call-blocking probability while maintaining the resource utilization. Summary of the design factors for the IoT-femtocell based applications is presented in Table 2.10.

2.7 Traffic Modeling of Femtocells in IoT Environments

Recently, many research works have been carried out by many researchers on the integration of femtocells with macrocellular networks to investigate the performance of the network. Although, Integration of femtocell technology with current macrocellular networks helps mobile network operators to reduce traffic loads of the macrocell layer, considerable amount of energy can be wasted if an appropriate traffic modeling strategy is not employed for the femtocells, which are deployed to provide services to the users in IoT environments. It is possible to classify these modeling approaches as static and dynamic.

TABLE 2.10 Summary of the Design Factors for the IoT-Femtocell Based Applications

References	*Primary Design Factors*					*Secondary Design Factors*			
	EE	*QoS*	*SE*	*Security*	*IM*	*Coverage/ Capacity*	*HM*	*RM*	*Cost*
[7]	–	X	–	–	–	X	–	–	–
[91]	X	–	–	–	X	–	–	X	–
[92]	X	X	–	–	–	–	–	–	–
[93]	X	–	–	–	–	–	–	–	–
[94]	X	–	–	–	X	–	X	–	–
[95]	X	–	–	–	X	–	–	–	–
[98]	X	–	–	–	–	X	–	–	–
[99]	–	–	X	–	–	X	–	–	–
[63]	–	X	–	–	–	X	–	–	–
[64]	–	X	–	–	–	–	X	–	–
[60]	X	X	–	–	–	X	–	–	–
[58]	X	–	X	–	–	–	–	–	–
[31,59]	X	–	–	–	–	–	–	–	–
[61,62]	–	X	–	–	–	–	–	–	–
[65]	–	X	X	–	–	–	X	–	–
[66]	–	X	–	–	–	X	–	–	–
[96]	X	–	–	–	–	X	–	–	–
[97]	X	–	X	–	–	–	–	–	–
[112,113]	–	–	–	X	–	–	–	–	–
[114,52]	–	–	–	X	–	–	–	–	–
[134]	X	X	–	–	–	–	–	–	X
[136]	–	X	–	–	–	–	X	–	X
[137,138]	X	–	–	–	–	–	–	–	X
[139]	–	X	–	–	–	–	–	–	X

–, not considered; X, considered; EE, energy efficiency; SE, spectral efficiency; QoS, quality of service; IM, interference management; HM, handover management; RM, resource management.

2.7.1 Models for Static Scenarios

In static modeling approaches, a fixed number of mobile terminals is assumed to communicate with a fixed number of BSs. Static traffic models do not consider the mobility of mobile terminals in terms of the arrivals and departures [179]. In addition, the call-level or packet-level dynamics in terms of packet arrival and call duration are not considered either in static traffic models. For example, a generic form approximation of the energy/spectral efficiency trade-off was derived in Ref. [180] for the uplink of coordinated multi-point system using static traffic model. The authors show the efficiency of the proposed model for the deployment of small cells such as femtocells compared to non-cooperative systems.

The energy/spectral trade-off for the uplink of coordinated multi-point system in Ref. [180] was defined as follows:

$$C_J = \frac{S}{N_0}\left[\frac{f^{-1}(\bar{S})}{\zeta} + \frac{tKP_c + bP_{sp} + cP_{bh}(\bar{S})}{N}\right]^{-1}, \tag{2.5}$$

where S is the achievable spectral efficiency, N_0 is the Additive White Gaussian Noise (AWGN) power spectral density, f^{-1} is the inverse function of the ergodic per-cell spectral efficiency of the uplink channel, K is the number of user terminals, P_c and ζ are the circuit power and amplifier efficiency of each user terminal respectively, N is the noise power, b is the cooling parameter, c is the ratio of the number of backhaul links to the number of BSs, and P_{sp} and P_{bh} are the BS signal processing power and additional backhauling induced power, respectively, for supporting coordinated multi-point system. In Ref. [181], performance evaluation of small cell networks such as femtocells was investigated and modelled using a Generalized Stochastic Petri Nets (GSPN) technique with channel breakdowns and BS breakdowns disciplines considering only static users in the system. For the channel breakdown, each channel can fail independently regardless of the state of the other channels. For the BS breakdown discipline, all the channels of the BS fail at the same time. The authors then derive a number of performance measures from the model such as failure probability of the channels, mean number of busy channels, and blocking probability.

We characterize the failure parameters in this domain by exploiting the synergy between fault detection and fault tolerance (FT) in 5G/IoT. The FT parameters are leveraged based on Markovian models via the coverage factor and the femtocell failure rate.

1. Coverage Factor: The coverage factor c is defined as the probability that the faulty active femtocell in a 5G/IoT paradigm is correctly diagnosed, disconnected, and replaced by a good inactive spare femtocell. The c estimation is critical in an FT 5G/IoT model and can be determined by:

$$c = c_k - c_c, \tag{2.6}$$

where c_k denotes the accuracy of the fault detection in diagnosing faulty femtocells and c_c denotes the probability of an unsuccessful replacement of the identified faulty femtocell with the good spare one. Whereas c_c depends on the femtocell switching circuitry and is usually a constant, c_k's estimation is challenging because different fault detection approaches have different accuracies. We analyzed the related work in the literature and observe that the accuracy of a fault detection algorithm depends on the average number of femtocells neighbors' k and the cumulative probability of femtocell failure p. Accordingly, c_k can be modelled as follows:

$$c_k = \frac{\left(k\left(1-p\right)\right)}{k\left(\frac{k}{M(p)}\right)^{\left(\frac{1}{M(p)}\right)} + \left(1-\frac{k}{M(p)}\right)^k}, \tag{2.7}$$

where $c_k < 1$ and $M(p)$ is a function of p representing an adjustment parameter that may correspond loosely to the desired average number of neighboring femtocells required to achieve a good fault detection accuracy for a given p.

2. Femtocell Failure Rate: The failure rate can be represented by exponential distribution with a failure rate of λ_s over the period t_s [182]. The failure rate curve approximation by piecewise exponential distributions is analogous to a curve approximation by piecewise straight-line segments. Consequently, the Cumulative Distribution Function (CDF) for the femtocell with an exponentially distributed failure rate can be represented by:

$$F_s(t_s;\lambda_s) = p = 1 - \exp(-\lambda t_s) \tag{2.8}$$

where p denotes the cumulative probability of a femtocell failure and t_s signifies the time over which p is specified.

2.7.2 Models for Dynamic Scenarios

Unlike static traffic models, in dynamic traffic models, the spatial and temporal fluctuations of the traffic load are captured which are discussed in the following subsections.

1. *Traffic Spatial Fluctuation Models:* It is well known that traffic characteristics can be quite different even among densely located BSs. Therefore, different traffic models have been recommended in the literature to obtain the spatial fluctuations in call traffic load. For instance, stochastic geometry has been used quite popularly for modeling of the spatial distribution of small

cells such as femtocells [183,184]. In this case, BSs are deployed according to homogeneous Poisson Point Process (PPP) with intensity λ_n in the Euclidean plane, and distribution of mobile terminals is according to various independent stationary point processes with intensity λ_m. With a stationary PPP, the distance between a mobile terminal and its serving BS is distributed regardless of the exact location of the mobile terminal. In Ref. [184], the probability density function of D_m where the traffic spatial variability among different cells is captured was given as follows:

$$fD_m(d) = 2\pi\lambda_n d \exp(-\lambda_n \pi d^2), \quad d > 0 \tag{2.9}$$

In Ref. [185], a mixed spatial traffic model was proposed based on the available K-tier HetNet of a macrocell and small cells. In this model, each tier has specific density, path loss exponent, Signal-to-Interference Ratio (SIR) coverage threshold, and transmission power. The authors also consider both uniform and nonuniform UEs at the same time for a more realistic scenario. In this model, each UE is connected to a BS that provides the highest downlink SIR. Moreover, UE locations are surrounded by various macro and small cells, and the downlink SIR is derived by considering a number of parameters such as the location of any interfering BS, UE downlink SIR with BS, the transmit power of the BSs, the magnitude squares of the channel gains, path loss exponent, and spatial PPP with various densities.

2. *Traffic Temporal Fluctuation Models:* Traffic temporal fluctuation models can be considered in two different time-scales: long-scale traffic models and short-scale traffic models [179]. In long-scale traffic models such as the one presented in Ref. [186], traffic fluctuations were captured over the days of the week, which is helpful for mobile network operators to investigate energy-efficient solutions. For instance, according to [186], traffic is much higher during daytime than that during night time or traffic during weekdays, holidays, and pick hours is much lower than that of a normal weekday. In order to model such a behavior, an activity parameter ψ_t can be used to specify the percentage of active subscribers over time t [179]. Therefore, by assuming p users per km^2 as the population density, N as the number of operators in which each can carry $\frac{1}{N}$ of the total traffic volume, and s_k as the fraction of the subscribers with an average data rate r_k for terminal type k such as smart-phone and tablet, the traffic demand in bit per second per km^2 can be calculated by the following equation:

$$A(t) = \frac{P}{N}\psi(t)\sum_k s_k r_k, \tag{2.10}$$

where ψ_t and r_t can be obtained from historical traffic statistics.

Short-scale traffic model is another time-scale in traffic temporal fluctuation models that models packet arrivals and departures of the mobile terminals [179]. In Refs. [187,188], following this approach, packet arrivals and departures are modelled using the concept of queuing theory and Markov chain. It is quite popular to use Markov chain for the analysis of different types of communication systems including femtocells since the arrival processes with an infinite population do not show intercorrelation and service times usually depend on the size and type of incoming packets. For example, in Ref. [187], arrival of packets were modeled using a Poisson process with rate λ and departure of packets follow exponential distribution with rate μ. The state transition probability of having h_i mobile terminals within cell i at time slot $t+1$ given that g_i mobile terminals were available at time slot t is given by the following equation:

$$\Pr\left\{M_i(t+1)=h_i \middle| M_i(t)=g_i\right\}, \tag{2.11}$$

where h_i and g_i are members of M and M is the state set of the number of packets in a given cell i. Similarly, a hybrid wireless cellular HetNet consisting of a macrocell and several femtocells were considered and modeled in Ref. [188] using two-dimensional Markov processes. The authors model femtocells as fault-tolerant wireless communication systems where mobility of the mobile users, multiple channels for the femtocells as well as failure and repair behavior of the channels are considered for more realistic performance measures in IoT environment. All the state probabilities (P_{ij}) are calculated using spectral expansion solution approach which can be used to calculate a number of important performance measures such as mean queue length (MQL) and throughput (γ) using the following equations:

$$\text{MQL}=\sum_{j=0}^{L} j\sum_{i=0}^{N} P_{i,j} \tag{2.12}$$

$$\gamma=\sum_{j=0}^{L}\sum_{i=0}^{N} i\mu P_{i,j}, \tag{2.13}$$

where i, j, L, N, and μ are the number of available channels, number of packets in the system, maximum number of packets in the system including the ones in service, maximum number of channels in the system, and total service rate of completed packet departures in the cell, respectively.

According to the objectives of the system, these traffic models can be utilized for performance evaluation and traffic modeling of the system under study. For example, stochastic geometry model presented in Ref. [183] provides precise and tractable information for hybrid wireless cellular systems in order to simplify the modeling techniques. This approach is quite powerful when used in the networks modeled with PPP with Rayleigh fading which results in general closed-form expressions. However, by generalizing the network models, the tractability of the network is reduced.

On the other hand, instead of comparing the system for a limited time, steady state analysis of traffic temporal fluctuation models considers the system when $t \rightarrow \infty$. In other words, regardless of the observation period, it identifies the effects of different parameters which can be used to control the data transmission throughput, the obtained bit error probability, and the mobile terminal radio interface state in order to balance energy saving and QoS [179].

All these categories are achieved via at least one of the following tools: analytical, simulation, or experimental test bed tools. Table 2.11 summarizes the tools used for the classification. Analytical modeling simulates behaviors of a system using mathematical concepts and language. For example, studies such as Refs. [53,91,189,190] used analytical techniques to evaluate the performance of the systems under study. In Ref. [53], closed-form expressions were used to investigate SE and EE of MFemtocell network. The authors also discussed the SE for multi-user system-level MFemtocells using opportunistic scheduling schemes. In Ref. [91], models based on exponential path loss and fast fading were utilized to foresee EE in static femtocells in comparison with the macrocells. Similarly, a HetNet composed of a single macrocell and several femtocells has been considered in Ref. [189]. The authors used an M/M/1 queuing model and Matrix Geometric Method (MGM) to analyze performance of the femtocell in terms of a number of important metrics such as average system delay and power savings. In addition to [189], the authors in Ref. [190] analyzed the performance characteristics of a finite capacity femtocell network in terms of a number of QoS parameters such as average packet delay,

TABLE 2.11 Tools Used for Modeling Traffics in Femtocells

References	*Analytical*	*Simulation*	*Experimental Test-bed*
[6,53,91,94]	X	X	–
[98,62,65,191]	–	X	–
[7]	X	–	–
[192,193,194]	–	–	X
[63,64,195]	X	X	–
[196]	–	X	X
[197]	X	X	–

–, not considered; X, considered.

packet-blocking probability, and buffer utilization for different sizes. The system is modeled as an M/M/1/K queue and the results presented show that the mentioned QoS parameters are highly dependent on traffic intensity as well as the buffer size.

In addition to the analytical models, studies in Refs. [6,91] utilized simulation to simulate the operations of a real-world system over time [198]. Simulation provides fairly accurate results but this technique requires high computation times [199]. Compared to simulation, analytical modeling approach is computationally more efficient [198,199]. Benchmarking (or experimental test-bed) is another technique which is performed through actual measurements and it is only possible when the system under study, or a prototype is already implemented [198]. Other efforts such as studies in Refs. [193,194] have used experimental test bed for the performance evaluation of the femtocell networks. The main problem related to the employment of experimental test beds in performance evaluation of the system is the difficulty in extrapolation of obtained results to different scenarios.

2.8 Femtocell Deployment Strategies

In order to provide services to the end users in a network having FBSs, it is important to determine the proper location of the FBS for a femtocell-based cellular network [22]. The deployment of femtocells brings some changes in the architecture of the current macrocellular networks and creates new design challenges. One of the main challenges is the problem of interference in telecommunication systems [200]. Therefore, it is vital to employ proper strategies and algorithms for the deployment of femtocells in the current macrocellular networks. These strategies can be classified as random, deterministic, and hybrid shown in Figure 2.10.

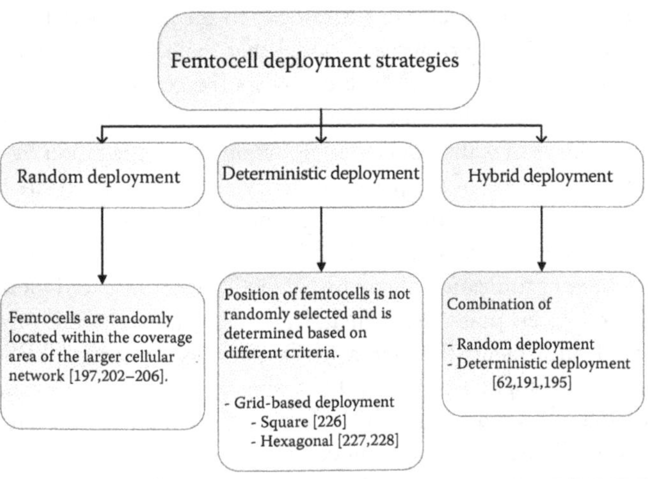

FIGURE 2.10 Femtocell deployment strategies.

2.8.1 Random Deployment of Femtocells

In random deployment strategy, femtocells are randomly located within the coverage area of the larger cellular network [197]. For example, in the case of Home Base Stations (HBSs), they are randomly placed within the coverage area of a macrocell so that higher spectrum efficiency and better coverage can be provided in the areas that are not completely covered by the macrocell [200]. However, it is needed to apply interference cancellation or avoidance techniques in order to make sure there will be no disruption of services in the vicinity of a femtocell [197]. In random deployments, the position of femtocells can be chosen in a purely random approach or based on the weighted random deployment where the distribution of femtocells is not uniform. Random deployment strategy is used in studies such as Refs. [197,201–205].

2.8.2 Deterministic Deployment of Femtocells

Unlike random deployment strategy, in deterministic deployment, position of femtocells is not randomly selected and is determined based on different criteria such as path loss [206–209], SIR statistics [210–214], managing the transmitting power and radio RM between the FAP and the outdoor cell-site [215–217], combination of path loss, SINR statistics, and cell overlapping [218], geometrical segmentation of macrocell, heuristic levels of traffic intensity and user distribution [219], resource allocation scheme [220], and interference and its impact on capacity, coverage, and handover [221–224]. These deployments can be accomplished through grid-based strategies [225–227]. For example, the authors in Ref. [221] proposed a dynamic greedy algorithm for deterministic placement of femtocells that can be adjusted and optimized while considering the best coverage for peak demanding spots. In Ref. [206], a model was developed for the optimization of BSs location for indoor wireless communications. The authors utilize a propagation model that includes parameters for the path loss and then define a cost function, which shows the coverage of the system under study. In Ref. [207], the authors used a model to analyze indoor wave propagation, considering the effects of walls, roofs, and floors on the propagation by using different terms in the expression of the path loss. They also investigated the possibility of optimization for indoor radio coverage using genetic algorithm in order to specify the locations of access points. The study in Ref. [208] reviewed different techniques such as discrete gradient optimization algorithm, genetic algorithm, and global optimization technique for access point deployment in wireless networks and proposed a new technique based on a Heuristic approach. Optimal placement of FBSs for indoor environment was investigated in Ref. [209]. Using femtocell path model, the authors optimize the QoS of associated users by locating femtocell at recommended location. They also analyze EE aspect of the femtocell placement.

The problem of finding an optimal BS placement for CDMA systems was investigated in Ref. [210]. Using a heuristic algorithm, each mobile user is assigned to a BS based on the smallest path loss considering the SIR in both uplink and downlink. A similar study was presented in Ref. [211] using Particle Swarm Optimization (PSO) technique while considering SIR statistics. The authors in Ref. [212] proposed a new hybrid algorithm to determine an optimal placement of BSs in indoor wireless communication systems considering SIR statistics in both uplink and reverse links. In Ref. [213], a combination of a heuristic algorithm and brute force search was used to propose a new algorithm to find an optimal placement for BSs from a set of potential BS sites in indoor CDMA networks. Both forward and reverse link SIR constraints are considered in the proposed approach for the placement problem. Research work carried out in Ref. [214] studied optimal locations to place a femtocell access point in an indoor environment while considering SIR statistics. The objective of the study is to improve the throughput and mean capacity of the system.

In studies [215–217], the authors investigate the problem of BS placement in femtocells networks by managing the transmitting power and radio RM between the FAP and the outdoor cell-site. The study in Ref. [218] investigated the problem of FAPs placement in LTE networks with the objective of improving the performance of automatic traffic sharing algorithms. This approach is evaluated by considering and simulating classical traffic sharing algorithms in an office environment with different femtocell position plans. In Ref. [219], the authors proposed a deployment algorithm for femtocells in multi-tier wireless cellular networks based on geometrical segmentation of macrocell, heuristic levels of traffic intensity, and distribution of the users with the objective of satisfying the existing traffic demand of the network. In Ref. [220], a dynamic algorithm for transmission optimization on dense femtocell deployments was proposed which specifies resource allocation based on satisfaction of the femtocell users and congestion level of the network. The authors in Ref. [222] investigated the effects of interference on inbound handover at a specified deployment placement of a femtocell. In Ref. [223], the authors proposed an algorithm that adjusts a femtocell coverage optimally based on user equipment's handover requests and the coverage size of the femtocell is monitored depending on whether or not a user equipment is permitted to the femtocell. The research work presented in Ref. [224] proposed an approach based on coverage adaptation for femtocell deployments. This approach uses information on mobility of the passing and indoor users for the optimization of the femtocell coverage.

2.8.3 Hybrid Deployment of Femtocells

In the hybrid deployment of femtocells, both deployments types are employed where random available femtocells are utilized as hotspots in addition to those that have been deterministically deployed at the beginning based on the cellular network operational conditions such as Siemens e-mobility project [228]. Hybrid deployment of femtocells are also utilized in m-health monitoring systems for medical scenarios, where femtocells are deployed deterministically inside a home or a hospital in addition to those that are randomly deployed inside an ambulance to transmit medical data to the hospital [62,191,195]. In general, hybrid femtocell deployments take into account density of mobile users, traffic volume, and coverage in the network models. It is also important to provide same or better QoS when EE is subjected. For example, the cross-tier interference in the deployment of femtocells can degrade the system performance significantly. The resource allocation problem in both the uplink and the downlink for two-tier networks comprising spectrum-sharing femtocells and macrocells had been investigated in Ref. [229]. The authors aim to maximize the capacity for both delay-sensitive users and delay-tolerant users subject to the delay-sensitive users' QoS constraint and an interference constraint imposed by the macrocell. The sub-channel and power allocation problem is modeled as a mixed-integer programming problem. Accordingly, an iterative sub-channel and power allocation algorithm considering heterogeneous services and cross-tier interference is proposed for the problem using the sub-gradient update.

2.9 Open Research Issues

In spite of numerous types of research activities and rapid progress that have been observed in recent years, deployment and performance enhancements in femtocells still harbor many open issues, which are still awaiting to be resolved. The main objectives of femtocell applications include providing coverage while maximizing throughput, extending lifetime, and preventing connectivity degradation. These goals can be achieved by employing energy management and efficient CAC. Additionally, there are restrictions on the utilized femtocells BSs and supported content that impose additional challenges. Therefore, there are open research issues that should further be explored in the field of femtocell performance assessments and planning. In this section, a broad range of research issues is outlined for future investigation.

2.9.1 Data Traffic Model

The data traffic and processing model significantly affects the performance of the femtocells especially with regard to energy consumption and QoS. In order to realistically model (can be analytical models or simulations) the behavior of the network, the traffic models should be studied, specified, and validated for femtocells as well as the systems they are interacting. In this regard, it is possible to create profiles for different strategies such as power allocation techniques and radio RM strategies by analyzing these interactions over time and investigate the behavior of network traffic in details. The performance of the femtocell should also be analyzed by various measures such as cell and user throughput, packet loss, and call-blocking/dropping probabilities since an appropriate data traffic model can support realistic and accurate performance evaluation.

2.9.2 Femtocell BS Deployment

Planning and placement strategies of FBSs are still studied extensively since improper placement strategies lead to interference and subsequently degrade performance of the entire network. Building dimensions, plan structures, and floor or wall partitions are important factors that must be taken into account in the placement and planning of FBSs [3]. In addition, the distance between FAP and the macro site as well as FAP density have an effect on coverage quality and interference.

2.9.3 Femtocell BS Capabilities

Several types of FBSs can have different role or capability according to the application, which is related to the capacity of the BS in terms of computation, communication, power. and UE support. Most of the applications use homogeneous BS, which have identical capabilities, or are produced by the same manufacturer. However, in practice, the network is considered heterogeneous, because some/all of the connected UEs have different capabilities and running varying applications. Therefore, in this context, research into the FBS capabilities would be beneficial in order to satisfy the users' demands.

2.9.4 Link Quality Estimators (LQEs)

Maintaining good link quality especially for high-speed vehicles on the move is still a challenging issue. Currently wireless links in femtocells are unreliable and unpredictable since they rely mainly on the internet as a backhaul in the majority of the surveyed scenarios in this paper such as Refs. [53,62]. Therefore, it is necessary to quantify a metric for communications between the femtocell BSs and UEs. This metric can be obtained through a Link Quality Estimator (LQE). A path is considered to be good when the Received Signal Strength (RSS) of the femtocells satisfies the link quality threshold. Therefore, investigating into LQE metric and proposing a link-quality-protection strategy would be beneficial in order to establish a better link quality for the associated femtocell users.

2.9.5 Mobility

Mobility is one of the key challenges in femtocells utilized in IoT applications, since the problem of frequent handover raises through such scenarios. The continuous change of UE coverage leads to an important issue regarding optimization, as well as achieving improvements in energy saving and bandwidth utilization. Generally, femtocells are utilized by users in indoor environments. Therefore, no specific mobility management is needed. However, it is necessary to have mobility management and handover procedures for dense deployment of femtocells [230]. In the case of dense deployment of femtocells, mobility management is one of the main challenges since it would be very difficult for a femtocell to keep track of all the neighboring femtocells due to the fact that every FBS may have a large number of mobile neighbors that dynamically change the network topology. Therefore, research for an efficient mobility management scheme would enhance the performance in terms of energy saving and bandwidth utilization.

2.9.6 Scalability

A typical HetNet consist of overlaying macrocell sites and many femtocells [231]. Therefore, it really is a challenging issue to keep up to date information on the femtocells. Scalability is one of the main design attributes in femtocells, and must be encompassed by the deployment and data-traffic modeling. The deployment strategy should be scalable enough to enable it to work with growing numbers of UEs, and it should continually assure the correct behavior of the application. Furthermore, it must be adapted to scalability changes in a transparent way (i.e., without requiring intervention of the user). Therefore, research for solutions applied on FBS management would be beneficial in order to have systems that are adapted to various transmission environments and available resources.

2.9.7 Served Content

Exchanged femtocell contents that can vary from a plain text message to video streaming can produce huge differences in QoS. In the past, voice and plain text messages were the only concerns of mobile operators. However, nowadays, with significant growth in the number of mobile connected devices and various types of mobile applications, different service types are available to the users (i.e., VoIP, IPTV, Skype, web browsing, online video, and video streaming) [232]. Each of these services requires different levels of QoS. Therefore, it is crucial to carry out research on the factors contributing on the perceived QoS in femtocell networks since these factors may be different for each service type. In addition, research on the network architecture would be beneficial to assure that the current architecture can support aforementioned service types in a way to satisfy the mobile users.

2.9.8 Quality Assurance

QoS specification can be used to provide various priorities to guarantee a certain level of performance to a data flow in accordance with requests from the application. QoS guarantee is considered as a must in femtocells, especially when the cellular network capacity is limited. For example, real-time streaming multimedia applications require fixed bit rate and are delay sensitive. Therefore, femtocells offering QoS guarantee for multimedia traffic should be flexible to support different application-specific QoS requirements (such as energy efficiency, end-to-end delay, reliability, delay jitter, and bandwidth consumption) in the heterogeneous traffic environment. Improvements in QoS can be obtained for small number of UEs being served by a femtocell. However, femtocells are recently being deployed in outdoor environments and public places such as shopping malls and airports where large number of UEs are associated with a femtocell. Therefore, available resources may not be adequate to achieve QoS requirements of each user. In addition, QoS management becomes more important in dense deployment of femtocells where each femtocell may have limited resources available. Therefore, in order to fulfill QoS requirements for each user, an optimized and efficient radio RM solution is required that considers many factors such as interference and limited radio resources.

2.9.9 Cognitive Radio (CR)

CR is based on Dynamic Spectrum Access (DSA), which has been proved to yield a promising approach for communications. CR has positive impacts on interference mitigation, power consumption level, and network lifetime and it should be utilized in several IoT-based femtocells. Capabilities such as sensing the spectrum, determining the vacant band and improving the overall utilization, may be exploited by the FBS to employ spectrum allocation and process resource constraints of low-end UE. The importance of CR lies under the fact that there is a trade-off between the power consumption and bandwidth, meaning that in order to reduce power consumption, we need to have more bandwidth [233]. In other words, we need to have dynamic and optimal spectrum management. This is where CR plays an important role. In addition, research into structures and techniques based on CR would be beneficial to propose less expensive and/or complex schemes compared to schemes presented in Refs. [234,235] in order to reduce power consumption while maintaining QoS.

2.9.10 Security

One of the other key challenges in femtocell networks is security. In open access mode, since everyone can connect to the network, providing efficient security measures is much more important because private information of the users have to be protected. There are many security issues associated with femtocell networks. For instance, private information of a subscriber which is transmitted over the backhaul internet connection may be hacked which breaks privacy and confidentiality laws [236]. Femtocells are also subjected to DoS attacks. For example, a hacker can make the link between FAP and the core network overloaded so that the femtocell users cannot get services [113]. Security is also needed in the case of closed access mode to prevent unauthorized users to have access to the network and resources. Due to these security related issues and with increase in the number of femtocell deployments, it is quite important to carry out extensive research in this area to provide sufficient levels of security for femtocell users.

2.9.11 Interference Management (IM)

Interference is one of the most important technical challenges for the dense deployment of femtocells. Two types of interference are observed in general: cross-tier interference (femto-to-macro) and co-tier interference (femto-to-femto) [237]. Interference is an important factor in the deployment of femtocells, which has the potential to limit the overall performance of the network and degrades QoS and network capacity. Therefore, research for IM algorithms and techniques such as interference cancellation and/or avoidance is of great importance to assure acceptable level of QoS.

2.9.12 Spectrum Management and Synchronization

Deployment of femtocells in the current macrocellular networks needs additional spectrum resources. Otherwise, the deployment can cause interference in the existing cellular networks [238]. Spectrum allocation is important to provide a broad range of services and deliver social and economic benefits. Therefore, the efficient use of spectrum is an important goal for many operators. Research for techniques and approaches that consider dynamic spectrum management methods would be beneficial in order to mitigate interference and enhance the capacity performance of the network. Frequency synchronization is another important factor in wireless communication systems since it minimizes interference and is necessary for IM. Synchronization is also required to have proper handovers between BSs in order to have good level of QoS [103]. An error in timing and synchronization can result in Inter Symbol Interference (ISI) in Orthogonal Frequency Division for Multiple Access (OFDMA) systems [239]. The problem of synchronization of the femtocell networks can be overcome by the use of backhaul Asymmetric Digital Subscriber Line (ADSL) or the use of GPS within the femtocells [103]. However, more research needs to be carried out towards timing and frequency synchronization in order to develop intelligent algorithms to overcome this challenge.

2.9.13 Mobile Video in m-Health Networking

With the rapid increase in the number of mobile and HD video devices and the rapid development in the infrastructure of the network used for video streaming, new techniques are required to assess the video QoE. In a few years, 90% of the internet traffic will contain video traffic, which will be watched by more than a billion people [240]. In addition, in a shared environment, medical applications will use the same resources as those of other bandwidth hungry applications [195]. Therefore, new wireless technologies such as 5G and femtocells are crucial to overcome video streaming-related challenges such as EE, SE, seamless coverage, high mobility, and different requirements of QoS and QoE. 5G and femtocells have great potential to benefit medical video streaming which is a well-known example of m-health applications. Therefore, extensive research is required to develop a model that can accurately describe QoE and how it is affected by QoS parameters in the field of m-health networking.

2.10 Concluding Remarks

Femtocells have proved to be a promising solution for mobile operators to enhance coverage and capacity, and provide high quality services to mobile users at low cost while maintaining good levels of QoS. In this article, through a comprehensive investigation on various IoT-femtocell based applications available in the literature, we discussed objectives and constraints of such systems and provided a classification of design factors that can be employed for various systems with potential of employing these infrastructures. Moreover, we investigated modeling and deployment strategies of femtocells in the IoT era, which provide a good foundation for researchers who are interested to gain an insight into femtocell and IoT technologies. Although various studies in existing literature consider surveys that cover various aspects of the femtocell and IoT enabling technologies, to the best of our knowledge this work is the first one to consider the use of femtocells and their applications in traffic modeling and deployment in the 5G/IoT environment.

References

1. "Cisco visual networking index (VNI): mobile data traffic forecast, 2016–2021," http://www.cisco.com, 2017.
2. "The Voice of 5G for the Americas, Cellular technologies enabling the internet of things–4G Americas, white paper," http://5gamericas.org, 2015.
3. V. Chandrasekhar, J. G. Andrews, and A. Gatherer, "Femtocell networks: A survey," *IEEE Communications Magazine*, vol. 46, no. 9, pp. 59–67, 2008.
4. J. Boccuzzi and M. Ruggiero, *Femtocells: Design & Application*. McGraw Hill Professional, New York, 2010.
5. "Small Cell market status, Informa and Small Cell forum, white paper," http://www.smallcells.com, 2012.
6. F. Haider, C.-X. Wang, H. Haas, D. Yuan, H. Wang, X. Gao, X.-H. You, and E. Hepsaydir, "Spectral efficiency analysis of mobile femtocell based cellular systems," in *IEEE 13th International Conference on Communication Technology (ICCT)*, 2011, pp. 347–351.
7. M. H. Qutqut, F. M. Al-Turjman, and H. S. Hassanein, "MFW: Mobile femtocells utilizing WiFi: A data offloading framework for cellular networks using mobile femtocells," in *IEEE International Conference on Communications (ICC)*, 2013, pp. 6427–6431.
8. M. Yavuz, F. Meshkati, S. Nanda, A. Pokhariyal, N. Johnson, B. Raghothaman, and A. Richardson, "Interference management and performance analysis of UMTS/HSPA+ femtocells," *IEEE Communications Magazine*, vol. 47, no. 9, pp. 102–109, 2009.
9. L. G. Garcia, K. I. Pedersen, and P. E. Mogensen, "Autonomous component carrier selection: Interference management in local area environments for LTE-advanced," *IEEE Communications Magazine*, vol. 47, no. 9, pp. 110–116, 2009.
10. X. Li, L. Qian, and D. Kataria, "Downlink power control in co-channel macrocell femtocell overlay," in *IEEE 43rd Annual Conference on Information Sciences and Systems (CISS)*, 2009, pp. 383–388.

11. K. Katzis and H. Ahmadi, "Challenges implementing Internet of Things (IoT) using cognitive radio capabilities in 5G mobile networks," in *Internet of Things (IoT) in 5G Mobile Technologies*. Springer, 2016, pp. 55–76.
12. A. Al-Fuqaha, M. Guizani, M. Mohammadi, M. Aledhari, and M. Ayyash, "Internet of things: A survey on enabling technologies, protocols, and applications," *IEEE Communications Surveys & Tutorials*, vol. 17, no. 4, pp. 2347–2376, 2015.
13. M. W. Woo, J. Lee, and K. Park, "A reliable IoT system for personal healthcare devices," *Future Generation Computer Systems*, vol. 78, part 2, pp. 626–640, 2017.
14. H. Al-Hamadi and R. Chen, "Trust-based decision making for health IoT systems," *IEEE Internet of Things Journal*, vol. 4, no. 5, pp. 1408–1419, 2017.
15. C. Brewster, I. Roussaki, N. Kalatzis, K. Doolin, and K. Ellis, "IoT in agriculture: Designing a Europe-wide large-scale pilot," *IEEE Communications Magazine*, vol. 55, no. 9, pp. 26–33, 2017.
16. J. Wu, S. Guo, J. Li, and D. Zeng, "Big data meet green challenges: Greening big data," *IEEE Systems Journal*, vol. 10, no. 3, pp. 873–887, 2016.
17. D. Evans, "The internet of things: How the next evolution of the internet is changing everything," *CISCO White Paper*, vol. 1, no. 2011, pp. 1–11, 2011.
18. G. Zhang, M. Chu, and J. Li, "Interference coordination based on random fractional spectrum reuse in femtocells toward Internet of Things," *Personal and Ubiquitous Computing*, vol. 20, no. 5, pp. 667–679, 2016.
19. G. Intelligence, "Understanding 5G: Perspectives on future technological advancements in mobile," *GSMA Intelligence Understanding 5G*, pp. 3–15, 2014.
20. S. Ashraf, Y.-P. E. Wang, S. Eldessoki, B. Holfeld, D. Parruca, M. Serror, and J. Gross, "From radio design to system evaluations for ultra-reliable and low-latency communication," in *Proceedings of 23th European Wireless Conference*. VDE, 2017, pp. 1–8.
21. R. Atat, L. Liu, H. Chen, J. Wu, H. Li, and Y. Yi, "Enabling cyber-physical communication in 5G cellular networks: Challenges, spatial spectrum sensing, and cyber-security," *IET Cyber-Physical Systems: Theory & Applications*, vol. 2, no. 1, pp. 49–54, 2017.
22. S. Mahmud, G. Khan, H. Zafar, K. Ahmad, and N. Behttani, "A survey on femtocells: Benefits deployment models and proposed solutions," *Journal of Applied Research and Technology*, vol. 11, no. 5, pp. 733–754, 2013.
23. L. Atzori, A. Iera, and G. Morabito, "The internet of things: A survey," *Computer Networks*, vol. 54, no. 15, pp. 2787–2805, 2010.
24. R. Khan, S. U. Khan, R. Zaheer, and S. Khan, "Future Internet: The Internet of things architecture, possible applications and key challenges," in *IEEE 10th International Conference on Frontiers of Information Technology (FIT)*, 2012, pp. 257–260.
25. Z. Sheng, S. Yang, Y. Yu, A. Vasilakos, J. McCann, and K. Leung, "A survey on the ietf protocol suite for the internet of things: Standards, challenges, and opportunities," *IEEE Wireless Communications*, vol. 20, no. 6, pp. 91–98, 2013.
26. J. K. Bare, "Comparison of the performance and capabilities of femtocell versus Wi-Fi networks," PhD dissertation, Monterey: CA, Naval Postgraduate School, 2012.
27. F. M. Al-Turjman, M. Imran, and S. T. Bakhsh, "Energy efficiency perspectives of femtocells in Internet of Things: Recent advances and challenges," *IEEE Access*, vol. 5, pp. 26808–26818, 2017.

28. E. Bou-Harb, C. Fachkha, M. Pourzandi, M. Debbabi, and C. Assi, "Communication security for smart grid distribution networks," *IEEE Communications Magazine*, vol. 51, no. 1, pp. 42–49, 2013.
29. H. Zhang, Y. Nie, J. Cheng, V. C. Leung, and A. Nallanathan, "Sensing time optimization and power control for energy efficient cognitive small cell with imperfect hybrid spectrum sensing," *IEEE Transactions on Wireless Communications*, vol. 16, no. 2, pp. 730–743, 2017.
30. Z. Fan, P. Kulkarni, S. Gormus, C. Efthymiou, G. Kalogridis, M. Sooriyabandara, Z. Zhu, S. Lambotharan, and W. H. Chin, "Smart grid communications: Overview of research challenges, solutions, and standardization activities," *IEEE Communications Surveys & Tutorials*, vol. 15, no. 1, pp. 21–38, 2013.
31. M. Hindia, T. Rahman, H. Ojukwu, E. Hanafi, and A. Fattouh, "Enabling remote health-caring utilizing IoT concept over LTE-femtocell networks," *PloS one*, vol. 11, no. 5, p. e0155077, 2016.
32. J. Wan, C. Zou, S. Ullah, C.-F. Lai, M. Zhou, and X. Wang, "Cloud-enabled wireless body area networks for pervasive healthcare," *IEEE Network*, vol. 27, no. 5, pp. 56–61, 2013.
33. Z. H. Hashmi, "Adaptive and efficient resource management for emerging wireless networks," PhD dissertation, Vancouver, University of British Columbia, 2013.
34. S. R. Saunders, S. Carlaw, A. Giustina, R. R. Bhat, V. S. Rao, and R. Siegberg, *Femtocells: Opportunities and Challenges for Business and Technology*. John Wiley & Sons, New York, 2009.
35. J. Manyika, M. Chui, J. Bughin, R. Dobbs, P. Bisson, and A. Marrs, *Disruptive Technologies: Advances That Will Transform Life, Business, and the Global Economy*. McKinsey Global Institute, San Francisco: CA, vol. 180, 2013.
36. "Commercial Building Automation Systems, Navigant Consulting Research, Boulder, CO, USA," https://www.navigantresearch.com, 2013.
37. E. Yaacoub, "Green 5G femtocells for supporting indoor generated IoT traffic," in *Internet of Things (IoT) in 5G Mobile Technologies*. Springer, 2016, pp. 129–152.
38. A. Rachedi, M. H. Rehmani, S. Cherkaoui, and J. J. Rodrigues, "IEEE access special section editorial: The plethora of research in Internet of Things (IoT)," *IEEE Access*, vol. 4, pp. 9575–9579, 2016.
39. I. Bisio, F. Lavagetto, M. Marchese, and A. Sciarrone, "Smartphone-centric ambient assisted living platform for patients suffering from co-morbidities monitoring," *IEEE Communications Magazine*, vol. 53, no. 1, pp. 34–41, 2015.
40. Y. Saleem, N. Crespi, M. H. Rehmani, R. Copeland, D. Hussein, and E. Bertin, "Exploitation of social IoT for recommendation services," in *IEEE 3rd World Forum on Internet of Things (WF-IoT)*, 2016, pp. 359–364.
41. E. Ferro and F. Potorti, "Bluetooth and Wi-Fi wireless protocols: A survey and a comparison," *IEEE Wireless Communications*, vol. 12, no. 1, pp. 12–26, 2005.
42. P. McDermott-Wells, "What is bluetooth?" *IEEE Potentials*, vol. 23, no. 5, pp. 33–35, 2004.
43. G. V. Crosby and F. Vafa, "Wireless sensor networks and LTE-A network convergence," in *IEEE 38th Conference on Local Computer Networks (LCN)*, 2013, pp. 731–734.
44. P. Kinney et al., "Zigbee technology: Wireless control that simply works," in *Communications Design Conference*, vol. 2, 2003, pp. 1–7.

45. R. S. Kshetrimayum, "An introduction to UWB communication systems," *IEEE Potentials*, vol. 28, no. 2, pp. 9–13, 2009.
46. S. Singh and Y. Singh, "The emerging technology—Li-Fi using femtocells," *International Journal of Scientific Research and Reviews*, vol. 3, pp. 115–120, 2014.
47. F. J. Mendieta and I. N. Hipólito, "Trends of the optical wireless communications," 2011.
48. I. Ashraf, F. Boccardi, and L. Ho, "Sleep mode techniques for small cell deployments," *IEEE Communications Magazine*, vol. 49, no. 8, 2011.
49. S. R. Hall, A. W. Jeffries, S. E. Avis, and D. D. Bevan, "Performance of open access femtocells in 4G macrocellular networks," in *Wireless World Research Forum*, vol. 20, 2008.
50. D. Chambers, "Using a femtocell API to create innovative new applications?" 2009.
51. H. Jung, J. H. Lee, C. Park, Y. Im, T. Kwon, and Y. Choi, "A femtocell-based testbed for evaluating future cellular networks," in *Proceedings of International Conference on Future Internet Technologies (CFI)*, 2008.
52. F. Van Den Broek and R. W. Schreur, "Femtocell security in theory and practice," in *Nordic Conference on Secure IT Systems*. Springer, 2013, pp. 183–198.
53. F. Haider, C.-X. Wang, B. Ai, H. Haas, and E. Hepsaydir, "Spectral/energy efficiency tradeoff of cellular systems with mobile femtocell deployment," *IEEE Transactions on Vehicular Technology*, vol. 65, no. 5, pp. 3389–3400, 2016.
54. D. Feng, C. Jiang, G. Lim, L. J. Cimini, G. Feng, and G. Y. Li, "A survey of energy-efficient wireless communications," *IEEE Communications Surveys & Tutorials*, vol. 15, no. 1, pp. 167–178, 2013.
55. J. Zhang and G. De la Roche, *Femtocells: Technologies and Deployment*. John Wiley & Sons, New York, 2011.
56. D. De and A. Mukherjee, "Femtocell based economic health monitoring scheme using mobile cloud computing," in *IEEE International Advance Computing Conference (IACC)*, 2014, pp. 385–390.
57. A. Mukherjee and D. De, "Femtocell based green health monitoring strategy," in *IEEE 31st URSI General Assembly and Scientific Symposium (URSI GASS)*, 2014, pp. 1–4.
58. A. Maciuca, D. Popescu, M. Strutu, and G. Stamatescu, "Wireless sensor network based on multilevel femtocells for home monitoring," in *IEEE 7th International Conference on Intelligent Data Acquisition and Advanced Computing Systems (IDAACS)*, vol. 1, 2013, pp. 499–503.
59. A. Maciuca, G. Stamatescu, D. Popescu, and M. Struţu, "Integrating wireless body and ambient sensors into a hybrid femtocell network for home monitoring," in *IEEE 2nd International Conference on Systems and Computer Science (ICSCS)*, 2013, pp. 32–37.
60. E. Mutafungwa, "Applying MTC and femtocell technologies to the Continua Health reference architecture," in *International Conference on Grid and Pervasive Computing*. Springer, 2011, pp. 105–114.
61. E. Mutafungwa, Z. Zheng, J. Hämäläinen, M. Husso, and T. Korhonen, "Exploiting femtocellular networks for emergency telemedicine applications in indoor environments," in *IEEE 12th International Conference on e-Health Networking Applications and Services (Healthcom)*, 2010, pp. 283–289.
62. I. U. Rehman, N. Y. Philip, and R. S. Istepanian, "Performance analysis of medical video streaming over 4G and beyond small cells for indoor and moving vehicle (ambulance) scenarios," in *EAI 4th IEEE International Conference on Wireless Mobile Communication and Healthcare (Mobihealth)*, 2014, pp. 211–216.

63. M. F. Feteiha, M. H. Qutqut, and H. S. Hassanein, “Outage probability analysis of mobile small cells over LTE-A networks,” in *International Wireless Communications and Mobile Computing Conference (IWCMC)*, 2014, pp. 1045–1050.
64. H. Zhang, H.-C. Wu, and L. Guo, “Multimedia services scheduling optimization using femtocell on high-speed trains,” in *IEEE Wireless Communications and Networking Conference (WCNC)*, 2015, pp. 464–469.
65. R. Raheem, A. Lasebae, and J. Loo, “Performance evaluation of LTE network via using fixed/mobile femtocells,” in *IEEE 28th International Conference on Advanced Information Networking and Applications Workshops (WAINA)*, 2014, pp. 255–260.
66. B. Naudts, J. Spruytte, J. Van Ooteghem, S. Verbrugge, D. Colle, and M. Pickavet, “Internet on trains: A multi-criteria analysis of on-board deployment options for on-train cellular connectivity,” in *IEEE 16th International Telecommunications Network Strategy and Planning Symposium (Networks)*, 2014, pp. 1–7.
67. A. Cimmino, T. Pecorella, R. Fantacci, F. Granelli, T. F. Rahman, C. Sacchi, C. Carlini, and P. Harsh, “The role of small cell technology in future smart city applications,” *Transactions on Emerging Telecommunications Technologies*, vol. 25, no. 1, pp. 11–20, 2014.
68. S. Talari, M. Shafie-khah, P. Siano, V. Loia, A. Tommasetti, and J. P. Catalão, “A review of smart cities based on the internet of things concept,” *Energies*, vol. 10, no. 4, p. 421, 2017.
69. A. E. Al-Fagih, F. M. Al-Turjman, W. M. Alsalih, and H. S. Hassanein, “A priced public sensing framework for heterogeneous IoT architectures,” *IEEE Transactions on Emerging Topics in Computing*, vol. 1, no. 1, pp. 133–147, 2013.
70. G. T. Singh and F. M. Al-Turjman, “A data delivery framework for cognitive information-centric sensor networks in smart outdoor monitoring,” *Computer Communications*, vol. 74, pp. 38–51, 2016.
71. T. Gea, J. Paradells, M. Lamarca, and D. Roldan, “Smart cities as an application of internet of things: Experiences and lessons learnt in Barcelona,” in *IEEE 7th International Conference on Innovative Mobile and Internet Services in Ubiquitous Computing (IMIS)*, 2013, pp. 552–557.
72. J. Jin, J. Gubbi, S. Marusic, and M. Palaniswami, “An information framework for creating a smart city through internet of things,” *IEEE Internet of Things Journal*, vol. 1, no. 2, pp. 112–121, 2014.
73. H. Amer, N. Salman, M. Hawes, M. Chaqfeh, L. Mihaylova, and M. Mayfield, “An improved simulated annealing technique for enhanced mobility in smart cities,” *Sensors*, vol. 16, no. 7, p. 1013, 2016.
74. A. Celik and A. E. Kamal, “Green cooperative spectrum sensing and scheduling in heterogeneous cognitive radio networks,” *IEEE Transactions on Cognitive Communications and Networking*, vol. 2, no. 3, pp. 238–248, 2016.
75. A. Celik and A. E. Kemal, “Multi-objective clustering optimization for multi-channel cooperative spectrum sensing in heterogeneous green CRNS,” *IEEE Transactions on Cognitive Communications and Networking*, vol. 2, no. 2, pp. 150–161, 2016.
76. S. K. Goudos, P. I. Dallas, S. Chatziefthymiou, and S. Kyriazakos, “A survey of IoT key enabling and future technologies: 5G, mobile IoT, semantic web and applications,” *Wireless Personal Communications*, vol. 97, no. 2, pp. 1645–1675, 2017.
77. 3GPP, “3G Home NodeB study item technical report, TR25.820,” 2008.
78. “3rd Generation Partnership Project 2 (3GPP2), Femtocell systems overview for CDMA2000 wireless communication systems,” http://www.scf.io, 2014.

79. 3GPP2, "System requirements for femtocell systems, S.R0126-0," 2008.
80. R. Want, "An introduction to RFID technology," *IEEE Pervasive Computing*, vol. 5, no. 1, pp. 25–33, 2006.
81. R. Frank, W. Bronzi, G. Castignani, and T. Engel, "Bluetooth low energy: An alternative technology for VANET applications," in *IEEE 11th Annual Conference on Wireless On-demand Network Systems and Services (WONS)*, 2014, pp. 104–107.
82. R. Want, "Near field communication," *IEEE Pervasive Computing*, vol. 10, no. 3, pp. 4–7, 2011.
83. I. W. Group et al., "IEEE standard for local and metropolitan area network—Part 15.4: Low-rate wireless personal area networks (LR-WPANs)," *IEEE Standard*, vol. 802, pp. 4–2011, 2011.
84. S. Aust, R. V. Prasad, and I. G. Niemegeers, "IEEE 802.11 ah: Advantages in standards and further challenges for sub 1 GHz Wi-Fi," in *IEEE International Conference on Communications (ICC)*, 2012, pp. 6885–6889.
85. L. Alliance, "A technical overview of LoRa and LoRaWAN," white paper, November, 2015.
86. B. B. Olyaei, J. Pirskanen, O. Raeesi, A. Hazmi, and M. Valkama, "Performance comparison between slotted IEEE 802.15. 4 and IEEE 802.1 lah in IoT based applications," in *IEEE 9th International Conference on Wireless and Mobile Computing, Networking and Communications (WiMob)*, 2013, pp. 332–337.
87. J. de Carvalho Silva, J. J. Rodrigues, A. M. Alberti, P. Solic, and A. L. Aquino, "LoRaWAN—Low power WAN protocol for Internet of Things: A review and opportunities," in *IEEE 2nd International Multidisciplinary Conference on Computer and Energy Science (SpliTech)*, 2017, pp. 1–6.
88. A. Rico-Alvarino, M. Vajapeyam, H. Xu, X. Wang, Y. Blankenship, J. Bergman, T. Tirronen, and E. Yavuz, "An overview of 3GPP enhancements on machine to machine communications," *IEEE Communications Magazine*, vol. 54, no. 6, pp. 14–21, 2016.
89. "Japan gears up for 2020 Olympics with further 5G trials at 4.5GHz and 28GHz," http://www.telecomtv.com, 2016.
90. C.-X. Wang, F. Haider, X. Gao, X.-H. You, Y. Yang, D. Yuan, H. Ag-goune, H. Haas, S. Fletcher, and E. Hepsaydir, "Cellular architecture and key technologies for 5G wireless communication networks," *IEEE Communications Magazine*, vol. 52, no. 2, pp. 122–130, 2014.
91. F. Cao and Z. Fan, "The tradeoff between energy efficiency and system performance of femtocell deployment," in *IEEE 7th International Symposium on Wireless Communication Systems (ISWCS)*, 2010, pp. 315–319.
92. Y. Hou and D. I. Laurenson, "Energy efficiency of high QoS heterogeneous wireless communication network," in *IEEE 72nd Vehicular Technology Conference Fall (VTC 2010-Fall)*, 2010, pp. 1–5.
93. M. Jada, M. Hossain, J. Hämäläinen, and R. Jäntti, "Impact of femtocells to the WCDMA network energy efficiency," in *IEEE 3rd International Conference on Broadband Network and Multimedia Technology (IC-BNMT)*, 2010, pp. 305–310.
94. I. Ashraf, L. T. Ho, and H. Claussen, "Improving energy efficiency of femtocell base stations via user activity detection," in *IEEE Wireless Communications and Networking Conference (WCNC)*, 2010, pp. 1–5.
95. J. Gambini and U. Spagnolini, "Wireless over cable for energy-efficient femtocell systems," in *IEEE GLOBECOM Workshops (GC Wkshps)*, 2010, pp. 1464–1468.

96. H. Leem, S. Y. Baek, and D. K. Sung, "The effects of cell size on energy saving, system capacity, and per-energy capacity," in *IEEE Wireless Communications and Networking Conference (WCNC)*, 2010, pp. 1–6.
97. J. B. Rao and A. O. Fapojuwo, "An analytical framework for evaluating spectrum/energy efficiency of heterogeneous cellular networks," *IEEE Transactions on Vehicular Technology*, vol. 65, no. 5, pp. 3568–3584, 2016.
98. H. Sherazi, R. Iqbal, A. Gilani, and M. Chaudary, "Energy efficient femtocell based architecture for low coverage areas in WiMAX," in *IEEE 6th International Conference on Computing, Communication and Networking Technologies (ICCCNT)*, 2015, pp. 1–6.
99. F. Haider, M. Dianati, and R. Tafazolli, "A simulation based study of mobile femtocell assisted LTE networks," in *IEEE 7th International Wireless Communications and Mobile Computing Conference (IWCMC)*, 2011, pp. 2198–2203.
100. "Continua Health Alliance," http://www.pchalliance.org/, 2017.
101. A. Khalifah, N. Akkari, and G. Aldabbagh, "Dense areas femtocell deployment: Access types and challenges," in *IEEE 3rd International Conference on e-Technologies and Networks for Development (ICeND)*, 2014, pp. 64–69.
102. H. Claussen, "Performance of macro-and co-channel femtocells in a hierarchical cell structure," in *IEEE 18th International Symposium on Personal, Indoor and Mobile Radio Communications (PIMRC)*, 2007, pp. 1–5.
103. T. Zahir, K. Arshad, A. Nakata, and K. Moessner, "Interference management in femtocells," *IEEE Communications Surveys & Tutorials*, vol. 15, no. 1, pp. 293–311, 2013.
104. S.-J. Wu, "A new handover strategy between femtocell and macrocell for LTE-based network," in *IEEE 4th International Conference on Ubi-Media Computing (U-Media)*, 2011, pp. 203–208.
105. V. Group, "Open and semi-open access support for UTRA Home NB, 3GPP-TSG RAN, Technical Report, R2–085280," 2008.
106. X. Chen, J. Wu, Y. Cai, H. Zhang, and T. Chen, "Energy-efficiency oriented traffic offloading in wireless networks: A brief survey and a learning approach for heterogeneous cellular networks," *IEEE Journal on Selected Areas in Communications*, vol. 33, no. 4, pp. 627–640, 2015.
107. F. Han, S. Zhao, L. Zhang, and J. Wu, "Survey of strategies for switching off base stations in heterogeneous networks for greener 5G systems," *IEEE Access*, vol. 4, pp. 4959–4973, 2016.
108. A. A. Abdulkafi, T. S. Kiong, J. Koh, D. Chieng, and A. Ting, "Energy efficiency of heterogeneous cellular networks: A review," *Journal of Applied Sciences*, vol. 12, no. 14, p. 1418, 2012.
109. T. Taleb and A. Ksentini, "QoS/QoE predictions-based admission control for femto communications," in *IEEE International Conference on Communications (ICC)*, 2012, pp. 5146–5150.
110. M.-S. Alouini and A. J. Goldsmith, "Area spectral efficiency of cellular mobile radio systems," *IEEE Transactions on Vehicular Technology*, vol. 48, no. 4, pp. 1047–1066, 1999.
111. O. Vermesan and P. Friess, *Internet of Things—From Research and Innovation to Market Deployment*. River Publishers, Aalborg, 2014.
112. C.-M. Chen, T.-Y. Wu, R. Tso, and M.-E. Wu, "Security analysis and improvement of femtocell access control," in *International Conference on Network and System Security*. Springer, 2014, pp. 223–232.

113. M. del Valle, B. Manikandan, and V. S. Sriram, "Securing the femtocells: Anonymity and location privacy," *Indian Journal of Science and Technology*, vol. 7, no. 4, pp. 46–51, 2014.
114. I. Syed and H. Kim, "A secure registration scheme for femtocell embedded networks," in *Multimedia and Ubiquitous Engineering*. Springer, 2013, pp. 103–109.
115. R. Raheem, A. Lasebae, M. Aiash, and J. Loo, "Interference management for co-channel mobile femtocells technology in LTE networks," in *IEEE 12th International Conference on Intelligent Environments (IE)*, 2016, pp. 80–87.
116. D. Xenakis, N. Passas, L. Merakos, and C. Verikoukis, "Handover decision for small cells: Algorithms, lessons learned and simulation study," *Computer Networks*, vol. 100, pp. 64–74, 2016.
117. G. Gódor, Z. Jakó, Á. Knapp, and S. Imre, "A survey of handover management in LTE-based multi-tier femtocell networks: Requirements, challenges and solutions," *Computer Networks*, vol. 76, pp. 17–41, 2015.
118. D. Xenakis, N. Passas, L. Merakos, and C. Verikoukis, "Mobility management for femtocells in LTE-advanced: Key aspects and survey of handover decision algorithms," *IEEE Communications Surveys & Tutorials*, vol. 16, no. 1, pp. 64–91, 2014.
119. H. Zhang, X. Wen, B. Wang, W. Zheng, and Y. Sun, "A novel handover mechanism between femtocell and macrocell for LTE based networks," in *IEEE 2nd International Conference on Communication Software and Networks (ICCSN)*, 2010, pp. 228–231.
120. H. Zhang, W. Ma, W. Li, W. Zheng, X. Wen, and C. Jiang, "Signalling cost evaluation of handover management schemes in LTE-advanced femtocell," in *IEEE 73rd Vehicular Technology Conference (VTC Spring)*, 2011, pp. 1–5.
121. S.-J. Wu and S. K. Lo, "Handover scheme in LTE-based networks with hybrid access mode femtocells," *Journal of Convergence Information Technology*, vol. 6, no. 7, pp. 68–78, 2011.
122. S. Wu, X. Zhang, R. Zheng, Z. Yin, Y. Fang, and D. Yang, "Handover study concerning mobility in the two-hierarchy network," in *IEEE 69th Vehicular Technology Conference (VTC Spring)*, 2009, pp. 1–5.
123. J.-M. Moon and D.-H. Cho, "Efficient handoff algorithm for inbound mobility in hierarchical macro/femto cell networks," *IEEE Communications Letters*, vol. 13, no. 10, pp. 755–757, 2009.
124. S. Deswal and A. Singhrova, "Handover algorithm for heterogeneous networks," in *IEEE 3rd International Conference on Computing for Sustainable Global Development (INDIACom)*, 2016, pp. 3358–3364.
125. D. Xenakis, N. Passas, L. Merakos, and C. Verikoukis, "Advanced mobility management for reduced interference and energy consumption in the two-tier LTE-advanced network," *Computer Networks*, vol. 76, pp. 90–111, 2015.
126. J.-M. Moon and D.-H. Cho, "Novel handoff decision algorithm in hierarchical macro/femtocell networks," in *IEEE Wireless Communications and Networking Conference (WCNC)*, 2010, pp. 1–6.
127. P. Xu, X. Fang, R. He, and Z. Xiang, "An efficient handoff algorithm based on received signal strength and wireless transmission loss in hierarchical cell networks," *Telecommunication Systems*, vol. 52, no. 1, pp. 1–9, 2013.
128. D. López-Pérez, A. Ladányi, A. Juttner, and J. Zhang, "OFDMA femtocells: Intracell handover for interference and handover mitigation in two-tier networks," in *IEEE Wireless Communications and Networking Conference (WCNC)*, 2010, pp. 1–6.

129. D. López-Pérez, A. Valcarce, Á. Ladányi, G. de la Roche, and J. Zhang, "Intracell handover for interference and handover mitigation in OFDMA two-tier macrocell-femtocell networks," *EURASIP Journal on Wireless Communications and Networking*, vol. 2010, p. 1, 2010.
130. Z. Becvar and P. Mach, "Adaptive hysteresis margin for handover in femtocell networks," in *IEEE 6th International Conference on Wireless and Mobile Communications (ICWMC)*, 2010, pp. 256–261.
131. A. Ulvan, R. Bestak, and M. Ulvan, "The study of handover procedure in LTE-based femtocell network," in *IEEE 3rd Joint IFIP Wireless and Mobile Networking Conference (WMNC)*, 2010, pp. 1–6.
132. M. S. Hung, J. Y. Pan, and Z. E. Huang, "Analysis of handover decision with adaptive offset in next-generation hybrid macro/femtocell networks," in *IEEE 10th International Conference on Intelligent Information Hiding and Multimedia Signal Processing (IIH-MSP)*, 2014, pp. 729–734.
133. T. M. Mutlu and B. Canberk, "A spatial estimation-based handover management for challenging femtocell deployments," in *IEEE International Black Sea Conference on Communications and Networking (BlackSeaCom)*, 2014, pp. 144–148.
134. K. S. B. Reguiga, F. Mhiri, and R. Bouallegue, "Handoff management in green femtocell network," *International Journal of Computer Applications*, vol. 27, no. 4, pp. 1–7, 2011.
135. D. Calin, H. Claussen, and H. Uzunalioglu, "On femto deployment architectures and macrocell offloading benefits in joint macro-femto deployments," *IEEE Communications Magazine*, vol. 48, no. 1, pp. 26–32, 2010.
136. D.-W. Lee, G.-T. Gil, and D.-H. Kim, "A cost-based adaptive handover hysteresis scheme to minimize the handover failure rate in 3GPP LTE system," *EURASIP Journal on Wireless Communications and Networking*, vol. 2010, no. 1, p. 750173, 2010.
137. F. Farias, M. Fiorani, S. Tombaz, M. Mahloo, L. Wosinska, J. C. Costa, and P. Monti, "Cost-and energy-efficient backhaul options for heterogeneous mobile network deployments," *Photonic Network Communications*, vol. 32, no. 3, pp. 422–437, 2016.
138. M. Arthi and P. Arulmozhivarman, "A flexible and cost-effective heterogeneous network deployment scheme for beyond 4G," *Arabian Journal for Science and Engineering*, vol. 41, no. 12, pp. 5093–5109, 2016.
139. R. Cai, W. Zhang, and P.-C. Ching, "Cost-efficient optimization of base station densities for multitier heterogeneous cellular networks," *IEEE Transactions on Wireless Communications*, vol. 15, no. 3, pp. 2381–2393, 2016.
140. M. Adedoyin and O. Falowo, "Self-organizing radio resource management for next generation heterogeneous wireless networks," in *IEEE International Conference on Communications (ICC)*. 2016, pp. 1–6.
141. M. Mehta, N. Rane, A. Karandikar, M. A. Imran, and B. G. Evans, "A self-organized resource allocation scheme for heterogeneous macro- femto networks," *Wireless Communications and Mobile Computing*, vol. 16, no. 3, pp. 330–342, 2014.
142. Z. Bharucha, A. Saul, G. Auer, and H. Haas, "Dynamic resource partitioning for downlink femto-to-macro-cell interference avoidance," *EURASIP Journal on Wireless Communications and Networking*, vol. 2010, no. 1, p. 143413, 2010.
143. R. An, X. Zhang, G. Cao, R. Zheng, and L. Sang, "Interference avoidance and adaptive fraction frequency reuse in a hierarchical cell structure," in *IEEE Wireless Communications and Networking Conference (WCNC)*, 2010, pp. 1–5.

144. F. Hu, K. Zheng, L. Lei, and W. Wang, "A distributed inter-cell interference coordination scheme between femtocells in LTE-advanced networks," in *IEEE 73rd Vehicular Technology Conference (VTC Spring)*, 2011, pp. 1–5.
145. Y. Bai, J. Zhou, L. Liu, L. Chen, and H. Otsuka, "Resource coordination and interference mitigation between macrocell and femtocell," in *IEEE 20th International Symposium on Personal, Indoor and Mobile Radio Communications*, 2009, pp. 1401–1405.
146. I. Guvenc, M.-R. Jeong, F. Watanabe, and H. Inamura, "A hybrid frequency assignment for femtocells and coverage area analysis for co-channel operation," *IEEE Communications Letters*, vol. 12, no. 12, pp. 880–882, 2008.
147. V. Chandrasekhar and J. G. Andrews, "Spectrum allocation in tiered cellular networks," *IEEE Transactions on Communications*, vol. 57, no. 10, pp. 3059–3068, 2009.
148. K. Sundaresan and S. Rangarajan, "Efficient resource management in OFDMA femto cells," in *Proceedings of the 10th ACM International Symposium on Mobile Ad Hoc Networking and Computing (ACM)*, 2009, pp. 33–42.
149. G. Cao, D. Yang, R. An, X. Ye, R. Zheng, and X. Zhang, "An adaptive sub-band allocation scheme for dense femtocell environment," in *IEEE Wireless Communications and Networking Conference (WCNC)*, 2011, pp. 102–107.
150. K. Zheng, F. Hu, L. Lei, and W. Wang, "Interference coordination between femtocells in LTE-advanced networks with carrier aggregation," in *IEEE 5th International ICST Conference on Communications and Networking in China (CHINACOM)*, 2010, pp. 1–5.
151. R.-T. Juang, P. Ting, H.-P. Lin, and D.-B. Lin, "Interference management of femtocell in macro-cellular networks," in *Wireless Telecommunications Symposium (WTS)*. IEEE, 2010, pp. 1–4.
152. S.-Q. Lee, B. H. Ryu, and N.-H. Park, "Call admission control for hybrid access mode femtocell system," in *IEEE 7th International Conference on Wireless and Mobile Computing, Networking and Communications (WiMob)*, 2011, pp. 512–516.
153. M.-S. Kim, H. W. Je, and F. A. Tobagi, "Cross-tier interference mitigation for two-tier OFDMA femtocell networks with limited macro-cell information," in *IEEE Global Telecommunications Conference (GLOBECOM)*, 2010, pp. 1–5.
154. M. Morita, Y. Matsunaga, and K. Hamabe, "Adaptive power level setting of femtocell base stations for mitigating interference with macrocells," in *IEEE 72nd Vehicular Technology Conference Fall (VTC 2010-Fall)*, 2010, pp. 1–5.
155. V. Chandrasekhar, J. G. Andrews, T. Muharemovic, Z. Shen, and A. Gatherer, "Power control in two-tier femtocell networks," *IEEE Transactions on Wireless Communications*, vol. 8, no. 8, pp. 4316–4328, 2009.
156. G. Cao, D. Yang, X. Ye, and X. Zhang, "A downlink joint power control and resource allocation scheme for co-channel macrocell-femtocell networks," in *IEEE Wireless Communications and Networking Conference (WCNC)*, 2011, pp. 281–286.
157. S.-M. Cheng, W. C. Ao, F.-M. Tseng, and K.-C. Chen, "Design and analysis of downlink spectrum sharing in two-tier cognitive femto networks," *IEEE Transactions on Vehicular Technology*, vol. 61, no. 5, pp. 2194–2207, 2012.
158. M. Mehta, O. G. Aliu, A. Karandikar, and M. A. Imran, "A self-organized resource allocation using inter-cell interference coordination (ICIC) in relay-assisted cellular networks," arXiv preprint arXiv:1105.1504, 2011.
159. Y.-S. Liang, W.-H. Chung, G.-K. Ni, Y. Chen, H. Zhang, and S.-Y. Kuo, "Resource allocation with interference avoidance in OFDMA femtocell networks," *IEEE Transactions on Vehicular Technology*, vol. 61, no. 5, pp. 2243–2255, 2012.

160. D. López-Pérez, A. Ladányi, A. Jüttner, and J. Zhang, "OFDMA femtocells: A self-organizing approach for frequency assignment," in *IEEE 20th International Symposium on Personal, Indoor and Mobile Radio Communications*, 2009, pp. 2202–2207.
161. F. Bernardo, R. Agustí, J. Cordero, and C. Crespo, "Self-optimization of spectrum assignment and transmission power in OFDMA femto-cells," in *IEEE 6th Advanced International Conference on Telecommunications (AICT)*, 2010, pp. 404–409.
162. A. Galindo-Serrano and L. Giupponi, "Distributed Q-learning for interference control in OFDMA-based femtocell networks," in *IEEE 71st Vehicular Technology Conference (VTC-Spring)*, 2010, pp. 1–5.
163. R. G. L. Narayanan, "Mobile video streaming resource management," in *Resource Management in Mobile Computing Environments*. Springer, 2014, pp. 461–480.
164. M. Salhi, S. Trabelsi, and N. Boudriga, "Mobility-assisted and QoS-aware resource allocation for video streaming over LTE femtocell networks," *ECTI Transactions on Electrical Engineering, Electronics, and Communications*, vol. 13, no. 1, pp. 42–53, 2015.
165. S. M. A. El-atty and Z. Gharsseldien, "Measuring QoS metrics in femto/macro cellular networks with CAC policy," *International Journal of Wireless Information Networks*, vol. 22, no. 3, pp. 240–251, 2015.
166. S. Khan, A. Ahmed, I. Ullah, and S. M. Zubair, "Call admission control based femtocell handover in LTE networks," in *IEEE International Conference on Computing, Electronic and Electrical Engineering (ICE Cube)*, 2016, pp. 196–201.
167. M. H. Ahmed, "Call admission control in wireless networks: A comprehensive survey." *IEEE Communications Surveys and Tutorials*, vol. 7, no. 1–4, pp. 50–69, 2005.
168. T. Sigwele, P. Pillai, A. S. Alam, and Y. F. Hu, "Fuzzy logic-based call admission control in 5G cloud radio access networks with preemption," *EURASIP Journal on Wireless Communications and Networking*, vol. 2017, no. 1, p. 157, 2017.
169. S.-Y. Kim, H.-Y. Lee, and S.-W. Ryu, "Analysis of call admission control for joint transmission-based LTE-advanced systems," *Journal of Korean Institute of Communications and Information Sciences*, vol. 38, no. 7, pp. 535–542, 2013.
170. M. Obaidat and N. Boudriga, "Modeling and simulation of ATM systems and networks," in *Applied System Simulation*. Springer, 2003, pp. 81–114.
171. Y. Zhang and E. Salari, "Modeling and analysis of a hybrid cac scheme in heterogeneous multimedia wireless networks," *International Journal of Handheld Computing Research (IJHCR)*, vol. 3, no. 1, pp. 23–36, 2012.
172. R. Singoria, T. Oliveira, and D. P. Agrawal, "Reducing unnecessary handovers: Call admission control mechanism between WiMAX and femtocells," in *IEEE Global Telecommunications Conference (GLOBE-COM)*, 2011, pp. 1–5.
173. T. F. Z. Badri, R. Saadane, M. Wahbi, and S. Mbarki, "Call admission control scheme for LTE femtocell-macrocell integrated system," in *IEEE International Conference on Multimedia Computing and Systems (ICMCS)*, 2014, pp. 1134–1139.
174. K. B. Ali, M. S. Obaidat, F. Zarai, and L. Kamoun, "Markov model-based adaptive CAC scheme for 3GPP LTE femtocell networks," in *IEEE International Conference on Communications (ICC)*, 2015, pp. 6924–6928.
175. P. Xia, V. Chandrasekhar, and J. G. Andrews, "Femtocell access control in the TDMA/OFDMA uplink," in *IEEE Global Telecommunications Conference (GLOBECOM)*, 2010, pp. 1–5.

176. C. Olariu, J. Fitzpatrick, P. Perry, and L. Murphy, "A QoS based call admission control and resource allocation mechanism for LTE femto-cell deployment," in *IEEE Consumer Communications and Networking Conference (CCNC)*, 2012, pp. 884–888.
177. M. Khan, "Fair admission control policies for femtocell operation," in *IEEE 4th International Conference on Ubiquitous and Future Networks (ICUFN)*, 2012, pp. 175–179.
178. S. Yi, G. Wang, and Y. Xia, "QoS-enabled dynamic resource management in multicell OFDMA-based systems," in *IEEE 73rd Vehicular Technology Conference (VTC Spring)*, 2011, pp. 1–5.
179. M. Ismail, W. Zhuang, E. Serpedin, and K. Qaraqe, "A survey on green mobile networking: From the perspectives of network operators and mobile users," *IEEE Communications Surveys & Tutorials*, vol. 17, no. 3, pp. 1535–1556, 2015.
180. O. Onireti, F. Héliot, and M. A. Imran, "On the energy efficiency-spectral efficiency trade-off in the uplink of CoMP system," *IEEE Transactions on Wireless Communications*, vol. 11, no. 2, pp. 556–561, 2012.
181. N. Gharbi, "Modeling and performance evaluation of small cell wireless networks with base station channels breakdowns," in *Proceedings of the Conference on Wireless and Mobile Communications*, 2012.
182. N. Johnson, S. Kotz, and N. Balakrishnan, *Continuous Univariate Distributions*. John Wiley & Sons, New York, 1995.
183. H. ElSawy, E. Hossain, and M. Haenggi, "Stochastic geometry for modeling, analysis, and design of multi-tier and cognitive cellular wireless networks: A survey," *IEEE Communications Surveys & Tutorials*, vol. 15, no. 3, pp. 996–1019, 2013.
184. Y. S. Soh, T. Q. Quek, M. Kountouris, and H. Shin, "Energy efficient heterogeneous cellular networks," *IEEE Journal on Selected Areas in Communications*, vol. 31, no. 5, pp. 840–850, 2013.
185. C. Li, A. Yongacoglu, and C. D'Amours, "Mixed spatial traffic modeling of heterogeneous cellular networks," in *IEEE International Conference on Ubiquitous Wireless Broadband (ICUWB)*, 2015, pp. 1–5.
186. E. Oh, B. Krishnamachari, X. Liu, and Z. Niu, "Toward dynamic energy-efficient operation of cellular network infrastructure," *IEEE Communications Magazine*, vol. 49, no. 6, 2011.
187. L. B. Le, D. Niyato, E. Hossain, D. I. Kim, and D. T. Hoang, "QoS-aware and energy-efficient resource management in OFDMA femtocells," *IEEE Transactions on Wireless Communications*, vol. 12, no. 1, pp. 180–194, 2013.
188. E. Ever, F. M. Al-Turjman, H. Zahmatkesh, and M. Riza, "Modelling green HetNets in dynamic ultra large-scale applications: A case-study for femtocells in smart-cities," *Computer Networks*, vol. 128, pp. 78–93, 2017.
189. W. Kumar, P. Kumar, and I. A. Halepoto, "Performance analysis of an energy efficient femtocell network using queuing theory," *Mehran University Research Journal of Engineering & Technology*, vol. 32, no. 3, pp. 535–542, 2013.
190. W. Kumar, S. Aamir, and S. Qadeer, "Performance analysis of a finite capacity femtocell network," *Mehran University Research Journal of Engineering & Technology*, vol. 33, no. 1, pp. 129–136, 2014.
191. I. U. Rehman, N. Y. Philip, and M. M. Nasralla, "A hybrid quality evaluation approach based on fuzzy inference system for medical video streaming over small cell technology," in *IEEE 18th International Conference on e-Health Networking, Applications and Services (Healthcom)*, 2016, pp. 1–6.

192. J. A. del Peral-Rosado, M. Bavaro, J. A. López-Salcedo, G. Seco-Granados, P. Chawdhry, J. Fortuny-Guasch, P. Crosta, F. Zanier, and M. Crisci, "Floor detection with indoor vertical positioning in LTE femtocell networks," in *IEEE Globecom Workshops (GC Wkshps)*, 2015, pp. 1–6.
193. A. Ulvan, M. Ulvan, and R. Bestak, "Integrated IMS-femtocell testbed," in *IEEE 36th International Conference on Telecommunications and Signal Processing (TSP)*, 2013, pp. 100–104.
194. P. Kulkarni, S. Gormus, W. H. Chin, and R. J. Haines, "Distributed resource allocation in small cellular networks—Test-bed experiments and results," in *IEEE 7th International Wireless Communications and Mobile Computing Conference (IWCMC)*, 2011, pp. 1262–1267.
195. I. U. Rehman and N. Y. Philip, "M-QoE driven context, content and network aware medical video streaming based on fuzzy logic system over 4G and beyond small cells," in *EUROCON IEEE International Conference on Computer as a Tool (EUROCON)*, 2015, pp. 1–6.
196. C. Park and H. S. Choi, "Optimization of downlink power control based on LTE," in *IEEE International Conference on ICT Convergence (ICTC)*, 2012, pp. 536–539.
197. A. Alexiou, C. Bouras, V. Kokkinos, K. Kontodimas, and A. Papazois, "Interference behavior of integrated femto and macrocell environments," in *Wireless Days (WD), IFIP*. IEEE, 2011, pp. 1–5.
198. I. Carson, D. M. Nicol, B. L. Nelson, J. Banks et al., *Discrete-Event System Simulation*. Prentice-Hall, Upper Saddle River: NJ, 2005.
199. A. M. Law, W. D. Kelton, and W. D. Kelton, Simulation Modeling and Analysis. McGraw-Hill, New York, vol. 2, 1991.
200. T. Omar, Z. Abichar, A. E. Kamal, J. M. Chang, and M. Alnuem, "Fault-tolerant small cells locations planning in 4G/5G heterogeneous wireless networks," *IEEE Transactions on Vehicular Technology*, vol. 66, no, 6, pp. 5269–5283, 2016.
201. S. He, Z. Lu, X. Wen, Z. Zhang, J. Zhao, and W. Jing, "A pricing power control scheme with statistical delay QoS provisioning in uplink of two-tier OFDMA femtocell networks," *Mobile Networks and Applications*, vol. 20, no. 4, pp. 413–423, 2015.
202. C. Bouras and G. Diles, "Resource management in 5G femtocell networks," in *IEEE 10th International Conference on Broadband and Wireless Computing, Communication and Applications (BWCCA)*, 2015, pp. 353–358.
203. R. Gonsalves, "Evaluation of outage probability in two-tier open access femtocell networks," in *IEEE International Conference on Circuits, Systems, Communication and Information Technology Applications (CSCITA)*, 2014, pp. 201–206.
204. Q. Zhang, Z. Feng, and W. Li, "Coverage self-optimization for randomly deployed femtocell networks," *Wireless Personal Communications*, vol. 82, no. 4, pp. 2481–2504, 2015.
205. Y. Lei and Y. Zhang, "Enhanced mobility state detection based mobility optimization for FEMTO cells in LTE and LTE-advanced networks," in *IET International Conference on Communication Technology and Application (ICCTA 2011)*, 2011, pp. 341–345.
206. Z. Ji, T. K. Sarkar, and B.-H. Li, "Methods for optimizing the location of base stations for indoor wireless communications," *IEEE Transactions on Antennas and Propagation*, vol. 50, no. 10, pp. 1481–1483, 2002.
207. L. Nagy, "Global optimization of indoor radio coverage," *AU-TOMATIKA: cˇasopis za automatiku, mjerenje, elektroniku, racˇunarstvo i komunikacije*, vol. 53, no. 1, pp. 69–79, 2012.

208. S. Rodd, A. Prof, and A. H. Kulkarni, "Optimization algorithms for access point deployment in wireless networks," *Journal of Computer Application*, vol. 2, no. 1, pp. 2–2, 2009.
209. A. Jain, K. Tawar, and A. Rathore, "Optimal placement of femtocell base station for indoor environment," in *IEEE International Conference on Advanced Computing and Communication Systems (ICACCS)*, 2013, pp. 1–5.
210. Y. Ngadiman, Y. H. Chew, and B. S. Yeo, "A new approach for finding optimal base stations configuration for CDMA systems jointly with uplink and downlink constraints," in *IEEE 16th International Symposium on Personal, Indoor and Mobile Radio Communications (PIMRC)*, vol. 4, 2005, pp. 2751–2755.
211. M. Talau, E. C. Wille, and H. S. Lopes, "Solving the base station placement problem by means of swarm intelligence," in *IEEE Symposium on Computational Intelligence for Communication Systems and Networks (CIComms)*, 2013, pp. 39–44.
212. L. Pujji, K. Sowerby, and M. Neve, "Development of a hybrid algorithm for efficient optimisation of base station placement for indoor wireless communication systems," *Wireless Personal Communications*, vol. 69, no. 1, pp. 471–486, 2013.
213. L. K. Pujji, K. W. Sowerby, and M. J. Neve, "A new algorithm for efficient optimisation of base station placement in indoor wireless communication systems," in *IEEE 7th Annual Communication Networks and Services Research Conference (CNSR)*, 2009, pp. 425–427.
214. S. Wang, W. Guo, and T. O'Farrell, "Optimising femtocell placement in an interference limited network: Theory and simulation," in *IEEE Vehicular Technology Conference (VTC Fall)*, 2012, pp. 1–6.
215. D. Fagen, P. A. Vicharelli, and J. Weitzen, "Automated wireless coverage optimization with controlled overlap," *IEEE Transactions on Vehicular Technology*, vol. 57, no. 4, pp. 2395–2403, 2008.
216. I. Ashraf, H. Claussen, and L. T. Ho, "Distributed radio coverage optimization in enterprise femtocell networks," in *IEEE International Conference on Communications (ICC)*, 2010, pp. 1–6.
217. J. Torregoza, R. Enkhbat, and W.-J. Hwang, "Joint power control, base station assignment, and channel assignment in cognitive femtocell networks," *EURASIP Journal on Wireless Communications and Networking*, vol. 2010, no. 1, p. 285714, 2010.
218. J. M. R. Avilés, M. Toril, and S. Luna-Ramírez, "A femtocell location strategy for improving adaptive traffic sharing in heterogeneous LTE networks," *EURASIP Journal on Wireless Communications and Networking*, vol. 2015, no. 1, p. 38, 2015.
219. E. Emelianova, S. Park, and S. Bahk, "Deployment algorithm for femtocells in multi-tiered wireless cellular network," in *IEEE International Conference on ICT Convergence (ICTC)*, 2012, pp. 131–136.
220. A. U. Ahmed, M. T. Islam, M. Ismail, and M. Ghanbarisabagh, "Dynamic resource allocation in hybrid access femtocell network," *Scientific World Journal*, vol. 2014, Article ID 539720, 7 pages, 2014. doi:10.1155/2014/539720
221. M. H. Qutqut, H. Abou-zeid, H. S. Hassanein, A. M. Rashwan, and F. M. Al-Turjman, "Dynamic small cell placement strategies for LTE heterogeneous networks," in *IEEE Symposium on Computers and Communication (ISCC)*, 2014, pp. 1–6.
222. S. Oh, H. Kim, B. Ryu, and N. Park, "Inbound mobility management on LTE-advanced femtocell topology using X2 interface," in *Proceedings of 20th IEEE International Conference on Computer Communications and Networks (ICCCN)*, 2011, pp. 1–5.

223. S. Jang, Y. Lee, J. Lim, and D. Hong, "Self-optimization of single femto-cell coverage using handover events in LTE systems," in *IEEE 17th Asia-Pacific Conference on Communications (APCC)*, 2011, pp. 28–32.
224. H. Claussen, L. T. Ho, and L. G. Samuel, "Self-optimization of coverage for femtocell deployments," in *IEEE Wireless Telecommunications Symposium*, 2008, pp. 278–285.
225. R. K. Saha and C. Aswakul, "A tractable analytical model for interference characterization and minimum distance enforcement to reuse resources in three-dimensional in-building dense small cell networks," *International Journal of Communication Systems*, vol. 30, no. 11, 2017.
226. T. D. Novlan, R. K. Ganti, and J. G. Andrews, "Coverage in two-tier cellular networks with fractional frequency reuse," in *IEEE Global Telecommunications Conference (GLOBECOM)*, 2011, pp. 1–5.
227. M. Herlich and H. Karl, "Energy-efficient assignment of user equipment to cooperative base stations," in *Proceedings of the Tenth International Symposium on Wireless Communication Systems (ISWCS)*. VDE, 2013, pp. 1–5.
228. "Siemens solutions for the e-mobility infrastructure management," https://www.siemens.com, 2016.
229. H. Zhang, C. Jiang, N. C. Beaulieu, X. Chu, X. Wen, and M. Tao, "Resource allocation in spectrum-sharing OFDMA femtocells with heterogeneous services," *IEEE Transactions on Communications*, vol. 62, no. 7, pp. 2366–2377, 2014.
230. L. Wang, Y. Zhang, and Z. Wei, "Mobility management schemes at radio network layer for LTE femtocells," in *IEEE 69th Vehicular Technology Conference (VTC Spring)*, 2009, pp. 1–5.
231. H. Klessig, M. Gunzel, and G. Fettweis, "Increasing the capacity of large-scale HetNets through centralized dynamic data offloading," in *IEEE 80th Vehicular Technology Conference (VTC Fall)*, 2014, pp. 1–7.
232. R. Trestian, "User-centric power-friendly quality-based network selection strategy for heterogeneous wireless environments," PhD dissertation, Dublin City University, 2012.
233. D. Grace, J. Chen, T. Jiang, and P. D. Mitchell, "Using cognitive radio to deliver green communications," in *IEEE 4th International Conference on Cognitive Radio Oriented Wireless Networks and Com- munications (CROWNCOM)*, 2009, pp. 1–6.
234. A. He, S. Srikanteswara, K. K. Bae, T. R. Newman, J. H. Reed, W. H. Tranter, M. Sajadieh, and M. Verhelst, "System power consumption minimization for multichannel communications using cognitive radio," in *IEEE International Conference on Microwaves, Communications, Antennas and Electronics Systems (COMCAS)*, 2009, pp. 1–5.
235. A. He, S. Srikanteswara, J. H. Reed, X. Chen, W. H. Tranter, K. K. Bae, and M. Sajadieh, "Minimizing energy consumption using cognitive radio," in *IEEE International Performance, Computing and Communications Conference (IPCCC)*, 2008, pp. 372–377.
236. C.-K. Han, H.-K. Choi, and I.-H. Kim, "Building femtocell more secure with improved proxy signature," in *IEEE Global Telecommunications Conference (GLOBECOM)*, 2009, pp. 1–6.
237. N. Saquib, E. Hossain, L. B. Le, and D. I. Kim, "Interference management in OFDMA femtocell networks: Issues and approaches," *IEEE Wireless Communications*, vol. 19, no. 3, 2012.

238. M. Feng, D. Chen, Z. Wang, T. Jiang, and D. Qu, "An improved spectrum management scheme for OFDMA femtocell networks," in *IEEE 1st International Conference on Communications in China (ICCC)*, 2012, pp. 132–136.
239. S. Huan, K. Linling, and L. Jianhua, "Interference avoidance in OFDMA-based femtocell network," in *IEEE Youth Conference on Information, Computing and Telecommunication (YC-ICT)*, 2009, pp. 126–129.
240. O. B. Maia, H. C. Yehia, and L. de Errico, "A concise review of the quality of experience assessment for video streaming," *Computer Communications*, vol. 57, pp. 1–12, 2015.

Chapter 3

Energy Efficiency Perspectives of Femtocells in IoT*

Fadi Al-Turjman
Antalya Bilim University

Muhammad Imran
King Saud University

Sheikh Tahir Bakhsh
King Abdulaziz University

Contents

* Fadi Al-Turjman, Muhammad Imran and Sheikh Tahir Bakhsh, "Energy Efficiency Perspectives of Femtocells in Internet of Things: Recent advances and challenges", *IEEE Access Journal*, 2017. doi: 10.1109/ACCESS.2017.2773834.

3.1 Introduction

The Internet of Things (IoT) is a novel paradigm where objects become part of the Internet. It paves the way for connecting actuators, sensors, mobile phones, Radio Frequency Identification (RFID) tags, and other objects to the Internet and these objects are uniquely defined; its status and position were known, accessible to the network, permitting the perception of the world [1,2]. Moreover, IoT promotes many new applications in various domains such as healthcare, environmental monitoring, and automotive and energy management in smart homes where it provides economic benefits [3]. Indeed, IoT is considered as "one of the six disruptive civil technologies with potential impact on US national power" by the US National Intelligence Council [4].

Figure 3.1 depicts some beneficiaries of a smart grid, which is one of the main IoT applications and one of the leading technological advancement today; it can monitor the energy supply and demand and efficiently adjusts power consumption in the network [5]. It can be used with smart meters; it provides real-time information to suppliers and consumers. Smart houses can communicate with a grid using these smart meters and enables consumers to manage their electricity usage [6]. In a

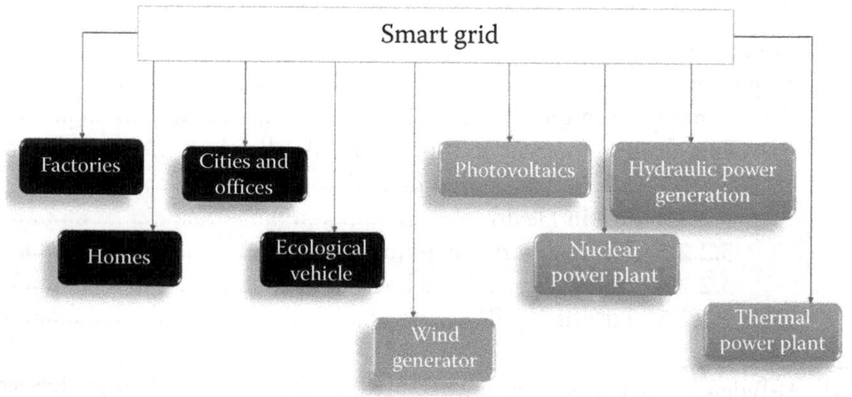

FIGURE 3.1 Some beneficiaries of a smart grid.

smart home, the consumer can access and operate appliances through energy management, using the network between appliances. In order to manage energy as mentioned, there is a need for a communication network in this layer such as "Home Area Network (HAN)" and "Neighborhood Area Network (NAN)." HANs have three basic components; it analyzes, measures, and collects energy usage of smart devices. On the contrary, multiple HANs are connected with NAN through local access points where data are carried to the utility. An infrastructure is needed for HANs to connect these elements. Femtocell can be used as a HAN communication mechanism like wireless LAN and Zigbee [7]. In addition, in Ref. [8], femtocell is used as an indoor or an outdoor network in smart grid wireless communications network, which is based on Radio-Over-Fiber (ROF) and cooperative relaying.

Energy consumption and carbon emission are some of the key challenges that need to be addressed; mobile networks are used in smart grids in order to provide energy efficiency. However, it is a fact that they consume energy and its amount cannot be regarded as too little. Moreover, carbon emission is the result of energy usage and as mentioned in Ref. [9], Information and Communication Technologies (ICT) are responsible for between 2% and 2.5% of total carbon emission, which is almost equivalent to global aviation industry [10]. It is expected that this rate will be doubled over the next decade. Mobile networks produce the 10% of ICT carbon emission and energy usage of Radio Access Network (RAN) is about 55%–60%. To illustrate, each macro base station (BS) use between 2.5 and 4 kW [8]. The number of these stations in each nation is about tens of thousands, which paves way for a lot of energy consumption and carbon emission.

Femtocell can be defined as a small cell, looking at its energy efficiency, which consumes 8–120 mW of energy; it is analyzed that it consumes less energy compared to a Wireless Fidelity (Wi-Fi) access point [11]. Therefore, it is an efficient communication solution in HANs from the consumer's perspective. They can track their energy usage and this can enable consumers to manage their electricity bills. Moreover, with the help of HAN, appliances can be automated and reduce electricity demand on the grid and also consumer energy bill [6].

Briefly comparing femtocell and macrocell, macrocell BSs work with high-energy consumption and most of this energy is used by driving high power radio frequency signal. However, it is estimated that only between 5% and 10% of this energy is used to create a useful signal. The reason behind that issue is that macrocells provide high area coverage and most of the space is unoccupied (i.e., without people that need radiated signals in these areas). Moreover, RF sections of BSs need high power, without efficiency. In other words, energy is used for just in case. Unlike macrocells, a femtocell is a solution to prevent these energy losses and it only provides local area coverage [12]. In other words, femtocell delivers power where there is a need and hence it requires less RF power to proffer high bandwidth since they are closer to the user. An average femtocell consumes 2 W while it is very high in the macrocell. The other advantage is that user equipment consumes less energy when it is connected to a femtocell since it is closer [13]. Thus, a femtocell is an efficient solution, which provides large coverage, longer battery, and better Quality of Service (QoS).

Accordingly, the main contributions in this article can be summarized as follows. A comprehensive background about the femtocells and their IoT-specific applications are outlined while emphasizing energy consumption and system capacity effects. Energy metrics and models for IoT-based femtocells have been investigated as well, and accurate ones are recommended. Furthermore, the femtocell base-station performance is inspected while discussing different deployment strategies in the literature. Key parameters that affect the femtocell power consumption under IoT setups have been reported. Moreover, femtocells' challenges and open research issues in the IoT era have been highlighted to open the door for further improvements in next generation networks such as the 5G cellular network.

This chapter is organized as follows. Section 3.2 provides a preliminary background of a femtocell and its role in IoT applications. Energy efficiency metrics are introduced in Section 3.3 and Section 3.4 includes how energy can be modeled. Section 3.5 gives some about energy efficiency schemes for deployment including the criteria and examples in the literature. In Section 3.6, we mentioned the factors that affect the energy consumption of femtocell base station (FBS), and in Section 3.7, we discuss some challenges facing small cell networks. In Section 3.8, we mentioned energy efficiency of user equipment when it is used under femtocell coverage. Open research issues are mentioned in Section 3.9 and concluding remarks are given in Section 3.10.

3.2 Background

3.2.1 What Is a Femtocell?

Femtocell connects mobile devices to the network through different wired and wireless technologies [14]. It is a low-cost Macrocellular Base Station (MBS), which provides radio access interface to a User Equipment (UE) [5]. It is a solution to offload from overloaded macrocells and increase the coverage area. They are specifically designed and used for increasing indoor coverage (e.g., home or small business), where there is a lack of cellular network or improving the QoS is desirable. It has advantages for both cellular operators and users. For the cellular operator, the advantage is to increase coverage and capacity. The coverage area is widened because the loss of signal is eliminated through buildings and capacity is increased by a reduction in the total number of UEs that uses the main cellular network. They use the Internet instead of using a cellular operator network. For customers, they have better service, improved coverage, and signal strength since they are closer to the BS. Moreover, using femtocells leads to increase UE battery life because of being close to femtocell [14].

A typical femtocell structure consists of five parts: Digital Subscriber Line (DSL) or cable router, femtocell device, cellular tower (macrocell), mobile operator network, and Internet Service Provider (ISP) Internet link [14] as shown in Figure 3.2. Femtocell does not require a cellular core network since it contains Radio Network Controller (RNC) and all other network elements. It acts as a Wi-Fi access point and it needs a data connection to the DSL or Internet connected to cellular operator core network [14]. Although it does not need to be under macrocell coverage area, there are many examples for deployment in macrocell coverage area in order to increase capacity and QoS where there is huge user demand, or there is a less coverage in buildings. Apart from this, it can be used in rural areas in order to provide cellular coverage where there is no macro coverage.

Although femtocell technology was first designed to use indoor, however, there are many outdoor applications of this technology as well. To illustrate, it can be deployed in transit systems such as bus and train. In this application, mobile users connect to femtocell instead of macrocells or satellite. There is a transceiver connected to Femtocell Access Point (FAP) through wired connection and to macrocell or satellite through wireless link [15]. Moreover, femtocells can be a good solution to increase coverage and capacity in public outdoor areas, especially in crowded areas.

The main idea behind femtocell is to bring cellular network closer to the user and it is a low-power and low-cost technology [15]. It is usually difficult for a macrocell to provide indoor service since there is a signal loss. The author in Ref. [16] says that macrocell is not efficient in delivering data indoors due to high penetration losses. Moreover, 50% of voice calls and 70% of data calls comes from indoor [15]. There is an estimation that 10% of active femtocell household deployment can offload 50% of the overall macrocellular tier load [17]. Thus, it increases revenue of cellular operators, and it is expected that there will be about 28 million units of femtocell by 2017 [18].

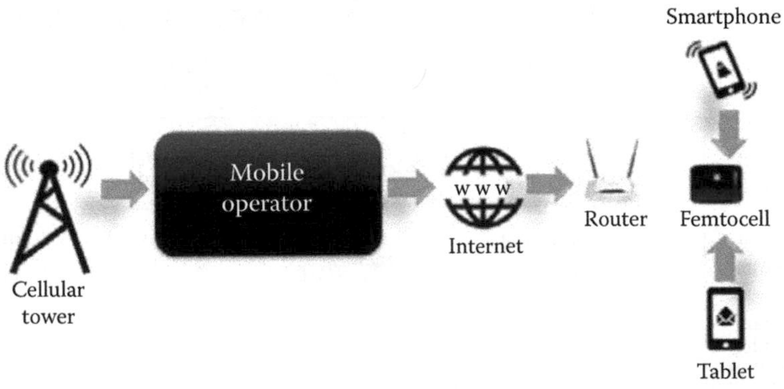

FIGURE 3.2 A typical femtocell network structure.

3.2.2 Femtocell in Internet of Things (IoT) Applications

IoT touches every facet of our lives, and it has potential to cover a wide range of applications (e.g., transportation and logistics, healthcare, smart environment, personal and social, and futuristic) that can positively influence the quality of life at different places (e.g., home, travelling, sick, and at work). Femtocell is preferential for operator networks as well as for industrial wireless sensor networks [1,19]. It describes communication capabilities of various objects with each other and elaborates information perceived. Most of these applications can be categorized and classified as shown in Figure 3.3.

Femtocell can be used as a communication mechanism of IoT, especially in the smart grid. For example, the authors in Ref. [7] introduced a femtocell-based communication mechanism in HAN and argued the security issues concerning femtocell utilization in the smart grid. Moreover, using femtocell in HAN as a cost-effective technology was also proposed in Refs. [20,21]. Healthcare IoT applications are another example for using a femtocell in IoT applications. In Ref. [22], they proposed an IoT-oriented healthcare monitoring system where sensors gather the data from the android application and then used LTE-based femtocell network in order to send the data with new scheduling technique.

In Ref. [23], they underline that femtocell is susceptible to man-in-the-middle attacks when it is used to fix shadow area problem. Moreover, they propose the interlock protocol to protect the confidential information. In Ref. [24], the benefits of using 56 femtocells for supporting indoor generated IoT traffic mentioned, and the fact that supporting the traffic produced from IoT is a major challenge for 5G and an important percentage of the traffic is generated indoor were also underlined. Thus, it can be said that femtocells can be used in most of the IoT applications

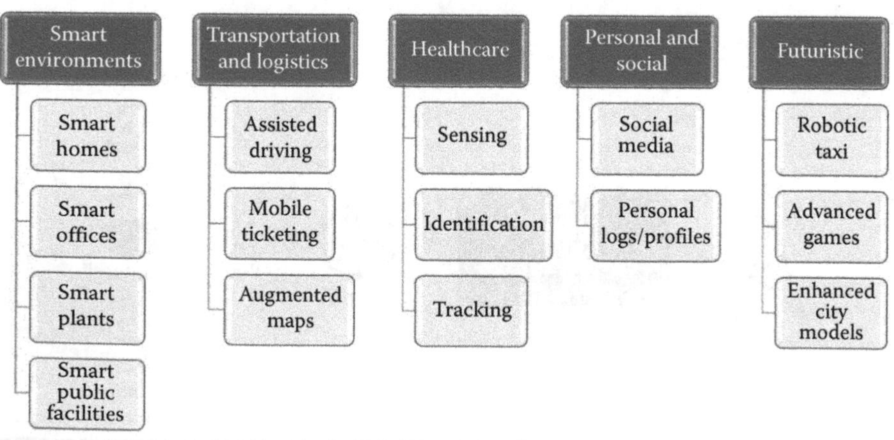

FIGURE 3.3 Application domains of IoT.

since there is a need for evolutionary telecommunication mechanisms. Indeed, it is expected that small cells such as femtocell will be central to 5G network architectures, both for human users and for IoT. Moreover, authors in Ref. [25] talk about a new method called network slicing for boosting up the performance of 5G so that it will be able to handle the high demand for information exchange with minimum energy.

3.2.2.1 IoT in Transportation and Logistics

As shown in Figure 3.3, there are many IoT applications in the field of transportation and logistics. Some of the listed areas include logistics, mobile ticketing, driving, monitoring environment, and improved maps. The authors in Ref. [16] concur with this statement by saying that IoT plays an important role in transport and logistics. They go on to point out some of the main areas where its application has made major strides, such as the ability of transportation and logistics companies to track their goods from origin to destination and have a real-time location of their property, using the bar codes and sensors planted on the goods. The recent technological advancement in vehicles increasing sensing, communication, and data processing capabilities, have opened up IoT to a new range of possibilities, where we can now track the exact location of the vehicle, track its path, and predict its next location. Moreover, the authors highlight an intelligent system called iDrive system, which monitors the conditions to enable better driving conditions. Additionally, the authors in Ref. [26] talks about the advancement of vehicular networks, which are able to make decision on their own and has autonomous control to cloud-assisted context-aware Vehicular Cyberphysical Systems (CVCs).

3.2.2.2 IoT in Healthcare

Figure 3.3 also shows some of the areas where IoT has been applied in healthcare. Some of these areas include sensing, tracking data collection identification and authentication. The authors in Ref. [16] say that IoT uncovers new opportunities in healthcare. The ability of IoT to sense, identify, and communicate has enabled the healthcare department to track and monitor all its objects such as people, equipment, and medicine among others. Moreover, due to the immense global connectivity of IoT, health care-related information such as logistics, therapy, diagnosis, medication, and the like can be managed, collected, and shared easily. The authors also note that by using personal computers and mobile phones, the healthcare system can be personalized. Moreover, the author in Ref. [27] talked about a synthesis of Wireless Body Area Networks (WBANs), which are largely adopted in the passive healthcare data collection, has limited storage capacity among a few other challenges, with Mobile Cloud Computing (MCC) that provides flexibility in massive computing and large storage spaces to allow for storage of data in the healthcare department.

3.2.2.3 IoT in Smart Environment

Figure 3.3 also features some of the areas in a smart environment where IoT is applied; these areas include relaxed homes, offices, and commercial places. The authors in Ref. [28] proposed a flexible low-cost home controlling and monitoring system where they use an android-based smartphone app to remotely control and monitor appliances in a smart home. The authors in Ref. [29] proposed a system for a smart factory aimed at improving safety in plants using an IoT-based WSN and RFID integrated solution. Moreover, the authors in Ref. [30] noted that wireless sensor network plays a major role in industrial monitoring and control; therefore, they proposed a new algorithm to assist the conventional orthogonal frequency division multiple access to maximize the subcarrier pairing, subcarrier allocation, and power allocation.

3.2.2.4 IoT in Personal and Social

Personal and social areas are some of the other fields where IoT has been employed and has made some improvements in these areas. Under this section, IoT has influenced fields such as social networking where the vast connection of smart devices has enabled users to connect and interact without worrying about the distance between them. Historical query is another field where anyone can retrieve or store information about anything from anywhere without distressing about losing that information. Moreover, IoT has made an impact in the security sector where the installation of smart security systems is used to guard property and prevent theft.

3.2.2.5 Futuristic IoT

The future of IoT is one that is very exciting because of the countless possibilities that it holds. Figure 3.3 highlights only a few of these possibilities i.e., robot taxi, city information model, and enhanced game rooms. In their conclusion, the authors in Ref. [31] talked about the web squared, which they say, is an evolution of the Web 2.0. In their chapter, they argued that this future model of the internet would help integrate the Web and sensing technologies by taking into consideration the information about the environment of the user, collected by sensors such as cameras and microphones and using the information to better the content provided to the user.

3.3 Energy Efficiency Metrics

There are three different Energy-Efficient (EE) metrics proposed at three different levels in literature: network, component, and a Base Station (BS). In the component level, millions of instructions per second can be calculated. Of course, the speed of communication has always been a critical issue that needs to be optimized at all cost, the author in Ref. [32] propose a handover system based on cell ID information which is able to effectively operate in a fast-moving environment such as Long-Term Evolution (LTE)—Advanced. Processor related energy consumption, whose formula is given in Eq. (3.1), and the Ratio of Output to Input (ROI) power amplifier can be used to calculate energy efficiency of power amplifier (PA) component. In BS level, EE metrics can be evaluated under two main categories. Bits per second per hertz per watt represent a trade-off between energy consumption and spectral efficiency (SE). SE and the transmission range of the BS are taken into consideration. In network level, obtained service relative to the consumed energy is evaluated by EE metrics which is power per area unit (watts per square meter) in order to evaluate the coverage energy efficiency. Table 3.1 gives the summary of the EE metrics.

$$P_{cpu} = P_{dyn} + P_{sc} + P_{leak}, \tag{3.1}$$

where P_{cpu} represents the CPU power consumption, P_{dyn} is the dynamic power consumption that varies based on the environment conditions, P_{sc} is the short-circuit power consumption, and P_{leak} denotes the power loss due to transistor current leakage.

Since femtocells structure is similar to macrocell, component level EE metrics are also suitable for a femtocell. However, for BS and network level, service difference between femtocell-supported and macrocell provided services and the

TABLE 3.1 Energy Efficiency Metrics

Energy Efficiency Metrics	*Levels*	*Descriptions*
ROI	Component level	Indicate energy savings of power amplifiers
MIPS/W or MFLOPS/W	Component level	Guages energy consumed during processing
bits/s/Hz/W	BS level	Represent trade-off between energy dissipation and spectral efficiency (SE)
(b*m)/s/Hz/W	BS level	Account for energy consumption, SE, and the transmission range of base stations
W/m^2	Network level	Assess energy efficiency with respect to coverage

interference between both of them should also be considered. In Ref. [33], an interference-aware pricing-based metric had been discussed, and energy factors had the dominant effect; they proposed an EE metric for femtocell and macrocell heterogeneous network that considers the service rate and power consumption in both femto-BS and macro-BS. In Ref. [17], uplink power control allocation was investigated through the circuit and transmit power to reduce energy consumption. The author in Ref. [34] states that there is an exponential increase in wireless data traffic due to the massive increase in wireless terminal equipment and wide usage of bandwidth-hungry applications of mobile Internet. Therefore, the conventional method of Macrocell Base Stations (MBS) deployment is no longer effective. It has low quality but preferred in 5G [35]. Smaller cells are used to curb this problem because, as noted in Ref. [34], smaller cells have low-power, low-cost, and small access points (e.g., microcell, picocell, and femtocell).

3.4 How Can Energy Be Modeled?

Generally, wireless network energy consumption is evaluated at two different levels (Table 3.2). The first one, embodied energy consumption of a femtocell calculates the total primary energy that is consumed for making the product. It is assumed as 162 MJ, same as a mobile terminal, and the average lifetime of a femtocell can also be assumed as 5 years. In other words, embodied the energy of femtocell per second is calculated as 1 W [36]. The second level is operational energy consumption, which is consumed during a system's lifetime and it changes depending on different configurations such as the age of the facility and load of the femtocell. The average operational energy consumption of a femtocell was considered 6 W in Ref. [36]. Femtocell power consumption mainly depends on radio frequency power amplifier and the power amplifier of the power supply.

In the literature, there are few models about power consumption of femtocells. On–off model is the most basic one, which can be used for theoretical analysis where FBS consumes unit and zero power in active and off mode, respectively. In reality, power consumption is different. In Ref. [37], they proposed a linear power

TABLE 3.2 Energy Consumption of a Femtocell BS at Different Levels

Energy Consumption Level	*Definition*	*Average Power*
Embodied energy	The total primary energy that is consumed for making the product.	162 MJ (1 W/s)
Operational energy	The aggregate energy consumed in a system's lifespan	6 W/s

model that considers traffic load, which is better than on–off model since traffic load is defined. In Ref. [38], they proposed a more detailed model, where the authors made an argument that the impact of traffic load on power consumption is insignificant so it can be omitted.

In Ref. [39], they proposed a simple analytic model to predict the FBS power consumption based on offered load and datagram size. The authors tried to fit the model to real experiment; the power consumption of a FBS is measured in idle and varied load. In the experiment, one femtocell is used, which supported up to four simultaneous end-user devices and it is connected to the campus network. The energy consumption of voice and File Transfer Protocol (FTP) was also predicted. In the experiment, radio energy is neglected. Since in a voice call, the downlink is active for a small period, while it remains most of the time during FTP download. It was expected that radio power consumption should be higher than voice calls.

In Ref. [40], the authors proposed power consumption model, it is based on a femtocell that consists of three interacting blocks: a microprocessor, Field Programmable Gate Array (FPGA), radio frequency transmitter, and power amplifier. They used energy efficiency, which is defined as power consumption needed to cover a certain area, in order to compare different technologies. They used ITU-R P.1238 propagation model with an indoor setup. They considered frequency, the floor penetration loss factor, the number of floors between the BS and terminal, and the distance power loss coefficient to calculate the range. Based on this model, they compared the energy consumption for different bit rates and different technologies. Moreover, they used this model in a deployment tool that allows designing energy efficient femtocell networks using a genetic algorithm. Table 3.3 summarizes the mentioned energy consumption models.

3.5 Energy Efficiency Schemes for Deployment

Femtocells can be deployed in a heterogeneous network with different combinations of macrocells, microcells, picocells, and femtocells and reduction in total power consumption can be obtained from these schemes. The amount of energy efficiency depends on different variables. However, how to create the combination is also a challenging issue, and there are some criteria to choose which cell to deploy. Although energy efficiency is not the first goal in some of the combinations modeled, energy efficiency is achieved. In general, cell deployment takes into account density of mobile users, traffic volume, and coverage in the network models. It is also important to provide same or better QoS when energy efficiency is subjected. Although femtocells have potential to improve coverage, it degrades the performance due to cross-tier interference. The network criteria mainly depend on the mobile user density, traffic, and coverage, in most of the previous studies. Mobile user density and traffic are considered to choose the correct cell size. Urban and rural areas are also important criteria in order to choose the cell. In the areas that

TABLE 3.3 Comparison of Different Energy Consumption Models

References	*Model Type*	*Input*
[37]	Practical	Static power consumption (idle mode power consumption of power amplifier, base band transceiver units, feeder network and cooling system) and dynamic power consumption (load on the base station and backhaul power consumption of base station)
[38]	Analytical (Log-normal distribution)	Radio frequency output power at maximum load, minimum load and in sleep mode and the dependency of the required input power on the traffic load
[39]	Analytical	Datagram size (byte), offered load (Mbps), baseline power consumption when the femtocell is idle (Watts)
[40]	Practical	Power consumption (Watts) of the microprocessor, the FPGA and the power amplifier (input power of antenna and the efficiency of the power amplifier which is the ratio of radio frequency output power to electrical input power)

have very low signal can be deployed with smaller cells, such as femtocell so that coverage can be increased and there will be a reduction in total power consumption.

There are different models in the literature to analyse energy efficiency using heterogeneous network that includes femtocell and macrocell. The authors [24] considered a different size of cells depending on mobile user density; the authors try to reduce power consumption without effacing coverage and QoS. An analytical model is developed for five different schemes; and it is concluded that it consumes less power compared to macrocell network. In the first schema, the femtocell-based network is used instead of macrocells and it is observed that power consumption is reduced by 82.72% and 88.37%. In the second schema, the area is divided into three parts as urban, suburban, and rural. The mobile user density, data traffic and required coverage are considered. The scheme covers urban areas by femtocells, suburban areas with macrocells, and rural areas with portable femtocells. Because of this simulation, they succeed between 78.53% and 80.19% reduction in power consumption. In the third schema, the femtocells are allocated to the congested urban area, picocells to less congested urban areas, the power consumption is reduced between 9.19% and 9.79% [35,41]. In the fourth schema, they allocated microcells, picocells, and femtocells to the border region and macrocell to remaining region.

The reduction in power consumption is between 5.52% and 5.98% for this combination. In the last schema, femtocells are allocated between the boundaries of the macrocell, where the signal is not enough for a call. The result of the last model was 1.94%–2.66% reduction in power consumption and shrunk macrocell coverage.

Bell Labs recommended a hybrid network, both femtocell and macrocell, with open access femtocell, where all subscribers can connect to these femtocells like any BS [8]. The implementation area was 10 × 10 km urban area of New Zealand, and the population was about 200,000 people that means 65,000 homes and the 95% of the population uses the mobile equipment. The authors deployed a different number of femtocells that can serve up to eight users in 100 × 100 m^2 with 15 W energy consumption. It consumes higher energy due to an open access model. In this technique, macrocells are permanently used with 2.7 kW energy and femtocells are randomly deployed. Energy consumption depends on the use of the network: voice call and data connections. It is analyzed femtocells can reduce energy consumption up to 60 for data connections, while it has no impact for a voice call. The authors used femtocell to offload capacity and macrocells to enhance coverage.

Another one was from Ofcom (the UK telecoms company) and Plextek as a consultant [42,43]. They analyzed two approaches: first, the femtocells are deployed to 8 million households, which are almost 25% of the UK population. It is analyzed that each femtocell daily consumed 7 W and annual energy consumption was 490 GWh. In the second approach, they modeled macrocell network in order to provide indoor coverage, same with the first approach and they would need 30,000 BSs in order to provide coverage. It was concluded that macrocells take 40 times more energy compared to femtocell for indoor signals. To provide same coverage, macrocell annual energy consumption was 700 GWh for each operator. For indoor coverage in the UK, energy consumption ratio was 7:1 over using macrocells.

3.6 What Affects the Power Consumption of a Femtocell BS?

Access type for the femtocell plays a key role in the amount of a femtocell energy consumption. Femtocell access typesets the rules about who can connect to a FBS. It can be categorized in three types: open, closed, and hybrid [44]. In the literature, most of the studies about access type examine interference, QoS, and handover issues. On the other hand, although there is no work directly compares the energy efficiency related the access type, energy consumption can be compared with examining different models in the literature.

In open access, sources are shared among the users and all users can connect the network in public places without any restriction. On the other hand, closed access, only authorized users to connect the network. There are different service levels among users, and they are mainly used in small buildings. A Hybrid access is a combination of both the techniques where outside users can access a femtocell. However, the outside user group must register with the system, and the system provides limited services depending on the management policies. The comparison of the access types is given in Table 3.4.

Although there is no work that compares the power consumption based on different access mode, it can result as open access consumes more energy. In Ref. [8], the power consumption of a femtocell is assumed as 15 W although it is assumed as 6 W in Ref. [36]. The reason for the high power consumption is open access. Bit rate is another important factor that affects the energy consumption of a femtocell. It can be defined as a number of bits that are conveyed or processed per unit of time. As it reaches higher rates, the speed of connection becomes better since it is speed-based measurement. Cellular operators try to increase the bitrates since it is important in the market. On the other hand, bit rate has an impact on energy consumption. In general, as the bit rate increases, the energy consumption also increases. Moreover, energy consumption for different bit rates is not same on different wireless technologies.

As mentioned, energy consumption of femtocell is not stable for different wireless technology standards such as Worldwide Interoperability for Microwave Access (WiMAX), High Speed Packet Access (HSPA), and LTE, which are rivals in the sector. WiMAX can be used for transferring data across an Internet service provider (ISP) network, as a fixed wireless broadband Internet access, replacing satellite Internet service or as a mobile Internet access [45]. HSPA is another wireless technology standard, which is enhanced version of 3G [46]. LTE is considered as 4G and provides better capacity and speed [47]. They provide different bitrates and energy consumption of a femtocell is different for these technology standards. On the other hand, it is hard to say which one provides better energy efficiency since it changes with different bitrate ranges.

TABLE 3.4 Comparative Performance Analysis of Access Types

	Open	*Closed*	*Hybrid*
Deployment	Public spaces	Residential area	Enterprise
High user densities	No	Yes	Yes
Owner preference	No	Yes	Yes
Number of handovers	High	Small	Medium
QoS	Low	High	High
Interference between femtocell and macrocell	Increase	Decrease	Decrease

The authors [40] compared the energy efficiency of a FBS with different wireless technologies, i.e., WiMAX, LTE, and HSPA. Based on this model, it is found that femtocell consumes nearly 10 W for a range between 9 and 130 m, WiMAX is the most energy efficient technology for bit rates more than 5 Mbps. LTE consumed the least energy for bit rates between 2.8 and 5 Mbps. They used this model in a deployment tool that allows designing energy efficient femtocell networks by using a genetic algorithm and they concluded WiMAX as the most energy efficient one in this scenario.

Network type also affects the energy consumption of femtocell. To illustrate, its power consumption is different for voice call and data transfer. Moreover, there are different data transfer protocols such as FTP and User Datagram Protocol (UDP). Datagram size and offered load have also impacts on the consumption. In Ref. [39], they studied the effects of network type, datagram size, and offered load. The result of the analysis conducted by Bell Labs [8] indicates that the network efficiency is significantly dependent on the type of data connection (voice and/or data). When femtocells mainly used for a voice call, there was no big saving on total energy consumption. On the other hand, when femtocells used for data connections, femtocells were able to reduce total energy consumption up to 60%.

The last factor that has an impact on femtocell power consumption is sleep mode. Since femtocell provides very small coverage compared to the macrocell and very few users are connected to it that makes it idle oftentimes, particularly in the case of indoor deployment. However, it consumes energy even if it serves no user. In this case, it is better to switch off the BS so this can be implemented with sleep mode. Sleep mode is energy efficient and feasible, but the decision is also important [48]. There are a few works in the literature that analyze the impact of sleep mode. In Ref. [40], they examined the sleep mode to reduce power consumption, and it reduces the power consumption supporting up to eight users and it led to 24% of power consumption in the network. Table 3.5 gives the summary of the section.

TABLE 3.5 Factors That Affect Power Consumption of a Femtocell BS

Factor	*Energy Consumption*
Access mode	Higher in open access, less in close access
Bit rate	Increases with higher bit rate
Wireless technology standards	Depends on bit rate range
Network type	Higher in data transfer
Sleep mode	Less energy consumed if sleep mode is an option

3.7 User Equipment Power Consumption

One of the main ideas of the femtocell is to become closer to the user equipment, which is an approach to enhance capacity and decrease energy depletion of both cellular network and user equipment [15]. However, since the coverage area of the femtocell is not wide, the number of handovers is very high in a femtocell network. Moreover, user equipment uses most of its energy for the handover process [49]. Another issue about handovers is that they decrease QoS and network capacity. Thus, the handover decision algorithm is an important issue in femtocells. Although there are many studies about handover decision, only a few of them depends on energy efficiency.

In Ref. [49], they studied about the energy efficiency of femtocell on user equipment battery. They worked on the fact that there is a reduction in power consumption of user equipment under femtocell coverage, compared to macrocell connection and they proposed a handover decision algorithm that aims to reduce user equipment power consumption while maintaining QoS. The suggested algorithm enhances the strongest cell handout policy using an adaptive handout hysteresis margin. Although there is a need for increased LTE network signaling in the proposed algorithm, they derived power consumption and also interference. The result is compared with different algorithms in the literature and a strongest cell based handout decision algorithm. It was analyzed that the proposed algorithm reduced 85% user equipment energy consumption compared to femtocell deployment in LTE network.

3.8 Challenges

Even though small cell cognitive network such as Femtocell is a promising technique, there are still many challenges that must be addressed, for instance, interference mitigation, spectrum access, and QoS provisioning [50]. Providing reliable services for applications, which demand low-energy consumption and latency within the IoT context is a challenging issue. It is well known that some wireless network applications require deterministic systems with a reliable and low latency aggregation service guarantees. Since the IEEE 802.15.4e standard is considered as the backbone of the IoT using WSNs, the existing Low Latency Deterministic Network (LLDN) has been used to fulfill the major requirements of these wireless network applications. In turn, research groups for this standard have conducted studies on improvement of QoS-related concerns including energy efficiency [51]. Test-beds, simulation studies, and analytical modeling approaches are being employed for this purpose effectively.

Moreover, the authors in Ref. [13] poses some questions on a few of the challenges faced by femtocell; for instance, they ask how will femtocell provide timing and synchronization, given that femtocell require synchronization to align incoming signals, and hence minimizing multiple access interference and guarantee a tolerable carrier off-set. Another challenge discussed by the authors is how backhaul will provide acceptable QoS; IP-based backhaul need to provide QoS for traffic that is sensitive to delays, additionally, microcells systems provide a latency guarantee of 15 ms. However, current backhaul networks do not have such protection against unnecessary delays. Another key challenge is how a femtocell will adjust to its surroundings and assign spectrum with the existence of intra- and cross-tier interference. The authors in Ref. [52] talked about an existing problem in small cell cognitive networks of power control and sensing time optimization; this problem has been evaded by many researchers, yet it has real effects in the operation of the network. Moreover, the author in Ref. [53] said that fairness and spectrum sensing errors were ignored in most research. The author in Ref. [54] defines Heterogeneous Cloud Small Cell Network (HCSNet) using small cells and cloud computing, while the author in Ref. [19] defines Ultra Dense Cloud Small Cell Network (UDCSNet) as a combination of cloud computing and massive deployment of small cells. They will play a major role in a 5G mobile communication network in the quest of trying to satisfy the massive data traffic generated by the ever-growing number of users. However, problems such as co-channel interference and handover management that came about due to massive deployment of small cells must be considered [54].

The author in Ref. [55] outlines a problem that arises in small cell networks when trying to achieve capacity growth through network densification; this attempt will be faced with the challenge of severe inter-cell interference. This problem arises due to the limited licensed spectrum for cellular networks. Therefore, researchers are looking at using unlicensed spectrum bands, including the 2.4 and 5 GHz to curb this problem.

3.9 Open Research Issues

Recently, popular technology solutions for various cellular network applications have emerged as a significant advancement for Internet and mobile networks. IoT is a novel paradigm where connected entities become part of these infrastructures, and the advancements in IoT make it quite popular where the traditional telecommunication systems facilitate basic communications between these entities. IoT has converged technologies in terms of sensing, computing, information processing, networking, and controlling intelligent technologies. Among the key technologies converged is the femtocell due to its low-energy wireless communications and cost effectiveness. Femtocells are composed of BS and numerous low-cost resources, in terms of communication, storage, and computation facilities. Wealth of various approaches have been proposed and designed for considering the collaborative nature of femtocells in the existing literature. However, there are key open research problems yet to be solved in the IoT era.

Since in general enabling technologies have restricted authentication privileges for mobile users, different strategies are introduced for the extension of user authentication over IoT-based environments. Commercialization of remote applications and security issues in femtocells have gained much attention of the researchers to satisfy the security properties of authentication and key agreement protocols. In general, the development of security protocols is more challenging and should also consider mitigation of the computation and communication cost. Moreover, considering the femtocell and the macrocell energy consumption for heterogeneous networks is another key challenge. Where energy consumption models, considering the aforementioned IoT setups with different access types, are mandatory for more accurate estimations. Furthermore, covering a large area with the optimized deployment of macrocell and many overlapping femtocells needs a careful consideration in order to realize the femtocells in the IoT era.

3.10 Concluding Remarks

Energy efficiency of femtocell networks becomes indispensable with the growing deployment of femtocells. Moreover, it can be used in order to manage energy efficiency when it is used in smart grids. In this survey, we mentioned energy efficiency of femtocell networks in IoT, considering energy metrics, energy consumption models, energy efficiency schemes for deployment, factors that affect the energy consumption of femtocell networks and energy efficiency of user equipment under femtocell coverage. The energy efficiency of the femtocell in IoT still needs to be investigated.

References

1. BI Research, High inventory and low burn rate stalls femtocell market in 2012, (July 5, 2012 [November 13, 2012]).
2. L. Coetzee, and J. Eksteen, The internet of things—promise for the future? An introduction, *Proceedings of IST-Africa Conference*, Gaborone, Botswana, May 2011.
3. F. Al-Turjman, Information-centric sensor networks for cognitive IoT: an overview. *Annals of Telecommunications*, vol. 72, no. 1, pp. 3–18, 2017.
4. National Intelligence Council, Disruptive civil technologies: six technologies with potential impacts on US interests out to 2025, April 2008. Available https://www.hsdl.org/?abstract&did=485606.
5. European Commission, Smart grid and meters, July 07 2017. Available http://ec.europa.eu/energy/en/topics/markets-and-consumers/smart-grids-and-meters.
6. SmartGrid.Gov, What is smart grid? n.d. Available https://www.smartgrid.gov/the_smart_grid/smart_grid.html.
7. E. Bou-Harb, C. Fachkha, M. Pourzandi, M. Debbabi, and C. Assi, Communication security for smart grid distribution networks. *IEEE Communications Magazine*, vol. 51, no. 1, pp. 42–49, 2013.

8. Z. Feng, and Z. Yuexia, Study on smart grid communications system based on new generation wireless technology. *International Conference on Electronics, Communications and Control,* Ningbo, China, pp. 1673–1678, 2011.
9. R. Baines, Femtocells - reducing power consumption in mobile networks, November 23, 2016. Available http://www.low-powerdesign.com/article_baines_092811.html.
10. F. Al-Turjman, H. Hassanein, and M. Ibnkahla, Towards prolonged lifetime for deployed WSNs in outdoor environment monitoring. *Elsevier Ad Hoc Networks Journal,* vol. 24, no. A, pp. 172–185, January 2015.
11. A. Banote, V. Ubale, and G. Khaire, Energy efficient communication using femtocell – a review. *International Journal of Electronics, Communication & Instrumentation Engineering Research and Development (IJECIERD),* vol. 3, no. 1, pp. 229–236, 2013.
12. HSPA or LTE? That is the question, *RCR Wireless News,* May 9, 2014. Available http://www.rcrwireless.com/20140509/hetnet-news/hspa-lte.
13. V. Chandrasekhar, J. Andrews, and A. Gatherer, Femtocell networks: a survey. *IEEE Communications Magazine,* vol. 46, no. 9, pp. 59–67, 2008.
14. R. Saeed, (Ed.), *Femtocell communications and technologies: Business opportunities and deployment challenges.* IGI Global, 2012. United States of America
15. M. Chowdhury, S. Q. Lee, B. H. Ru, N. Park, and Y. M. Jang, Service quality improvement of mobile users in vehicular environment by mobile femtocell network deployment, in *International Conference on ICT Convergence (ICTC),* pp. 194–198, 2011. Seoul, South Korea
16. M. Qutqut, F. Al-Turjman, and H. Hassanein, HOF: a history-based offloading framework for LTE networks using mobile small cells and Wi-Fi, *Proceedings of the IEEE Local Computer Networks (LCN),* Sydney, Australia, pp. 77–83, 2013.
17. D. Xu, W. He, and S. Li, Internet of things in industries: a survey. *IEEE Transactions on Industrial Informatics,* vol. 10, no. 4, pp. 2233–2243, 2014.
18. S. R. Hall, A. W. Jeffries, S. E. Avis, and D. D. N. Bevan, Performance of open access femtocells in 4G macrocellular networks, *Wireless World Research Forum 20 (WWRF 20),* Ottawa, Canada, 2008.
19. J. Zhao, W. Zheng, X. Wen, X. Chu, H. Zhang, and Z. Lu, Game theory based energy-aware uplink resource allocation in OFDMA femtocell networks. *International Journal of Distributed Sensor Networks,* vol. 10, no. 3, pp. 245–252, 2014.
20. H. Zhang, Y. Nie, J. Cheng, V. Leung, and A. Nallanathan, Sensing time optimization and power control for energy efficient cognitive small cell with imperfect hybrid spectrum sensing. *IEEE Transactions on Wireless Communications,* vol. 16, no. 2, pp. 730–743, 2017.
21. Z. Fan, P. Kulkarni, S. Gormus, C. Efthymiou, G. Kalogridis, M. Sooriyabandara, and W. Chin, Smart grid communications: overview of research challenges, solutions, and standardization activities. *IEEE Communications Surveys & Tutorials,* vol. 15, no. 1, pp. 21–38, 2013.
22. M. Hindia, T. Rahman, H. Ojukwu, E. Hanafi, and A. Fattouh, Enabling remote health-caring utilizing IoT concept over LTE-femtocell networks. *PloS One,* vol. 11, no. 5, p. e0155077, 2016.
23. T. Cho, and G. Jeon, A method for detecting man-in-the-middle attacks using time synchronization one time password in interlock protocol based internet of things. *Journal of Applied and Physical Sciences,* vol. 2, no. 2, pp. 37–41, 2016.
24. Z. Hashmi, Adaptive and efficient resource management for emerging wireless networks, *Electronic Theses and Dissertations (ETDs),* 2008. doi: 10.14288/1.0073687

25. H. Zhang, N. Liu, X. Chu, K. Long, A. Aghvami, and V. C. M. Leung, Network slicing based 5g and future mobile networks: mobility, resource management, and challenges, 2017. arXiv preprint arXiv: 1704.07038.
26. J. Wan, D. Zhang, S. Zhao, L. T. Yang, and J. Lloret, Context-aware vehicular cyber-physical systems with cloud support: architecture, challenges and solutions. *IEEE Communications Magazine*, vol. 52, no. 8, pp. 106–113, 2014.
27. J. Wan, C. Zou, S. Ullah, C. Lai, M. Zhou, and X. Wang, Cloud-enabled wireless body area networks for pervasive healthcare. *IEEE Network*, vol. 27, no. 5, pp. 56–61, 2013.
28. R. Piyare, Internet of things: ubiquitous home control and monitoring system using android based smart phone. *International Journal of Internet of Things*, vol. 2, no. 1, pp. 5–11, 2013.
29. M. Petracca, S. Bocchino, A. Azzarà, R. Pelliccia, M. Ghibaudi, and P. Pagano, WSN and RFID integration in the IoT scenario: an advanced safety system for industrial plants. *Journal of Communications Software and Systems*, vol. 9, no. 1, pp. 104–113, 2013.
30. H. Zhang, H. Xing, J. Cheng, A. Nallanathan, and V. Leung. Secure resource allocation for OFDMA two-way relay wireless sensor networks without and with cooperative jamming. *IEEE Transactions on Industrial Informatics*, vol. 12, no. 5, pp. 1714–1725, 2016.
31. L. Atzori, A. Iera, and G. Morabito. The internet of things: a survey. *Computer Networks*, vol. 54, no. 15, pp. 2787–2805, 2010.
32. D. Su, X. Wen, H. Zhang, and W. Zheng, A self-optimizing mobility management scheme based on cell ID information in high velocity environment, *Proceedings of the 2nd International Conference on ICCNT*, pp. 285–288, 2010. Bangkok, Thailand.
33. Y. Hou, and D. I. Laurenson, Energy efficiency of high QoS heterogeneous wireless communication network, *IEEE VTC-Fall*, Ottawa, Canada, 2010.
34. H. Zhang, C. Jiang, N. Beaulieu, X. Chu, X. Wang, and T. Quek. Resource allocation for cognitive small cell networks: a cooperative bargaining game theoretic approach. *IEEE Transactions on Wireless Communications*, vol. 14, no. 6, pp. 3481–3493, 2015.
35. H. Zhang, C. Jiang, N. Beaulieu, X. Chu, X. Wen, and M. Tao, Resource allocation in spectrum-sharing OFDMA femtocells with heterogeneous services, *IEEE Transactions on Communications*, vol. 62, no. 7, pp. 2366–2377, 2014.
36. M. W. Arshad, A. Vastberg, and T. Edler, Energy efficiency gains through traffic offloading and traffic expansion in joint macro pico deployment, *Proceedings of the IEEE WCNC*, Shanghai, China, 2012.
37. A. De Domenico, R. Gupta, and E. Calvanese Strinati, Dynamic traffic management for green open access femtocell networks, *IEEE Vehicular Technology Conference*, Yokohama, Japan, 6–9 May 2012.
38. R. Riggio, and D. J. Leith, A measurement-based model of energy consumption in femtocells, *IEEE/IFIP Wireless Days*, Dublin, Ireland, 2012.
39. M. Deruyck, D. De Vulder, W. Joseph, and L. Martens, Modelling the power consumption in femtocell networks, *IEEE Wireless Communications and Networking Conference (WCNC 2012): Workshop on Green Communications*, Paris, France, pp. 30–35, April 2012.
40. A. Mukherjee, S. Bhattacherjee, S. Pal, and D. De, Femtocell based green power consumption methods for mobile network. *Computer Networks*, vol. 57, no. 1, pp. 162–178, 2013.

41. H. Zhang, H. Lui, C. Jiang, N. Beaulieu, X. Chu, A. Nallanathan, and X. Wen, A practical semi-dynamic clustering scheme using affinity propagation in cooperative picocells, *IEEE Transactions on Vehicular Technology*, vol. 64, no. 9, pp. 4372–4377, 2015.
42. J. Zhang, P. Hong, H. Xue, and H. Zhang, A novel power control scheme for femtocell in heterogeneous networks, *Proceedings IEEE CCNC*, Las Vegas, NV, USA, pp. 802–806, January 2012.
43. A. Khalifah, N. Akkari, and G. Aldabbagh, Dense areas femtocell deployment: access types and challenges, *Third International Conference on e-Technologies and Networks for Development (ICeND)*, Beirut, Lebanon, 2014.
44. B. Mitchell, WiMax vs. LTE for mobile broadband, September 17, 2016. Available https://www.lifewire.com/wimax-vs-lte-for-mobile-broadband-818319.
45. L. Cassavoy, November 24. What is HSPA? 2014. Available https://www.lifewire.com/definition-of-hspa-578679.
46. M. H. Qutqut, F. M. Al-Turjman, and H. S. Hassanein, MFW: mobile femtocells utilizing WiFi: a data offloading framework for cellular networks using mobile femtocells, *Proceedings of the IEEE International Conference on Communications (ICC)*, Budapest, Hungary, pp. 5020–5024, 2013.
47. Y. Li, H. Celebi, M. Daneshmand, C. Wang, and W. Zhao, Energy-efficient femtocell networks: challenges and opportunities. *IEEE Wireless Communications*, vol. 20, no. 6, pp. 99–105, 2013.
48. M. Qutqut, H. Abou-zeid, H. Hassanein, A. Rashwan, and F. Al-Turjman, Dynamic small cell placement strategies for LTE heterogeneous networks, *in Proceedings of the IEEE Symposium on Computers and Communications (ISCC)*, Madeira, Portugal, pp. 1–6, 2014.
49. N. Xenakis, C. Passas, and C. Verikoukis, An energy-centric handover decision algorithm for the integrated LTE macrocell-femtocell network. *Elsevier Computer Communications*, vol. 35, no. 14, pp. 1684–1694, 2012.
50. H. Zhang, J. Chunxiao, M. Xiaotao, and C. Hsiao-Hwa, Interference-limited resource optimization in cognitive femtocells with fairness and imperfect spectrum sensing. *IEEE Transactions on Vehicular Technology*, vol. 65, no. 3, pp. 1761–1771, 2016.
51. Y. Al-Nidawi, H. Yahya, and A. H. Kemp, Tackling mobility in low latency deterministic multihop IEEE 802.15. 4e sensor network. *IEEE Sensors Journal*, vol. 16, no. 5, pp. 1412–1427, 2016.
52. H. Zhang, C. Jiang, X. Mao, and A. Nallanathan, Resource management in cognitive opportunistic access femtocells with imperfect spectrum sensing, in *Proceedings of the IEEE International Conference on Global Communications Conference (GLOBECOM)*, pp. 3098–3102, 2014.
53. H. Zhang, C. Jiang, J. Cheng, and V. Leung, Cooperative interference mitigation and handover management for heterogeneous cloud small cell networks. *IEEE Wireless Communications*, vol. 22, no. 3, pp. 92–99, 2015.
54. H. Zhang, X. Chu, W. Guo, and S. Wang, Coexistence of Wi-Fi and heterogeneous small cell networks sharing unlicensed spectrum. *IEEE Communications Magazine*, vol. 53, no. 3, pp. 158–164, 2015.
55. H. Zhang, Y. Dong, J. Cheng, M. Hossain, and V. Leung, Fronthauling for 5G LTE-U ultra-dense cloud small cell networks. *IEEE Wireless Communications*, vol. 23, no. 6, pp. 48–53, 2016.

Chapter 4

Modeling Green Femtocells in Smart Grids*

Fadi Al-Turjman
Antalya Bilim University

Contents

* **F. Al-Turjman**, "Modelling Green Femtocells in Smart-Grids", *Springer Mobile Networks and Applications*, 2017. doi: 10.1007/s11036-017-0963-1.

4.1 Introduction

Recently, heterogeneous wireless technologies have been developed rapidly to support different Radio Access Technologies (RATs) such as GSM, Long-Term Evolution (LTE)-Advanced, Wireless Fidelity (WiFi), etc. in order to connect mobile users to the Internet and form what we call Heterogeneous Networks (HetNets). In fact, mobile HetNets experience explosive growth in usage and energy consumption due to explosion of smart devices in massive volumes and energy-hungry mobile applications among which the smart-grid systems start to be the foremost ones [1,2]. Smart grid has been evolved recently in managing our gigantic electricity and energy demands in a sustainable, reliable, and economic manner, while utilizing already existing HetNets' infrastructures. Smart grids are energy networks that can automatically monitor energy flows and adjust to changes in energy supply and demands accordingly. It can be used with smart meters to enhance the quality of experience by providing information on real-time energy consumption. Smart homes can also communicate with the grid using these smart meters and enable consumers to manage their electricity usage. This can be achieved via wireless HetNets connecting different appliances at home. These home appliances and other electrical devices, which are heavily used during our daily life and are considered as smart devices, have also energy-saving issues. The trend in *e-Mobility* [3] (i.e., the examined use case in this study), for example, is focused mainly on minimizing the waste in electrical energy provided by the municipal utility. Smart grids in such scenarios can improve operating efficiencies, lower costs, shorten outages, and reduce peak demands and electricity consumption. This is due to automated meter readings and service connections and disconnections, which can significantly reduce the need to dispatch trucks and personnel while accomplishing these tasks [4]. Where reducing the truck rolls by 15,000 miles can save approximately 6.3 metric tons of CO_2 emissions. In addition, the integrated wireless communication networks in smart grids can be used for outage management, where utilities can ping smart meters to determine which customers do not have power and pinpoint outage locations for quicker and more efficient service restoration. In addition, it can accommodate future smart grid upgrades, automate other city services (such as water and gas metering), and deliver valuable returns for cities and taxpayers. Accordingly, the utilized HetNets by the municipal of the city is not only used to connect, communicate, and control most of the city's smart-grid equipment/devices, it is also used to avoid the global energy waste and reduce the carbon footprint. To realize these benefits, the utilized HetNets have to overcome several challenges in practice. These include integrating heterogeneous technologies and communication systems optimally for data backhauling, catering with the mobility factor, tolerating system failures to reduce electricity consumption and lower bills, and implementing new communication systems in a dynamic environment of rapidly evolving users and devices. Toward this end, there is a need for an efficient HetNet between the different heterogeneous appliances/devices at home and outside the home, such as the Home Area Network (HAN) and the Neighborhood Area Network (NAN). HAN consists of three components; it measures, collects, and analyzes energy usage from smart devices [5]. NAN connects multiple HAN to local/regional access points (i.e., access points within the range of instantaneous transmitters), where transmission lines carry the data to the municipal [6]. Such HetNets in the smart grid need a communication approach that connects the vast counts of consumers and suppliers in an energy-efficient way.

One promising solution for HetNets and cellular providers in this regard is the deployment of femtocells [2]. A femtocell is a cell that provides cellular coverage and is served using a FBS, which is short-range and low-power cellular base station typically deployed in indoor environments and/or outdoor rural and densely populated areas for enhanced reception of voice and data traffics [6,7]. It broadcasts using GSM signals and allows a cellular phone to connect to the Internet for better indoor coverage. It uses between 8 and 120 mW, which is much lower than a WiFi access point [4]. Thus, it can be used as a green communication method in HAN and HetNets in general, where consumers can save energy and money using their smart home systems. They can track their energy usage and this can give consumers the ability to manage their electricity bills. Moreover, with the help of the femtocell, appliances can be automated and remotely controlled to reduce unnecessary electricity demands on the grid. This can be generalized for all smart homes in a smart-city scenario for example, where mobile outdoor femtocells can be utilized to achieve ULS coverage. However, this increases the HetNets overload and makes our expectations unrealistic in proximity of the green planet vision [2].

In fact, RATs are responsible for more than 2.5% of the total carbon emission rates, which is more than the carbon emission rate of the global aviation industry [4,8]. This rate is expected to be doubled over the next decade [4], which means more energy waste since carbon emission is a straightforward result of energy usage in the world; especially, when we know that the main component of a HetNet, which is the Macro base station, consumes between 2.5 and 4 kW [4]. Since the number of these stations in every country is about tens of thousands and still growing more and more. This means many energy consumption demands and carbon emissions are forecasted. It is estimated that only between 5% and 10% of this energy will be used to create useful signals. This is because macrocells provide wide area coverage, and most of the area is empty space where transmitted signals are not utilized. Moreover, the radiated wireless signals need high power without efficiency, which means power is used for just in case scenarios where a user might exist. Thus, femtocell is a powerful candidate to prevent these energy losses. It provides local coverage while macrocell is providing huge area coverage. In other words, femtocell delivers power only where there is a need. Therefore, it consumes less energy to provide higher bandwidth since it is closer to user. The other advantage is that user equipment consumes less energy when it is connected to femtocell since it is closer, where it takes 40 times more energy to deliver signal to indoor from a macrocell compared to a femtocell. Thus, femtocell is more energy efficient in addition to providing more coverage, capacity, and better Quality of Service (QoS). Nevertheless, significant energy amounts can be wasted in data (re)transmission unless a reasonable load balance is applied between the deployed FBSs that are typically planned to serve huge counts of static/mobile users of the smart grid. This would not be achieved without a realistic case study analysis and an accurate analytical model that can predict the system performance in such setup.

There have been a few attempts in the literature to analytically model the energy consumption in a femtocell. On–off model is the most basic one, which can be used for theoretical analysis where femtocell base station (FBS) is assumed to consume unit power in active mode and zero power when it is off. However, it does not reflect actual power consumption. In Ref. [9], the authors proposed a linear power model that considers traffic load. This model is used for analysis and more accurate than on–off model since traffic load is considered. In Ref. [10], the authors proposed a simple analytic model to predict the FBS power consumption based on the offered load and datagram size. However, they neglected that radio energy must be consumed by the base station when making downlink transmission. In Ref. [11], energy consumption is modeled based on three main interactive components in the FBS: the microprocessor, the Field Programmable Gateway Array (FPGA), and the radio frequency transmitter for indoor applications. Nevertheless, it is mainly designed for static indoor applications and cannot predict outdoor energy consumptions for mobile FBS.

In this study, unlike the aforementioned studies, an analytical modeling approach is proposed, where varying workload, communication range, multiple FBS/user mobility related issues, as well as channel failures in the femtocell infrastructure are considered. Our modeling approach has been utilized in a quite useful offloading approach for discovering the operational space of various femtocell configurations/decisions. This approach is maintained mainly in two algorithms. It leads to more accurate QoS measurements in real-time applications (e.g., VoIP and video streaming), while considering the varying speed effect of the mobile FBS/user equipment on the HetNet performance of a smart grid and other QoS parameters. Unlike other approaches in the literature, our approach caters for failures and data packets, which can leave the system due to mobility and energy saving modes. In addition, the effect on the performance characteristics of the grid system such as Mean Queue Length (MQL), throughput, and delay has been investigated accordingly. In the following, the main contributions of this research are summarized.

- A queue-based analytical model is proposed while considering varying characteristics in the grid infrastructure.

- Femtocell predictions in energy consumptions with high accuracy and efficacy have been achieved and discussed.
- A novel offloading approach for discovering the operational space of various femtocell configurations/decisions is proposed.
- A hybrid wireless cellular HetNet consisting of a macrocell and several femtocells has been considered as a case study toward more energy efficient HetNets in the smart-grid era.
- Detailed analysis of the case study is given based on the queuing theory concept and results have been validated through extensive simulations.

The remainder of this chapter is organized as follows. Section 4.2 overviews the related attempts for HetNet modeling in the literature. Section 4.3 highlights the assumed and used system models. In Section 4.4, we propose our detailed grid model toward realizing green femtocell applications. Then, a real case study, namely the *e-mobiliy*, is examined and used to validate our proposed model via extensive simulation results in Section 4.5. Finally, we conclude this work in Section 4.6.

4.2 Related Work

There are several attempts in the literature toward modeling HetNets telecommunication systems for better performance assessments and energy consumption predictions. Modeling these systems using analytical models like queuing theory is a well-known approach in the literature [11–16]. Different performance metrics, depending on system and required analysis' studies, such as average number of requests in the system, average resource utilization, average power consumption, average waiting time, throughput, etc. have been investigated [14,15]. Such modeling systems and performance metrics can be classified into static versus dynamic models. By static and dynamic models, we refer to systems with/without mobile femtocells. In static models such as the ones presented in Refs. [17–19], performance characteristics of cellular networks have been investigated without considering mobility, and thus we call it static systems. Unlike static models, in the dynamic models, mobility is considered as one of the utmost important issues in the performance evaluation process [20–26].

4.2.1 Static Modeling

Several studies have been performed on analyzing and evaluating the performance of typical small cell deployments in static scenarios (i.e., without mobility considerations). For example, in Ref. [12], a set of algorithms had been proposed to reduce the energy consumption of a dense network and provide better QoS to the end-users. In Ref. [12], the authors study the energy consumption of a campus WiFi. A simple approximation queuing model is used to save energy in WiFi by considering cut-offs for the small cells according to user demands due to sleep modes, channel failures, mobility issues, etc. Presented results show that by using sleep modes for the small cell, a considerable amount of energy can be saved when the number of users connected to the network is small. The authors in Ref. [16] evaluated the performance advantage of using fixed small cells as relays by communicating with the macro base station to improve and extend the HetNet coverage. In Ref. [11], the authors proposed a simple analytical model to predict a static FBS power consumption based on a specific offered load and packet size. They assumed one femtocell that supports up to four simultaneous end-user devices in a campus network to predict energy consumption of voice and FTP messages. However, in their prediction, they neglected that radio energy consumption while performing downlink transmissions. The authors in Ref. [18] analyzed the behavior of Adaptive Modulation and Coding (AMC) systems with sleep mode using queuing theory. They were interested in evaluating energy consumption rates per packet, average delays, and packet loss. An admission control problem for a multi-service LTE radio network was addressed in Ref. [19]. A model for two resource-demanding video services: video conferencing and video on demand was proposed. Teletraffic and queuing theories were applied to obtain a recursive algorithm in order to calculate performance measures such as blocking probability and the mean bit rate. However, a limited number of researchers have studied and modeled mobile small cell deployments as an option in providing green systems.

4.2.2 Dynamic Modeling

Authors in Ref. [20] propose the idea of deploying small cells in vehicles to improve the uplink throughput for mobile users. Results show that mobile small cells can enhance QoS and maintain an acceptable level of Signal to Interference Noise Ratio (SINR). In Ref. [21], the authors proposed seamless multimedia service for mobile users in high-speed trains through deploying small cells onboard. The onboard small cells communicate with macrocells to facilitate the seamless handover. In Ref. [22], an integrated cellular/WiFi system is modeled for high mobility using a two-stage open queuing system with guard channel and buffering to obtain acceptable levels of QoS in heterogeneous environments. An exact analytical solution of the system is given using the spectral expansion solution approach that can be useful for vertical handover decision management. Similarly, the authors in Ref. [23] model an integrated cellular/WiFi HetNet in order to study specific performance characteristics such as MQL, blocking probability, and throughput. The system is modeled as a two-stage open queuing network and the exact solution is presented using the spectral expansion solution approach. Simulation is also employed to validate the accuracy of the proposed system. The authors in Ref. [23] presented a mathematical model for analytical study on complete and partial channel allocation schemes. By employing Markov models, results can be presented for performance measures such as MQL and blocking probability.

However, among the most significant issues in performance evaluation of such HetNets is mobility [25]. In Ref. [25], wireless cellular networks are modeled using a Markov reward model. An S-channel per cell in homogeneous cellular system and mobility related issues are considered. Performance characteristics of the system such as MQL and blocking probability are presented using an analytical model. In Ref. [26], the authors considered a network in which cells of different sizes have been deployed depending on mobile user density, traffic, and coverage such that power consumption can be minimized without compromising QoS. They developed analytical models of power consumption in five different ways and obtained the reduction in power consumption compared to macrocell networks. In the first one, they used femtocell-based network instead of the macrocell-based one in an area that is fully covered by femtocells only and accordingly obtained between 82.72% and 88.37% power consumption reduction. In the second way, they divided the area into three parts as urban, suburban, and rural areas. They considered mobile user density, mobile user traffic, and required coverage and they covered urban areas with femtocells, suburban areas with macrocells, and rural areas with mobile femtocells. Because of this setup, they achieved between 78.53% and 80.19% reduction in power consumption. In the third way, they allocated femtocells to densely populated urban area, picocells to sparsely populated urban areas, microcells to suburban areas, and mobile femtocells to rural areas and the reduction in power consumption rate was predicted to be between 9.19% and 9.79%. In the fourth way, they allocated microcells, picocells, and femtocells to the border region and macrocell to remaining region. The reduction in power consumption is between 5.52% and 5.98% for this setup. The last one, femtocells are allocated at the boundaries of macrocells, where the radio signal is not enough for making a call. Because of the last setup, a reduction in power consumption between 1.94% and 2.66% and a macrocell coverage shrink is achieved. Moreover, two different handoff schemes with/without preemptive priority procedures for integrated wireless mobile networks are proposed and analyzed in Ref. [27]. Service requests are categorized into four different types: as voice requests, data requests, voice handoff requests, and data handoff requests. A 2D Markov chain is used to model the system and analyze the HetNet performance in terms of average delay, blocking probability, and forced termination probability. Existing research efforts, however, do not target assessing the performance gains of mobile small cells. A quantitative performance analysis of such gain is definitely needed nowadays for better energy utilization and more green applications. Hence, the work in this chapter is proposed. Particularly, we considered the different velocity effect of modeled mobility, which is a typical case in smart-grid applications such as the *e-Mobility* project in Siemens [3].

4.3 System Models

In this study, mobile users may move to neighboring cells while they are either in the queue or being served in the system. It is typical in grid systems to experience some outage periods due to many different reasons including load balance and/or sleep modes for better energy consumption. These cut-offs and unavailability of an FBS may degrade the performance of the grid system. It is assumed that a single recover facility is available not only for every FBS, but also for every FBS channel to make the cut-off recovered again. Similar to previous studies in Refs. [7,20], each macrocell can be represented by a circle of radius *R* so that it is served by a base station placed at the center. The femtocells that are deployed within the coverage area of a macrocell are also represented by circles of radius *r* and are served by mobile/static FBSs.

4.3.1 Queuing Model

In order to satisfy energy requirements in grid-based HetNets applications, a Markovian discrete-time stochastic process M/M/N/L queuing model is assumed. To cope with the grid heterogeneity nature, we assume a priority-based approach in queuing the incoming requests. The queuing capacity of the system is denoted by W, and L represents the maximum capacity that includes the number of FBSs in the system (N). Similar to related studies in Refs. [27–29], arrivals to the system are assumed to follow a Poisson process with arrival rate σ, and service time exponentially distributed with rate μ. This system was proposed under a realistic assumption of a mobile FBS queue that can hold waiting requests as long as they are within the required communication range. The arrival and departure of the data packets are regulated under a finite queue size. Multiple mobile FBSs may move while they are inside the coverage area of a macrocell, and thus, user requests will handover to other neighboring cells with rate μ_{cd} while they are either in the queue or being served in the system. Moreover, cut-offs may also occur in the system. The cut-off rate of the FBS is assumed to be exponentially distributed and is denoted by ξ [27]. Following the cut-off, the failed

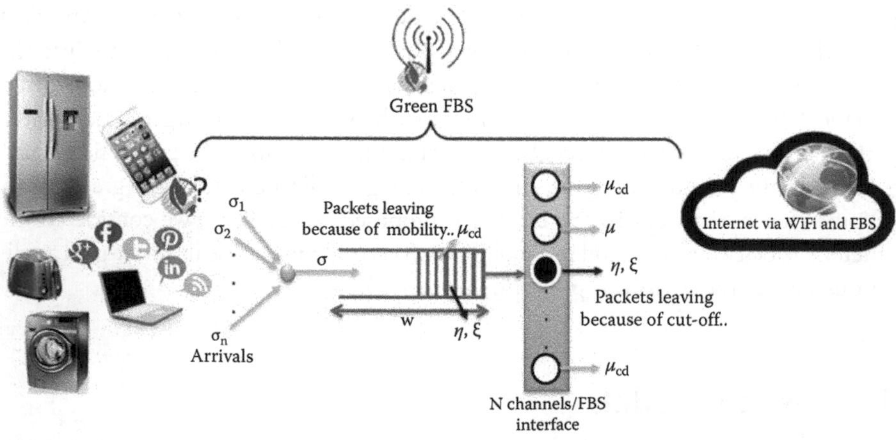

FIGURE 4.1 The queuing system considered with cut-off and mobility effects.

FBS is recovered with a recover rate η with exponential distribution as well. Figure 4.1 represents our FBS queuing system under this study. Symbols/notations used in this model are summarized also in Table 4.1. According to Ref. [15], the dwell time of a mobile FBS is the time that the mobile node spends in a given system. The dwell time is assumed to have an exponential distribution with a mean rate μ_{cd}. Thus, the service rate due to mobility, can be calculated by $\mu_{cd} = \frac{P \cdot E[v]}{\pi A}$, where $E[v]$ is the average expected velocity of the mobile FBS. P and A are the length of the perimeter and the area of the macrocell, respectively.

4.3.2 Communication Model

Practically, the signal level at distance d from a transmitter varies depending on the surrounding environment. These variations are captured through what we call log-normal shadowing model. According to this model, the signal level at distance d from a transmitter follows a log-normal distribution centered on the average power value at that point [30]. This can be formulated as follows:

$$P_r = K_0 - 10\rho \log(d) - \chi, \tag{4.1}$$

TABLE 4.1 Summary of Symbols

Symbol	*Definition*
r	Radius of the femtocell
V	Velocity of the mobile users
P	Perimeter of the femtocell
A	Area of the femtocell
N	Total number of channels in the FBS
W	Queue capacity of the cell
L	Maximum number of requests in the cell
σ	Total arrival rate of requests in the cell
μ	Total service rate of completed request departures in the cell
μ_{cd}	Mean service rate of handover requests in the cell
ξ	Cut-off rate of a server
H	Mean recover rate of the cut-off

where d is the Euclidian distance between the transmitter and receiver, ρ is the path loss exponent calculated based on experimental data, χ is a normally distributed random variable with zero mean and variance σ^2, i.e., $\chi \sim \mathcal{N}(0, \sigma^2)$, and K_0 is a constant calculated based on the mean heights of the transmitter and receiver.

4.3.3 Energy Consumption Model

Assuming d is the distance in meters between the transmitter in a smart-grid point of interest and the mobile FBS. The achievable transmission rate at the FBS can be approximated using Shannon's capacity equation with signal-to-noise ratio (SNR) clipping at 20 dB for practical modulation orders as follows:

$$R = B \log_2 \left(1 + \frac{P_r}{N_0 \cdot B} \right), \tag{4.2}$$

where R is the data rate in bit-per-second for a given received power P_r and system bandwidth B. The P_r is computed using Eq. (4.1), and N_0 is the background noise power spectral density (i.e., AWGN). Consequently, and based on [31], the total power consumption, P_{tot}, of an FBS can be formulated as $P_{tot} = P_{FPGA} + P_{mp} + P_t + P_a$, which is the total of power consumption parameters of the FPGA: the microprocessor, the transmitter, and the power amplifier, respectively. And thus, we define the energy consumed at a FBS by $\boldsymbol{E_{FBS}} = \boldsymbol{P_{tot}} * \boldsymbol{R} * (\boldsymbol{E_{TX}} + \boldsymbol{E_{RX}})$, where most of the energy consumption at the FBS is due to data communication indicated by E_{TX} for transmission energy and E_{RX} for energy consumed during the data reception process. And T represents the number of transmitted packets. Hence, expected energy consumption at a FBS with communication N channel/interface can be estimated as follows:

$$E(x) = \sum_{i=1}^{N} P_i \cdot i \cdot \mu \cdot E_p, \tag{4.3}$$

Where P_i, i, μ, and E_p are probability of having i channels available (sum of all probabilities in columns of the 2D state diagram shown in Figure 4.2), number of available channels, service rate, and energy consumption for each transmitted packet, respectively. In this work, we refer to one-hop neighbors' communication as the first tier of nodes. Since no other node can reach the macrocell station directly, traffic from every other node will have to be forwarded, in the last hop, by one of the FBS. If the spatial distribution of nodes is assumed to be uniform, then the traffic load is equally distributed. Each first tier node will forward hardly the same amount of traffic, and all first tier nodes' energy will be depleted at times very close to each other, after the network is first put into operation. Since the entire first tier nodes' energies are depleted at once, the FBS will be overwhelmed and plenty of handovers will be performed, in

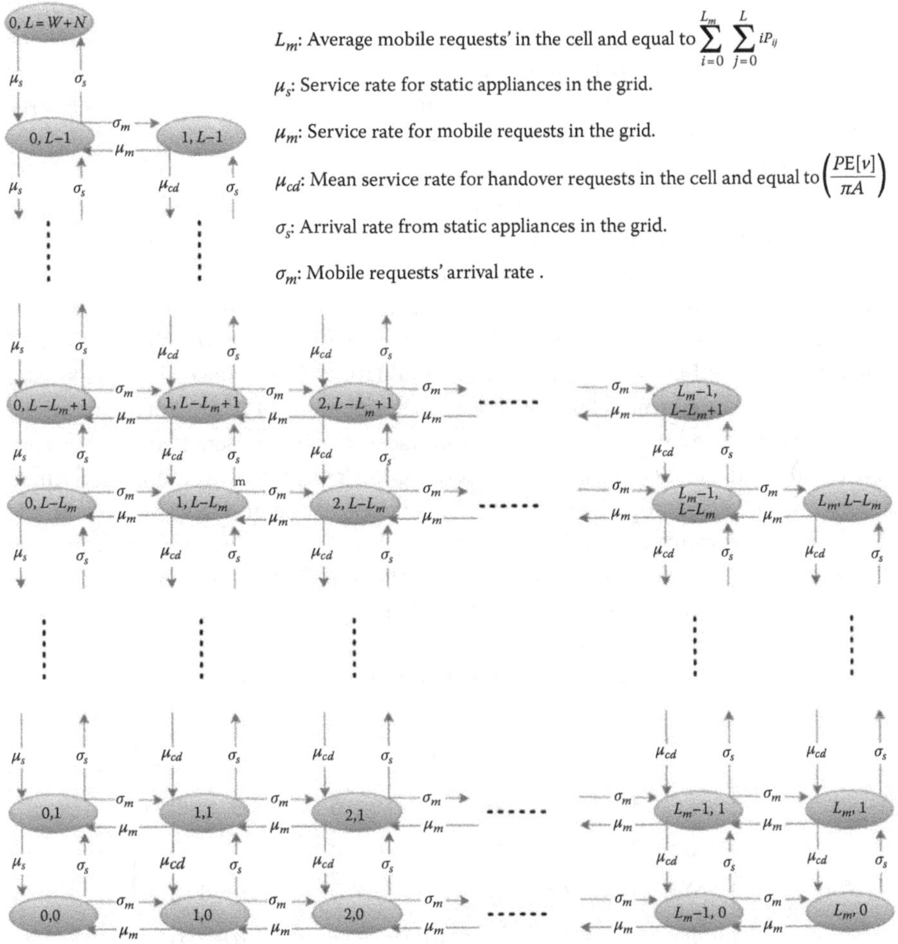

FIGURE 4.2 The 2D Markov-chain for the targeted grid system.

addition to experiencing peak energy-consumption periods. Increasing the number of nodes in the network accentuates this effect, since there is more traffic to forward and the first tier of nodes has no indicator to manage their energy budget. Hence, delegating the role of device scheduling to the FBS based on the proposed energy consumption model can lead to significant gain in terms of energy savings and QoS.

4.4 The Green FBS Model for Smart Grids

Future smart grids are expected to be a combination of macrocells and small cells such as femtocells [32]. FBSs are mostly expected to be static/mobile stations on busses, taxis, trains, smart homes, etc. in a grid application such as Siemens' *e-Mobility* [3] to provide better coverage and capacity to dramatically increasing mobile users. The overall system can be considered in Q-theory as a 2D Markov process. A pair of integer valued random variables, i and j, can be used to describe the state of the system at time t, where $i(t)$ represents the number of available communication channels/interfaces at time t, and $j(t)$ determines the number of existing requests/packets at time t. A 2D state diagram for the Q-system is considered in this study as depicted in Figure 4.2. We can assume the minimum value of i is 0, and the maximum is N representing the maximum number of channels in the system. For the random variable j, the minimum value is 0, and it can take values from 0 to L which is the total number of requests in the system at time t, including the one(s) in service (i.e., $W+N$). The Markov process is denoted by Z and is used for performance evaluation of the considered grid system in this chapter. We assume Z is irreducible with a state space of $N \times N$. Furthermore, we assume that the number of channels/FBS interfaces, $i(t)$, is represented in the horizontal direction and the total number of requests, $j(t)$, is represented in the vertical direction of a finite lattice strip. Thus, possible transitions of the grid system Z are purely lateral transitions from state (i, j) to state (k, j), one-step upward transitions from state (i, j) to state $(k, j+1)$, and one-step downward transitions from state (i, j) to state $(k, j-1)$. In this study, spectral expansion approach is employed, where A is the matrix of purely lateral transitions with zeros on the main diagonal, and one-step upward and one-step downward transitions are represented in matrices B and C, respectively.

$$A=\begin{pmatrix} 0 & \eta & 0 & 0 & 0 & 0 & 0 & 0 \\ \xi & 0 & \eta & 0 & 0 & 0 & 0 & 0 \\ 0 & 2\xi & 0 & \eta & 0 & 0 & 0 & 0 \\ 0 & 0 & 3\xi & 0 & \ddots & 0 & 0 & 0 \\ 0 & 0 & 0 & \ddots & 0 & \ddots & 0 & 0 \\ 0 & 0 & 0 & 0 & \ddots & 0 & \eta & 0 \\ 0 & 0 & 0 & 0 & 0 & (n-1)\xi & 0 & \eta \\ 0 & 0 & 0 & 0 & 0 & 0 & n\xi & 0 \end{pmatrix} \tag{4.4}$$

$$B=\begin{pmatrix} \sigma_s,\sigma_m & 0 & 0 & 0 & 0 & 0 & 0 \\ 0 & \sigma_s,\sigma_m & 0 & 0 & 0 & 0 & 0 \\ 0 & 0 & \sigma_s,\sigma_m & 0 & 0 & 0 & 0 \\ 0 & 0 & 0 & \ddots & 0 & 0 & 0 \\ 0 & 0 & 0 & 0 & \ddots & 0 & 0 \\ 0 & 0 & 0 & 0 & 0 & \sigma_s,\sigma_m & 0 \\ 0 & 0 & 0 & 0 & 0 & 0 & \sigma_s,\sigma_m \end{pmatrix}, \tag{4.5}$$

$$C = \begin{pmatrix} \min(0,j)\mu_s, \mu_m + j\mu_{cd} & 0 & 0 & 0 & 0 \\ 0 & \min(1,j)\mu_s, \mu_m + j\mu_{cd} & 0 & 0 & 0 \\ 0 & 0 & \ddots & 0 & 0 \\ 0 & 0 & 0 & \ddots & 0 \\ 0 & 0 & 0 & 0 & \min(n,j)\mu_s, \mu_m + j\mu_{cd} \end{pmatrix}, \quad 0 \le j \le L. \tag{4.6}$$

In this grid system, the transition rate matrices are always dependent on j, because the requests in the queue may leave the system due to the assumed user/FBS mobility.

Lemma 4.1:

Elements of matrix A depend only on the cut-off and recover rates of the FBS channels, ξ and η, respectively. The transition rate matrices A, B, and C are square matrices each of size $(N+1) \times (N+1)$, and given in Eqs. (4.4)–(4.6). The matrix C depends on the number of requests in the system for $j = 0, 1, \ldots, L$. Thus, the threshold M of the spectral expansion approach can be set to L.

If the number of requests in the system is less than the number of available channels, each request is served using a channel. On the other hand, if the number of requests is greater than the number of available channels, N requests with the highest priority are served first and the remaining can only handover to a neighboring cell with the service rate μ_{cd}. The spectral expansion solution approach is employed for the steady-state solution. Details of the spectral expansion solution approach can be found in Ref. [33]. Following spectral expansion solution, *Lemma 2* can be stated as follows:

Lemma 4.2:

The steady-state probabilities of the states can be expressed as:

$$P_{ij} = \lim_{t\to\infty}\left(Pi(t) = i,\ \ j(t) = j\right), 0 \le i \le N \text{ and } 0 \le j \le L. \tag{4.7}$$

From *Lemma 1* and *Lemma 2*, it is possible to obtain *Theorem 1*, as follows:

Theorem 1: Probability of being in a state *ij* is

$$P_{ij} = \sum_{l=0}^{N}\left(a_l\psi_l(i)\lambda^{j-M+1} + b_l\varphi_l(i)\beta^{L-j}\right), \quad M-1 \le j \le L, \tag{4.8}$$

where, λ and ψ are eigenvalues and left-eigenvectors of $Q(\lambda)$, respectively, and ψ is a row-vector defined as $\psi = \psi_0, \psi_1, \ldots, \psi_N$, λ is a row-vector defined as $\lambda = {}_0, \lambda_1, \ldots, \lambda_N$, and $\psi Q(\lambda) = 0$; $|Q(\lambda)| = 0$. On the other hand, β and φ are eigenvalues and left-eigenvectors of $\bar{Q}(\beta)$, respectively, and φ is a vector defined as $\varphi = \varphi_0, \varphi_1, \ldots, \varphi_N$, $\beta = \beta_0, \beta_1, \ldots, \beta_N$. ■

Consequently, the state probabilities in *Theorem 1* can be used to calculate important performance measures such as MQL, response time (R), and throughput (γ) as follows.

$$MQL_m = \sum_{i=0}^{L_m} \sum_{j=0}^{L} iP_{ij}, \quad MQL_s = \sum_{i=0}^{L_s} \sum_{j=0}^{L} jP_{ij} \tag{4.9}$$

$$=> MQL = MQL_m + MQL_s$$

$$\gamma_m = \sum_{i=0}^{L_m} \sum_{j=0}^{L} \mu_m P_{ij}, \quad \gamma_s = \sum_{i=0}^{L_s} \sum_{j=0}^{L} \mu_s P_{ij} \tag{4.10}$$

$$=> \gamma = \gamma_m + \gamma_s,$$

$$R_m = \frac{MQL_m}{\gamma_m}, \quad R_s = \frac{MQL_s}{\gamma_s}. \tag{4.11}$$

$$=> R = R_m + R_s$$

Based on the aforementioned formulas, the FBS simply can decide on whether to accept a new user in the queue or not, where a User Equipment (UE) should communicate with the FBS that has the best SNR. In fact, the UE_i will report the mobile FBS's SNR to the serving macrocell and send a Req_i to offload. Algorithm 1 represents this stage at the macro-BS. When the SNR trigger condition is satisfied, the serving macro-BS checks the status of the UE indicated in lines 2–5. If UE's status is active, the macro-BS will classify the UE_i traffic as indicated in line 6. As a result, the user category (C_i) will be associated with UE_i. The macro-BS waits a predefined residence time (*res_time*) and then sends the Req_i with its associated C_i to the mobile FBS as in lines 7–8. However, if UE_i status is other than active, the macro-BS ignores its request and keeps it associated with macro-BS. After receiving the decision from the mobile FBS, the macro-BS will switch (offload) the UE_i to the mobile FBS if it receives an *Accept* message from the FBS (lines 8–9). Where the *Decide()* function can be as detailed in Algorithm 2.

Based on Algorithm 2, once the mobile FBS receives the Req_i from an UE_i, it checks whether to accept or not (lines 2–12). If the FBS chooses to accommodate the UE_i, it informs the serving macro-BS to transfer the data session of UE_i, update the list of offloaded users (U_f) and their total counts T_{total}, (lines 3–4). However, if the FBS reaches its maximum threshold L, where L here is equal to U_{fmax} and there exists an UE_i connected with the same FBS but with lower priority than the new UE_j. In this case, the mobile FBS will transfer/return the user with the minimum priority (UE_{min}) to the macro-BS and accept the new UE_j (lines 6–9). Finally, if the WiFi signal strength degrades below a certain threshold, the mobile FBS will ask the macro-BS to take its list of UE (lines 14 and 15).

Algorithm 1: UE_i Categorization at the Macro-BS

Input: Req_i: a request by UE_i to switch to mob-FBS

1. Receive a Req_i from a UE_i
2. **If** UE_i is active **then**
3.   **If** the UE_i has a voice call **then**
4.   Ignore // i.e., keep connected to the macro-BS
5.   **Else**
6. C_i = Classify UE_i based on application types
7.   wait for *res_time*
8.   **If** (**Decide** (Req_i) == Accept) **then**
9.   Transfer UE_i to the FBS
10.   **Else**
11.   Ignore
12.   **Endif**
13.   **Endif**
14. **Else**
15. Ignore
16. Endif
17. **End**

Algorithm 2: Decision at the Mobile FBS

Decide (Req_i)

Input: Req_i: is a data user request from a UE_i associated with its C_i

Output: Accept/Reject: message sent to macro-BS to transfer/keep the UE_i

Initialize: U_f, U_{fmax}, T_{total}, UE_{min}

Receive a Req_i from the macro-BS of UE_i
If $U_f < U_{fmax}$ AND $T_{total} \leq L$, **then**
  Accept UE_i
  $U_f = U_f + UE_i$
  $T_{total} = T_{total} + 1$
Elseif $U_f = U_{fmax}$ AND ⟨I⟩∃⟨/I⟩ C_{i-1} ⟨I⟩⟨SUB⟩∈⟨/SUB⟩⟨/I⟩ U_f then
    Switch UE_{min} to macro-BS & Accept UE_i
    $U_f = U_f + UE_i$
    $T_{total} = T_{total} + 1$
    Else
      Reject UE_i
    Endif
Endif
If FBS is *cut-off* **then**
    Transfer $\{U_f\}$ back to the macro-BS
Endif
End

4.5 A Typical Case-Study in Smart Grids: *e-Mobility*

e-Mobility is a real case scenario proposed and instantiated by Siemens [3] for applied mobile FBS in smart grids not only as a mean of transportation but also to feedback electricity into the grid at the peak hours. It represents the concept of using electric powertrain technologies, in-vehicle information, and communication technologies and connected infrastructures to enable the electric propulsion of vehicles and fleets. Powertrain technologies include full electric vehicles and plug-in hybrids, as well as hydrogen fuel cell vehicles that convert hydrogen into electricity. *e-Mobility* efforts are motivated by the need to address corporate fuel efficiency and emission requirements, as well as market demands for lower operational costs. Electrical vehicles/trains in this case study are relying on mobile FBS in exchanging their energy status during the day in addition to other smart-home appliances, and thus, a heavy data traffic is expected to be generated. In such a comprehensive mobile HetNet model, an efficient energy-consumption policy is required to motivate the usage of femtocells in serving thousands of incoming requests per hour in a green framework. Moreover, such kind of a mobile HetNet model introduces plenty of challenges regarding the system's capacity and targeted QoS. Hence, we visualize a green HetNet-driven femtocell case study for *e-Mobility* in smart grid that tackles the aforementioned concerns.

We consider a set of femtocells, which are deployed inside the coverage area of a macrocell to provide sufficient users' capacity while maintaining adequate QoS in terms of throughput, MQL, response time, and energy consumption. Mobile users might be static/mobile and may use their smartphones and energy-hungry mobile applications such as mobile video streaming while they are commuting over the city road. Each FBS can be described via a set of hardware parameters as described above in Section 4.3. Typically, users are assumed to be uniformly distributed in the coverage area of their serving cell. The FBS parameters used in the proposed queuing system for this case study are summarized in Table 4.2 while assuming

TABLE 4.2 Specifications of FBS Parameters [31,34]

Parameter	*Value*
P (mW)	20
BW (MHz)	5
N_0 (W/Hz)	$4*10^{-21}$
p_{mp} (W)	3.2
p_{FPGA} (W)	4.7
p_{trans} (W)	1.7
p_{amp} (W)	2.4
FBS radius (m)	30
FBS velocity (km/h)	Low, medium, high
FBS channels	8
Expected cut-off rate per hour (ξ)	0.001
Expected arrival rate per hour (σ)	2000
Expected recover rate per hour (η)	0.5
ρ	4.8
δ^2	10
K_0	42.152
R	30 m

typical LTE-values that have been used in practice [34] and the assumed *e-Mobility* scenario is shown in Figure 4.3. In order to assess the proposed Q-model under this scenario, we consider the summarized performance metrics in Table 4.3.

4.5.1 Simulation Setups

Using MATLAB R2016a and Simulink 8.7, we simulate randomly generated HetNets to represent the targeted smart-grid environment in a smart city. A discrete event simulator is built on top of these MATLAB platforms, which considers practical aspects in the network physical layer for more realistic performance evaluations. Our Simulink simulator supports wireless channel temporal variations, node mobility, and cut-offs. Based on experimental measurements taken in a site of dense heterogeneous nodes [35], we adopt the described signal propagation model in Section 4.3, where we set the communication model variables as shown in Table 4.2, and χ to be a random variable that follows a log-normal distribution

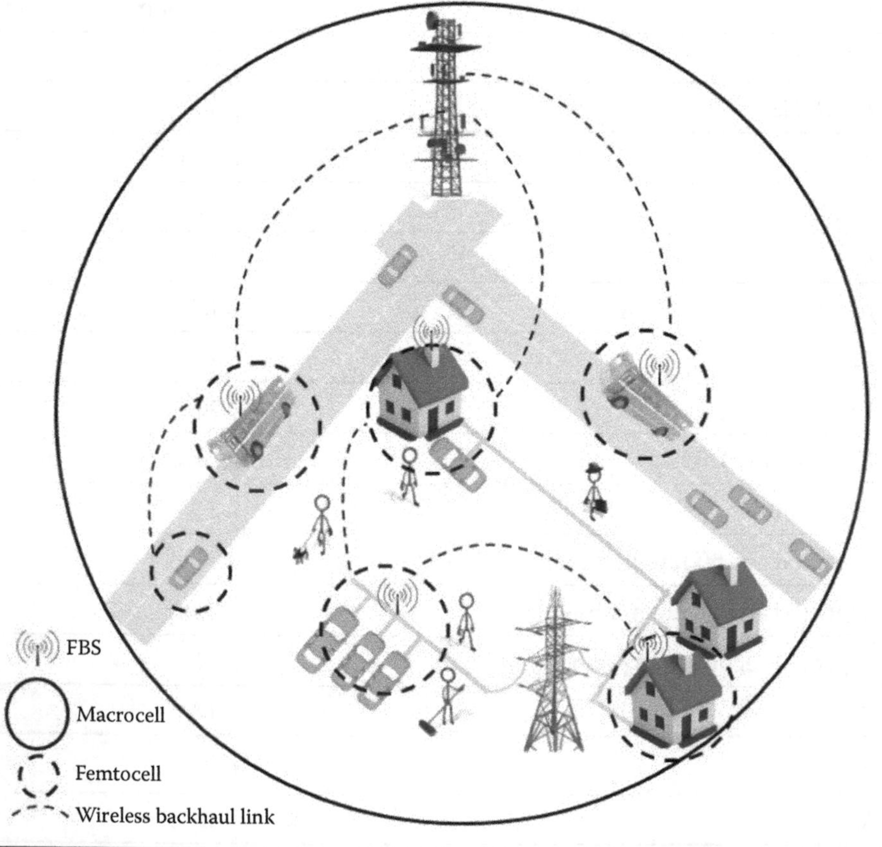

FIGURE 4.3 Mobile FBSs serving mobile/static users in a smart grid of FBSs.

TABLE 4.3 Performance Metrics and Parameters

Performance Metrics & Parameters	*Definitions*
Throughput (γ)	The average percentage of transmitted data packets that succeed in reaching the destination. This metric has been chosen to reflect the effectiveness of the mobile Femtocell Base Station (FBS) in a Heterogeneous Networks (HetNet) setup and it is measured in *"packets/h"*
Mean Queue Length (MQL)	The average number of the requests pending in the system, either waiting in the queue or being served. This metric represents the Quality of Service (QoS) from the FBS perspective and is measured in *"packets"*
Response time (R)	The time spent by a mobile/static user from arrival until departure and plays a significant role in performance evaluation since it incorporates all the delays involved per user request
Energy consumption (E)	The amount of energy consumed by a single FBS based on the arrived user requests and the FBS dynamic status change
Service rate (μ)	The number of served users per hour at a FBS. It is used to analyze the effect of traffic loads on the performance of the FBS system
FBS velocity	The velocity of a mobile FBS measured in kilometer per hour

function with mean zero and variance of δ^2. In this simulator, an event-based scheduling approach is taken into account, which depends on the events and their effects on the system state. The assumed event-based scheduling approach is typical for data gathering in wireless networks. It allows multiple, parallel indirect transmissions across multiple, adjacent clusters, with collision avoidance techniques. Since we are assuming priority-based queue, scheduled packets can be delayed in favor of the node with the highest rank, and packets having the same sender are ordered according to their arrival time. As for the stopping criterion, we assume a commonly used one, called relative precision [9]. It is employed in our simulations to be stopped at the first checkpoint when the condition $\beta < \beta_{max}$, where β_{max} is the maximum acceptable value of the relative precision of confidence intervals at $(1-\alpha)$ significance level. Accordingly, our achieved simulation results are within the confidence interval of 5% with a confidence level of 95%, where both default values for

β and α are set to 0.05. The simulation results obtained from our MATLAB code are presented comparatively with the analytical results from our queue model and validated to reflect the performance of the actual femtocell system.

4.5.2 Results and Discussions

Obtained results for the proposed case study are presented in this section under two subsections. First, impact of the mobility-speed factor on an FBS performance in terms of throughput and queue length is studied. Second, the impact of traffic load and service rate on the FBS energy consumption is considered and analyzed.

4.5.2.1 The Impact of Velocity on the FBS Performance

Based on the arrival rate (σ) that varies in this study from 2500 to 6000 users/h, performance metrics are compared for three different FBSs' mobility speed categories: (i) low speed FBSs such as stationary ones at smart homes and in pedestrian handheld smart devices with the velocity from 0 to 15 km/h, (ii) medium speed FBSs like those on top of busses with the velocity from 15 to 40 km/h, and (iii) high speed FBSs with the velocity above 40 km/h, such as those on top of electric cars/trains.

Figure 4.4 shows the effect of velocity of the mobile FBS on the MQL for various arrival rates. It is clear from the figure that when the system is congested like in densely populated areas such as airports and city centers, the MQL will also grow. This is because more users request service from the FBS at the same time. As the mobile FBS moves faster, the MQL decreases. This is because the service rate, μ_{cd},

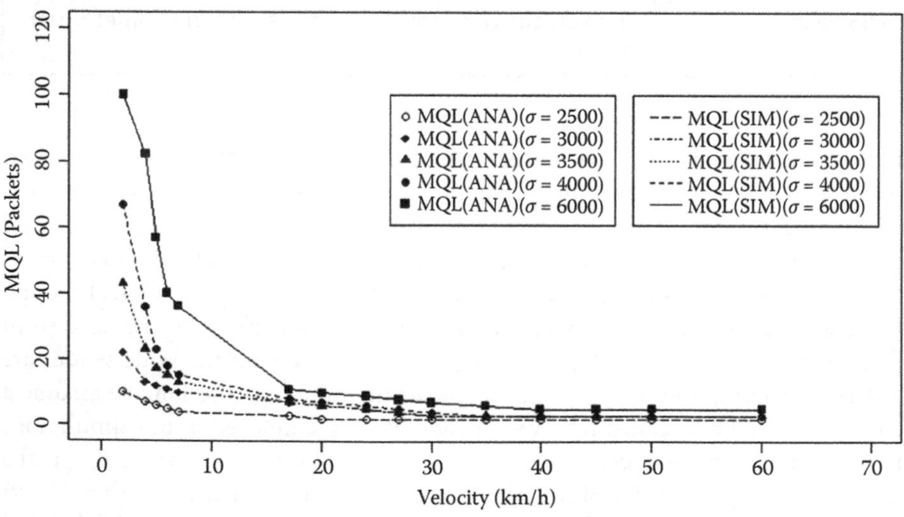

FIGURE 4.4 The effect of velocity of mobile FBSs on MQL.

is directly proportional to the expected velocity of mobile FBS. Therefore, as the velocity increases, the users' requests will leave the FBS queue sooner and the MQL will decrease. For instance, when the arrival rate σ is equal to 6000 the MQL is very close to queue capacity at velocity of 1 km/h. But, when the mobile FBS start moving faster at speed of 60 km/h, the MQL is equal to four requests.

In Figure 4.5, throughput of the system is presented as a function of average velocity of the mobile FBS for different values of arrival rates. The parameters are same as the parameters used in Figure 4.4. It is obvious that as arrival rate increases, more requests are served and throughput will increase too. It has been also observed that as the mobile FBS moves faster, the smart grid throughput decreases. This is because when the velocity increases and the mobile FBS moves faster, the femtocell users are removed away from the FBS before they are served. Therefore, the number of served requests decreases, and consequently, throughput of the grid decreases as well. It's worth pointing out that for both Figures 4.4 and 4.5, simulation results are also performed comparatively for validation purposes. The maximum discrepancy between the analytical results and simulation are 1.96%, and 0.07% for Figures 4.4 and 4.5 respectively which is less than the confidence interval 5%.

4.5.2.2 The Impact of Traffic Load on the Energy Consumption

Generally, when $d \leq R$, the single-hop communication is considered to be more energy efficient. Particularly, because of the low-path-loss exponents because the distance for one hop is close to the perfect value of SNR, and the start-up power overhead makes the multi-hop strategy inefficient for hop-distances less than d. The assumed reduction in energy consumption is considered as a central issue and should be utilized in terms of reducing the power overhead through a new strategy for HetNets, where FBS can be deployed everywhere to provide a single

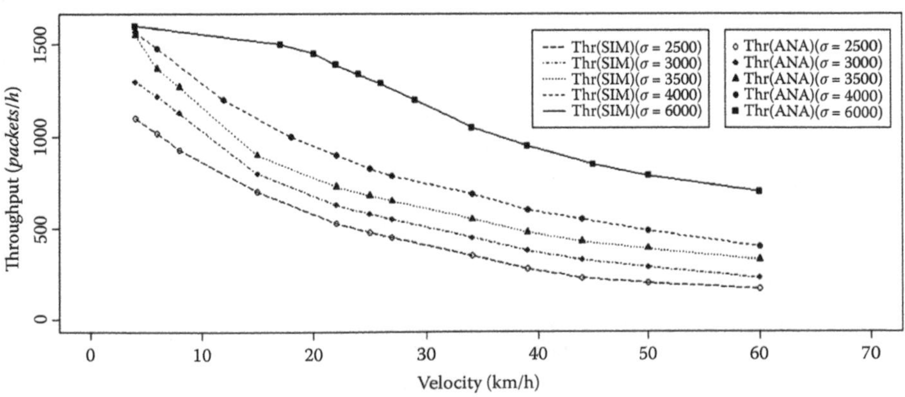

FIGURE 4.5 The effect of velocity of mobile FBSs on throughput.

hop communication in connecting heterogeneous nodes viz., sensors, Personal Digital Assistants (PDAs), tablets, etc. Thus, multiple channel FBS can dramatically decrease power consumption while relying on single-hop communications. Figure 4.6 represents the amount of energy consumed while MQL is increasing. Obviously, the increasing MQL parameter here causes undesired exponential energy increase. Thus, more attention shall be given for the queue length in HetNets. In Figure 4.7, the effect of how much dense is the environment in terms

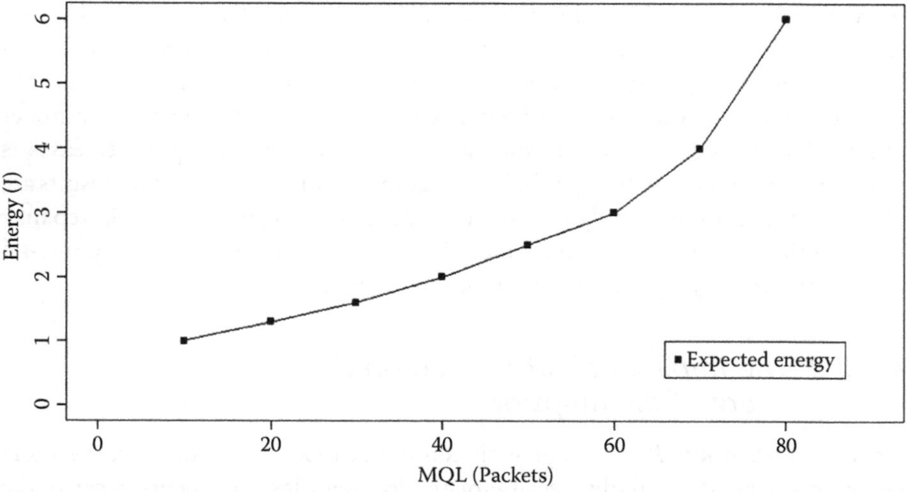

FIGURE 4.6 Energy spent per hour vs. the average MQL.

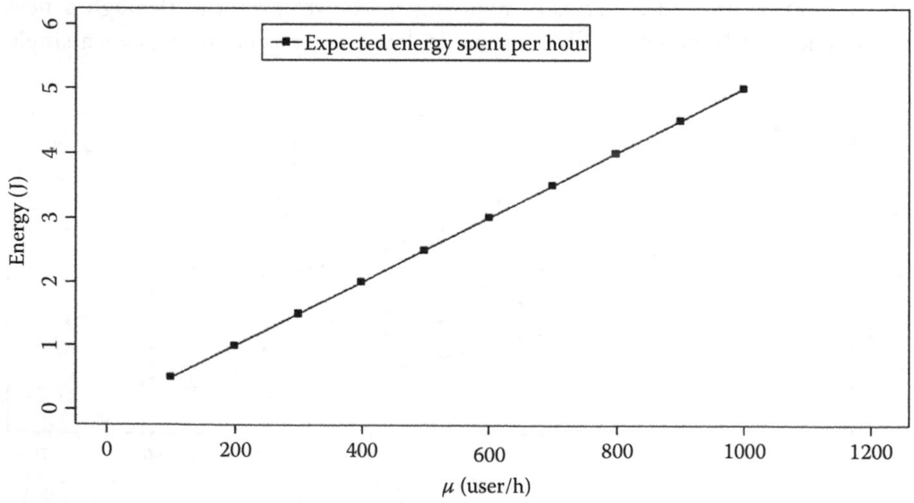

FIGURE 4.7 Energy spent per hour vs. the average service rate μ.

of mobile FBS users is studied against the average energy consumption, as well. The service rate μ is used to reflect the traffic load in the vicinity of an FBS. We notice that even with relatively large increments in user/arrivals counts, the FBS energy consumption is not that much affected. Unlike the effect of MQL, number of served users can barely affect the consumed FBS energy, where the increment in energy is linear instead of the observed exponential one in Figure 4.6. This can be returned to the mobility factor, where a great portion of these arrivals is leaving the queue due to varying communication range conditions (d shall be much less than R, which is the macrocell range). Nevertheless, as the service rate increases, the more energy is consumed of course. For example, at μ equal to 1000, the expected energy consumption is 5.3 J, which is 10 times higher when μ is equal to 100.

In Figure 4.8, we show the effect of the service rate against the response time (R). We note that as the service rate increases, the response time starts to decrease significantly. For instance, at μ equal to 1000, response time is 5.4 s, which is 60 times lower when μ is equal to 100 users/h. By merging Figures 4.7 and 4.8 in to one plot, we can examine the optimal service rate that guarantees the fastest response time against the lowest energy consumption. As it is shown in Figure 4.9, the optimum area is obtained at μ equal to 380 users/h at which the response time is equal to 1.8 s and the expected energy consumption is equal to 1.59 J. In addition, Figure 4.9 shows that there is a trade-off between energy consumption and the performance of the FBS as expected. It also shows that the proposed Q-model can play a key role in specifying the operative space, performance level as well as energy consumption in smart-grid systems.

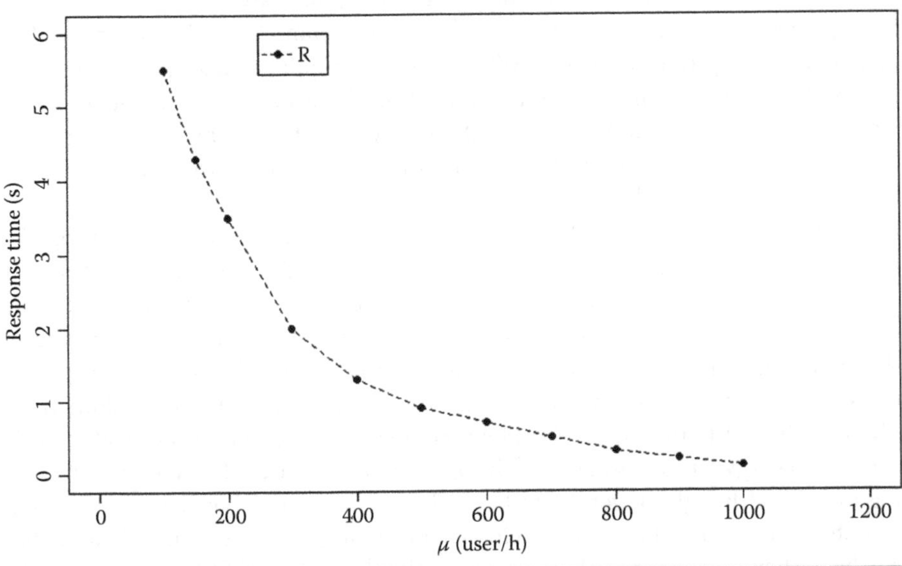

FIGURE 4.8 Response time versus the service rate.

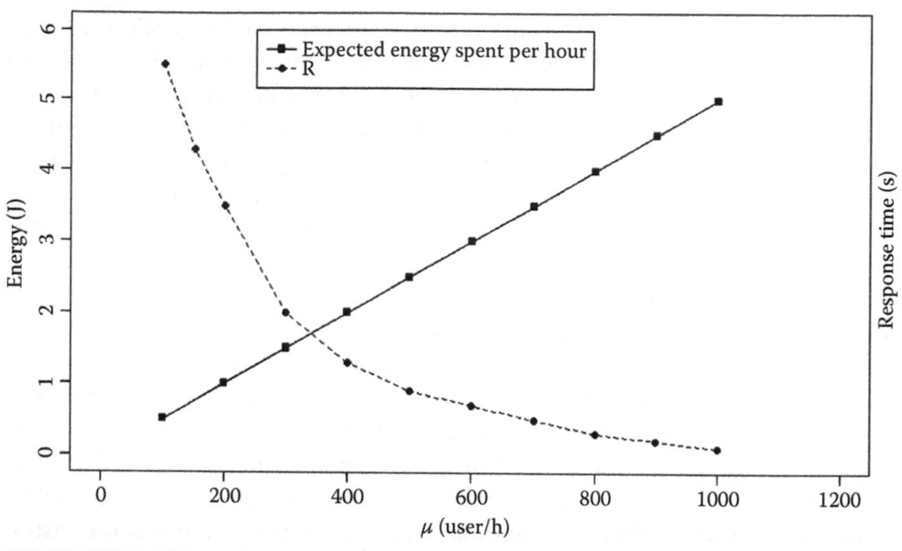

FIGURE 4.9 Response time and energy spent per hour as a function of the service rate.

Figure 4.10 depicts the energy consumption of the macrocell, with varying arrival rates in relation to the number of FBSs per cell. The energy saving trend in smart-grids reflects that there is some power saving achieved through the deployment of FBSs while experiencing high and low arrival rates (i.e., σ = 2500 and σ = 6000). The trend reflects that the energy saving increases from 6.9% to about 9.45% when the number of FBSs per cell increases from 10 to 20 when σ = 6000. The power saving trends for σ = 2000 reflects that quite smaller power saving is achieved through the introduction of the FBS. The lower power saving achieved while experiencing σ = 2000 can be explained by the efficient energy-consumption model proposed in this research in predicting and tolerating the extra incoming load and reacting accordingly.

To further investigate the increasing energy saving trend while increasing the count of FBSs, the energy saving is evaluated for two different service rates; μ is equal to 300, and 400 users/h. The results are depicted in Figure 4.11. It can be seen that the power saving is doubled for both μ = 300 and μ = 400. Therefore, the increasing number of FBSs is contributing more toward the power saving when the service rate value is closer to the optimum value. Thus, we can conclude from Figures 4.10 and 11 that the FBS deployment in smart grids with heavy traffic loads can be an energy efficient solution if the optimal service rate value is used. Where in Figure 4.11, although the total number of FBSs is the same, but the energy saving is quite better while experiencing a closer service rate μ to the optimized value that has been shown in Figure 4.9. Thus, the optimal number of FBS depends significantly on the usage of the optimal μ value.

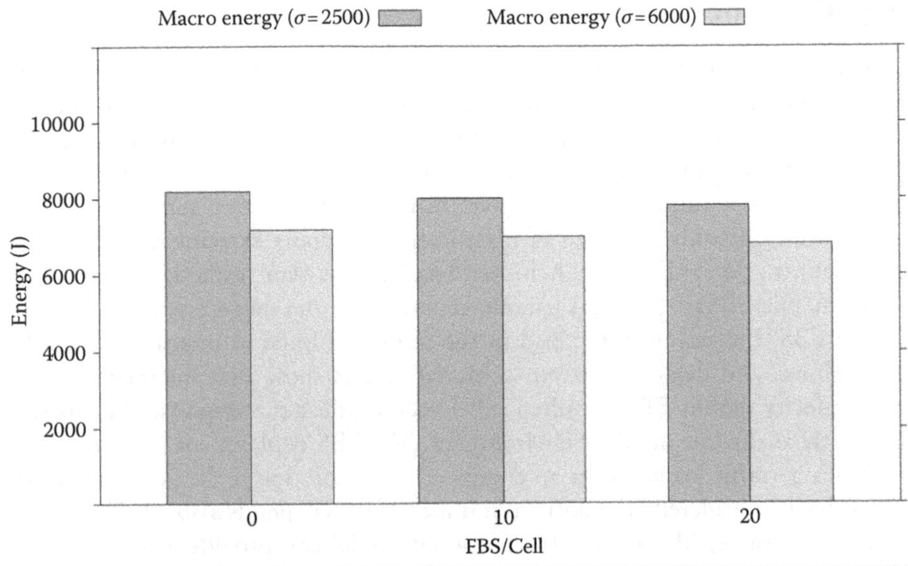

FIGURE 4.10 Energy consumption of the macrocell as a function of FBS per cell while the arrival rate is varying.

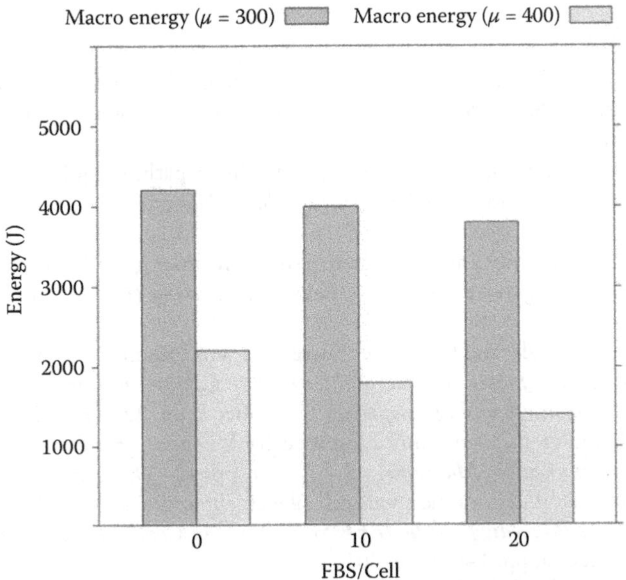

FIGURE 4.11 Energy consumption of the macrocell as a function of FBS per cell while the service rate is varying.

4.6 Conclusion

In this chapter, a hybrid HetNet consisting of a macrocell and several mobile FBSs is considered for a smart-grid application in presence of cut-offs and energy-saving assumptions. With the introduction of internet of things, we believe that the aid of highly available femtocells in such setups for the increasing traffic load will become very important. A new system model is presented for mobile FBSs, which are commonly deployed in macrocells to support extreme data traffic in green smart-grid applications. Achieved results show that traffic load and velocity of mobile FBSs are crucial parameters affecting the mean energy consumed by the FBS. Increasing traffic load in the femtocell leads to increases in MQL, throughput, and delay. Moreover, achieved results show that for medium and high velocity mobile FBSs, traffic load does not affect the response time as significantly as the low mobility environment. The FBS deployment in smart grids with heavy traffic loads can be an energy efficient solution if the optimal service rate value is considered. The optimal number of FBS depends also on the average service rate value. Meanwhile, the proposed model can provide a threshold for the mean energy consumption while expecting a specific response time. Such a model can be quite useful in specifying the operative space for FBSs. In general, this model can have a key impact on the next generation HetNets design and planning aspects.

References

1. CISCO, Cisco Visual Networking Index (VNI): VNI mobile forecast highlights, 2015–2020, 2016, http://www.cisco.com/assets/sol/sp/vni/forecast_highlights_mobile/index.html.
2. G. Singh, and F. Al-Turjman, Learning data delivery paths in QoI-aware information-centric sensor networks, *IEEE Internet of Things Journal*, vol. 3, no. 4, pp. 572–580, 2016.
3. M. Morte, e-Mobility and multiagent systems in smart grid, *Proceedings of the IEEE International Conference on Electric Power Engineering (EPE)*, Karlsruhe, pp. 1–4, Sept. 2016.
4. A. Banote, V. Ubale, and G. Khaire, Energy efficient communication using femtocell – a review, *International Journal of Electronics, Communication & Instrumentation Engineering Research and Development*, vol. 3, no. 1, pp. 229–236, 2013.
5. E. Bou-Harb, et al., Communication security for smart grid distribution networks, *IEEE Communications Magazine*, vol. 51, no.1, pp. 42–49, 2013.
6. M. Qutqut, et al., Dynamic small cell placement strategies for LTE heterogeneous networks, *in Proceedings of the IEEE Symposium on Computers and Communications (ISCC)*, Madeira, pp. 1–6, June, 2014.
7. M. Qutqut, et al., MFW: mobile femto-cells utilizing WiFi, *Proceedings of the IEEE International Conference on Communications (ICC)*, Budapest, pp. 5020–5024, June, 2013.

8. M. Z. Hasan, et al., A survey on multipath routing protocols for QoS assurances in real-time multimedia wireless sensor networks, *IEEE Communications Surveys and Tutorials*, 2017, doi: 10.1109/COMST.2017.2661201.
9. A. Law, Statistical analysis of simulation output data: the practical state of the art, *Proceedings of the IEEE Simulation Conference*, Washington, DC, pp. 77–83, Dec. 2007.
10. M. Arshad, A. Vastberg, and T. Edler, Energy efficiency gains through traffic offloading and traffic expansion in joint macro pico deployment, *Proceedings of the IEEE International Conference on Wireless Communications and Networking (WCNC)*, Paris, pp. 2203–2208, Sept. 2012.
11. R. Riggio, and D. Leith, A measurement-based model of energy consumption in femtocells, *Proceedings of the IEEE International Conference on Wireless Days*, Dublin, pp. 1–5, Nov. 2012.
12. M. Marsan, and M. Meo, Queueing systems to study the energy consumption of a campus WLAN, *Computer Networks*, vol. 66, no. 1, pp. 82–93, 2014.
13. A. Silva, M. Meo, and M. Marsan, Energy-performance trade-off in dense WLANs: a queuing study, *Computer Networks*, vol. 56, no. 1, pp. 2522–2537, 2012.
14. G. Ambene, and G. Anni, *Queuing Theory and Telecommunications*, Springer, New York, 2014.
15. G. Bolch, et al., *Queueing Networks and Markov Chains: Modeling and Performance Evaluation with Computer Science Applications*, John Wiley & Sons, New York, 2006.
16. T. Elkourdi, and O. Simeone, Femtocell as a relay: an outage analysis, *IEEE Transactions on Wireless Communications*, vol. 10, no. 12, pp. 4204–4213, 2011.
17. F. Al-Turjman, Information-centric sensor networks for cognitive IoT: an overview, *Annals of Telecommunications*, vol. 72, no. 1, pp. 1–16, 2016.
18. J. Gong, S. Zhou, and Z. Niu, Queuing on energy-efficient wireless transmissions with adaptive modulation and coding, *Proceedings of the IEEE International Conference on Communications (ICC)*, Kyoto, pp. 1–5, June 2011.
19. V. Borodakiy, et al., Modelling and performance analysis of pre-emption based radio admission control scheme for video conferencing over LTE, *Proceedings of the ITU Kaleidoscope Academic Conference*, St. Petersburg, pp. 53–59, June 2014.
20. M. Chowdhury, et al., Service quality improvement of mobile users in vehicular environment by mobile femtocell network deployment, *Proceedings of the International Conference on ICT Convergence (ICTC)*, Seoul, pp. 194–198, Sept. 2011.
21. O. Karimi, J. Liu, and C. Wang, Seamless wireless connectivity for multimedia services in high speed trains, *IEEE Selected Areas on Communications*, vol. 30, no. 4, pp. 729–739, 2012.
22. F. Saghezchi, A. Radwan, and J. Rodriguez, Energy-aware relay selection in cooperative wireless networks: an assignment game approach, *Ad Hoc Networks*, vol. 56, no. 1, pp. 96–108, 2017.
23. R. Baloch, et al., A mathematical model for wireless channel allocation and handoff schemes, *Telecommunication Systems*, vol. 45, no. 4, pp. 275–287, 2010.
24. Q. Zeng, and D. Agrawal, Modeling and efficient handling of handoffs in integrated wireless mobile networks, *IEEE Transactions on Vehicular Technology*, vol. 51, no. 6, pp. 1469–1478, 2002.
25. V. Sucasas, A survey on clustering techniques for cooperative wireless networks, *Ad Hoc Networks*, vol. 47, no. 1, pp. 53–81, 2016.

26. A. Mukherjee, et al., Femtocell based green power consumption methods for mobile network, *Computer Networks*, vol. 57, no. 1, pp. 162–178, 2013.
27. K. Trivedi, S. Dharmaraja, and X. Ma, Analytic modeling of handoffs in wireless cellular networks, *Information Sciences*, vol. 148, pp. 155–166, 2002.
28. I. El Bouabidi, et al., Design and analysis of secure host-based mobility protocol for wireless heterogeneous networks, *Journal of Supercomputing*, vol. 70, no. 1, pp. 1036–1050, 2014.
29. H. Beigy, and M. Meybodi, A learning automata-based adaptive uniform fractional guard channel algorithm, *Journal of Supercomputing*, vol. 71, no. 1, pp. 871–893, 2015.
30. F. Al-Turjman, and H. Hassanein, Towards augmented connectivity with delay constraints in WSN federation, *International Journal of Ad Hoc and Ubiquitous Computing*, vol. 11, no. 2, pp. 97–108, 2012.
31. M. Deruyck, et al., Modelling the power consumption in femtocell networks, *Proceedings of the IEEE Wireless Communications and Networking Conference*, Paris, pp. 30–35, April, 2012.
32. W. Wang, and G. Shen, Energy efficiency of heterogeneous cellular network, *Proceedings of the IEEE Vehicular Technology Conference*, Taipei, pp. 1–5, May, 2010.
33. R. Chakka, Spectral expansion solution for some finite capacity queues, *Annals of Operations Research*, vol. 79, pp. 27–44, 1998.
34. J. Zhang, et al., *Femtocells: Technologies and Deployment*, Wiley Online Library, New York, 2010.
35. G. Solmaz, M. Akbas, and D. Turgut, A mobility model of theme park visitors, *IEEE Transactions on Mobile Computing*, vol. 14, no. 12, pp. 2406–2418, 2015.

Chapter 5

Agile Medium Access in Smart-Cities Vehicular-IoT

Fadi Al-Turjman
Antalya Bilim University

Contents

5.1 Introduction

Every year, hundreds of thousands of people are killed in car accidents. For example, over 1.2 million people were killed in road traffic accidents around the world in 2016 [1]. The causes of road accident vary from one country to another. However, a common major reason for car accidents is driving under bad weather and traffic conditions. This behavior is denoted as speeding. A speeding driver might be driving below the posted speed limits. However, his speed could be much higher than what it should be. Hence, it is important to emphasize that 50% of fatal accidents occur at impact speeds less than 55 km/h [1]. Consequently, speed limits should be set and enforced according to weather, traffic, and road conditions. With the recent revolution in wireless telecommunications, several advanced solutions relying on wireless communication standers have been proposed to provide Intelligent Transportation Systems (ITS) in the Internet of Things (IoT) paradigm. For example, authors in Ref. [2] proposed an Automatic Speed Control (ASC) system to automatically adjust the vehicle speed in order to match the speed limit. The development of a smart box termed "Telematics," a tool similar to the so-called black box found in aircraft, have been investigated in collaboration with IBM's Engineering & Technology Services. Telematics can capture, analyze, and deliver relevant data via a wireless network. Using multiple microprocessors, plus a multitude of other sensors that can be attached to an automobile's carriage to, it is able to monitor the vehicle's speed for example, comparing it to the speed limit of the street. If the car speed is higher than the speed limit allowed by the traffic department, the box would talk to the driver and issue a verbal warning. Baró et al. [3] had proposed to use digital image processing to recognize the traffic signs on the street sides and generate certain signals to alert the driver or to control the car. Different versions of this system have been investigated intensively all over the world. Results in Ref. [4] have shown that this solution can reduce the accidents rate by 35%. It is believed that the speed control system will rely heavily on the standard of IEEE 802.16 to locate each vehicle and satisfy the demands of real-time services such as voice and video. The standard of IEEE 802.16 is designed and developed to offer specific services for wireless radio interface [5]. The Task Group (TG) of IEEE 802.16 has significant advantages, namely higher data rate, scalability, real-time serviceability, low-cost maintenance and cost upgrade [6]. However the IEEE 802.16d/e does not have an appropriate scheduler algorithm for the real-time service [5]. This research focuses its attention on IEEE 802.16 system to enhance the reliability of the emerging networks. Since IEEE 802.16 system cannot be distinctly specific in its resource allocation among the real time applications, the base station unfairly allocates and shares the resources, namely frequency spectrum and time slots with different types of traffic flows [5].

In IEEE 802.16, there are different Medium Access Control (MAC) scheduling services, such as Unsolicited Grant Service (UGS), real-time Polling Service (rtPS), and non-real time Polling Service (nrtPS) to provide better Quality of Services (QoSs). UGS and rtPS are the two schedulers for the real-time traffic. Each real-time scheduling mechanism has a parameter to quantify its bandwidth requirements, namely delay, minimum and maximum transmission rate [7]. However, these schedulers do not fulfill the requirements for the real-time services in smart cities. The suitability of Batch Markovian Arrival Process (BMAP) is analyzed in studies such as Ref. [8] for modelling of IP traffic, and it is shown that the BMAP model is a better candidate especially compared to other popular processes such as Markov Modulated Poisson Process. Thus, we propose a Real-Time-BMAP

(RT-BMAP) model for real-time services. The objective of RT-BMAP is to achieve the required QoS with the minimum delay. The major contributions of this chapter are as follows:

1. We present the Enhanced-real time Polling System (E-rtPS) with RT-BMAP and proactive resource allocation framework to resolve the real-time network traffic issues and solve the interference problem optimistically.
2. The proposed proactive resource allocation framework offers less computational overhead and yet again well-nigh performance in terms of resource allocation to maximize the transmission rate of each users.
3. The examination of real-time results shows that the proposed framework outperforms the existing IEEE 802.16 services in terms of throughput, session setup delay and packet reception ratio.

5.2 Related Work

In the Long-Term Evolution-Advanced (LTE-A), QoS Class Identifier (QCI) uses Guaranteed Bit Rate (GBR) and Non-Guaranteed Bit Rate (Non-GBR) to support different priorities and delay requirements [9]. According to the statistical report generated for mobile data traffic in 2013, the global usage of mobile data has grown dramatically to 83% [8]. The datum report says that the data traffic reached 1.5 exabytes at the end of 2013, which was a hike of 820 petabytes per month at the end of 2012. As reported in CISCO forum, the network applications on global networks grew the data traffic 63% in 2016, which was 7.2 exabytes per month in 2016 and 4.4 exabytes per month in 2015. In fourth generation (4G) networks, mobile data traffic has exponentially surged into 69%, which is four times more data traffic than third generation (3G) connection. Since the smartphone usage had raised for half-a-billion, the mobile device and connection were accounted of 89% for the usage of mobile data traffic. Therefore, in 2016, the smartphones generate 13 times more data traffic than the non-smartphones. The CISCO statistical survey reports that the average traffic rate per smartphone was 1,614 MB per month in 2016, whereas it was about 1,169 MB per month in 2015.

By 2021, the mobile networks will have three-quarters of smart mobile devices, which are 36.7% more devices than in 2016. Besides, the usage of mobile video service would increase 9-fold in the interval of 2016 and 2021. The review of CISCO 2016 reports that the mobile data traffic is ever growing exponentially in Middle East and Africa [about 96%] followed by Asia Pacific [about 71%], Latin America [about 66%], Central and Eastern Europe [about 64%], Western Europe [about 52%] and North America [about 44%] in 2016 illustrated in Figure 5.1.

Since the system resources are limited and sometimes this cause the violation of QoS, the base station should proactively allocate the available bandwidth to the

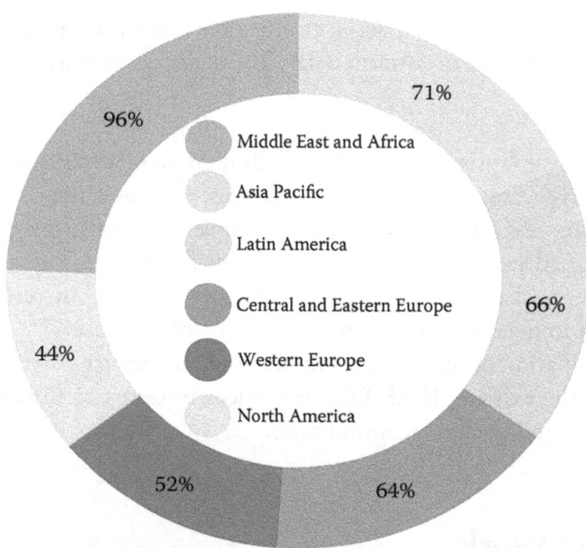

FIGURE 5.1 Mobile data traffic growth in 2016.

networking systems. IEEE 802.16 and LTE-A standard has not had any native scheduling algorithms for the effective usage of bandwidth. Therefore, it is an open challenge for individual venders to implement the scheduling mechanism [10].

5.3 Framework Description

In this section, we propose a detailed description for the proposed Vehicular-Cloud & IoT framework. System components and operations are described as follows.

5.3.1 System Components

The proposed system consists of the following main components, as shown in Figure 5.1.

- Management and Control Center (MCC): It specifies and controls the speed limits of every route according to the road, weather, and traffic conditions.
- Speed Limits Transmitters (SLT-x): It is used to inform the driver about the speed limit in the street he/she is driving through. The speed limits are transmitted as a wireless signal. The MCC controls the speed limits and might change them according to the road, traffic, and weather conditions. The transmitters are fixed at predetermined points the highway.

- Speed Limits Receiver (SLR-x): It receives the transmitted speed limit of a given road and displays it on a display that can be clearly seen by the driver. The speed limits might also be communicated to the driver's smartphone.
- Vehicle Speed Sensor: This is required to measure the speed of the vehicle accurately. The speed measurement system that already exists in the car can be used as well.
- In-vehicle Micro Controller: Its main task is to compare the actual vehicle speed and compare it to the road speed limit, which is received by the SLR-x. Based on the vehicle speed; the controller may generate an audio warning or may communicate such incidence to the MCC through the 4G/5G network. In addition, this information will be communicated to the driver's smartphone.
- 4G/5G Modem: This modem is used to send information about the vehicle speed to a central server, which is controlled and monitored by the Transportation Authority or any other governmental or private entity. This modem can be replaced by the smartphone of the driver during pilot testing stages.
- Drivers' Record Server (DRS): The DRS is a database that has the drivers' records. The DRS updates the record of a certain driver when the GSM modem of that vehicle sends a note to the DRS. The DRS can be accessible by third parties with the approval of the driver. Such third parties include parents, family members and insurance companies.

5.3.2 System Operations

A simplified system-level block diagram of the proposed vehicular-cloud system is presented in Figure 5.2. In normal operating conditions, the SLT-xs transmit the default speed limits and the in-vehicle SLR-x receives the speed limit and display it to the driver. The SLT-xs are connected to the MCC through the Public Switched Telephone Network (PSTN) and the GSM/4G/5G networks, which are connected to the Cloud data centers. The SLT-xs are controlled by the MCC, which is supposed to change the transmitted speed limits according to the road, traffic, and weather conditions. The MCC collects data from several sources, such as police patrols, weather stations and drivers' smartphones and then decides if the speed limit in a specific road needs to be changed. The vehicle speed is continuously compared to the received speed limit, which is also displayed to the driver. Therefore, the driver will always be aware of the speed limit of the road he/she is driving through; hence, which can be considered as an attractive alternative to the traditional speed limit signs located on streets sides. If the vehicle speed is less than the received speed, no action will be taken; however, this might be considered as a congestion indicator. If the vehicle speed limit exceeds the received one, a warning signal will be generated to alert the driver that he exceeded the speed limit. If the driver does not respond within a given period, the In-vehicle Micro

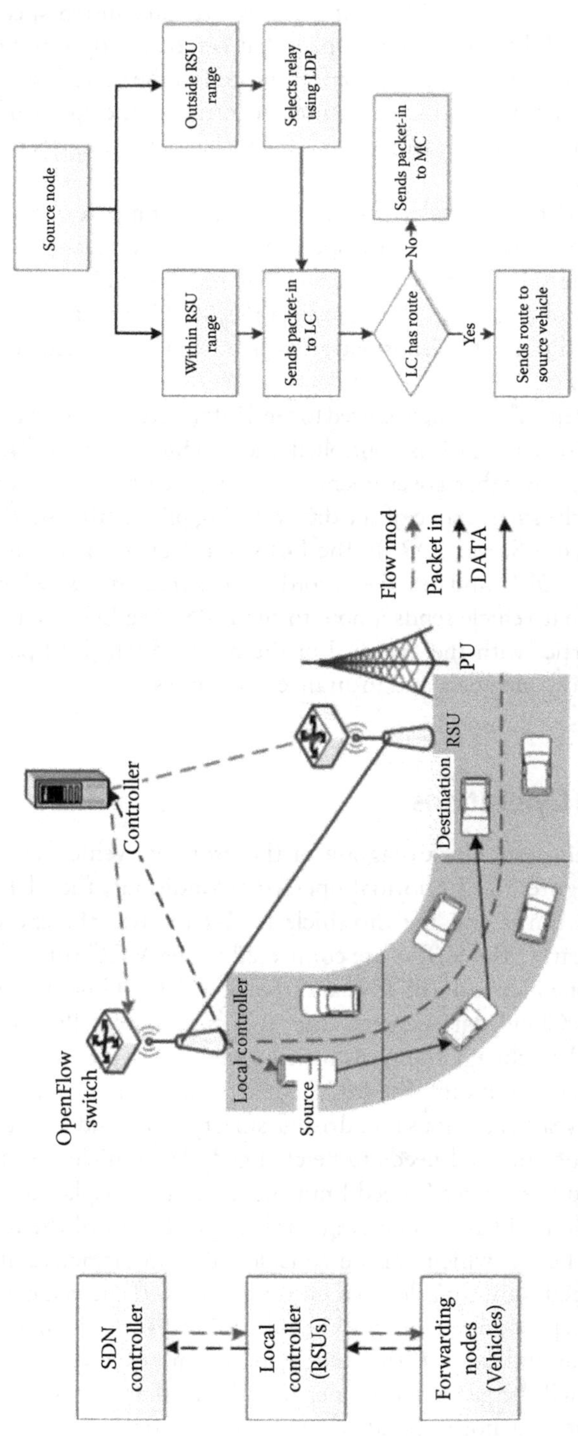

FIGURE 5.2 A simplified block diagram and flow chart for the vehicular-cloud in smart cities.

Controller sends a note (violation) through the 4G/5G modem to the DRS. The same information can also be communicated to the drivers' smartphones. The violation notice should contain information such as the car identity, date and time of the violation. The violation can also be recorded using various network-positioning techniques. If the vehicles are equipped with other types of sensors, then other types of driving violations can be considered as well, such as red light crossing and tail tracking. In such a case where a violation was recorded, the DRS can inform the driver instantaneously through his/her Mobile/e-mail/mail, or by some other IEEE 802.16 standard means. We remark that this study focused on the Voice over Internet Protocol (VoIP) service as an example of the exchanged real-time contents over the vehicular cloud.

Existing service categories supported by 802.16, UGS, and rtPS are designed and developed for the support of real-time (e.g., VoIP) communication flow. The existing and proposed real-time services can be concisely described as follows:

Service of UGS: This service is designed to provide fixed-size data packet flow to the real-time applications, such as E-Carrier(E1)/T-Carrier(T1)/Integrated Services Digital Network (ISDN) technologies, UCT IMS client and real-time services such as VoIP (without silence detection). To send the voice data packet, the base station (BS) usually asserts the fixed size to the Mobile Subscriber Station (MSS). When the real-time application integrates the voice codec with silence detector, the bandwidth consumption of the application should be less for the off period. Otherwise, it wastes the resource availability.

Service of rtPS: The service of rtPS is intended to provide variable-size data packet flow to the real time application like MPEG-Video Streaming periodically [5]. The BS puts up a request of bandwidth as polling and the MSS sends a report of bandwidth demand through the request of BS, thus the process of bandwidth request incurs an additional MAC overhead and queuing-delay. The voice connection process can be negotiated in the requisition process; and called as polling or bandwidth request process. As the rtPS service always relies on the bandwidth request process as a suitable grant size, it is able to transport the voice data more efficiently than the UGS service. However, this service causes connection delay and MAC overhead. In addition, the MSS uses a piggyback request to grant VoIP services, since the voice service is delay sensitive. In this study, we propose an improved real-time service support E-rtPS.

5.3.3 Registration Phase

The road trajectory has several RSUs that are in different positions from each other. We apply a tree-based approach to select the LC at each depth in this proposed scheme. The MC is considered a root node, and all RSUs are reflected as leaf nodes, as shown in Figure 5.2. These leaf nodes at different depths form several layers from the MC. RSUs at the same depth exchange control messages with each other and select one RSU as the LC. The registering phase is divided into the following four steps:

Step 1: The MC first sends a request message to all RSUs within communications range in order to get the network topology. This message is a control message, i.e., a Hello message. As shown in Figure 5.2, only nodes 2 and 3 are within communications range of the MC; therefore, these nodes first receive this message.

Step 2: Each RSU receiving the message calculates its delay using (1) below and compares it with the neighboring RSUs at the same depth. The RSU having the minimum delay at that depth announces itself as the LC:

The number of Hello messages in the previous few seconds can be calculated by RSUs as the vehicles periodically update each other and the nearby RSUs with their current state while moving. This parameter helps in improving network stability by avoiding sparse network conditions, thereby selecting the LC that has high connectivity. HopCount is calculated by the number of layers and RSUs that a message has to travel through to reach the MC (see Figure 5.2).

Step 3: The selected LC (i.e., node 3) advertises a reply message to the MC and all neighboring RSUs, as shown in Figure 5.2, with the updated network information of all its neighboring nodes (vehicles and RSUs). All the nodes receiving this message save the route to the MC in their flow table.

Step 4: The first three steps are repeated at every depth and, last of all, the MC establishes the global topology for all the LCs, RSUs, and vehicles in the network.

At the conclusion of this phase, the selection of LCs provides a localized global view at each depth from the MC. Hence, different types of controller in this proposed scheme are used to reduce the network burden from the single main controller and to reduce overall overhead and delay.

5.4 Enriched-rtPS (E-rtPS) with RT-BMAP

Since the services of UGS and rtPS do not support the method, of on/off to infer whether the state of Voice Activity Detector (VAD)/Silence Detector (SD) is active or not, the VoIP service is not considered to be efficient. In addition, the service of UGS incurs the resource wastage during off period whereas the service of rtPS incurs the MAC overhead and queuing-delay. To resolve these issues, the E-rtPS employs a VAD/SD for the voice codec. To act genuinely, the IP Realtime Subsystem-Network (IRS-N) uses Channel Quality Information Channel (CQICH) to demand bandwidth from available wireless channel resources. The generic MAC header of IEEE 802.16 has two reserved bits to do additive operation, and one of them is used to inform the BS about the state of MSS voice transition. This reserved bit is called as "Grant Size" (GS) bit. When the voice connection of the MSS is "on," the MSS assigns the GS bit to "1"; otherwise it assigns its GS bit to "0." Most importantly, the MSS informs the BS about its change in voice transition state without MAC overhead, since it imparts its change of transition using traditional generic MAC header. The generic MAC header only transmits the voice transition state between the MSS and BS; however, the bandwidth request process related to IEEE 802.16 system header is not transmitted. Instead, the bandwidth request process can be gained through the uplink resources, while the MSS and BS are monitoring the GS bit transmission.

5.4.1 Real Time-Batch Markovian Arrival Process (RT-BMAP)

The performance degradations (packet delay, MAC-Overhead and resource wastage) for UGS and rtPS services, are mainly caused because the BS assignment regards the uplink resource with the consideration of voice transition to the IRS-N. To infer the state of the voice/real-time transition, in this study, we integrate the technical feature of VAD/SD with the IRS client codec. While the feature of VAD/SD is integrated with the IRS client, the IRS-N can deduce whether the voice state transition is on/off. The deduction of such transition is usually performed in the higher-layer and the layer of MAC is used to find the primitives of convergence for sub-layers from the standard system of IEEE 802.16d/e. Since the IRS-N informs the BS regarding the state of the voice transition, the IRS-N necessitates the method of on/off to infer the current status of voice transition. We thus decide to utilize the reserved bits of IEEE 802.16d/e to deduce the status of voice transition. To probe the reserved bit realistically, we represent the bit as Status-Check (SC). While the voice state transition is going "on," the IRS-N represents the SC bit as "1"; otherwise, the IRS-N sets the SC bit as "0." The significant use of SC is to avoid the MAC-Overhead. The IRS-N employs the bandwidth request to hold the usage of uplink resource and the frame of voice codec incurred in the IRS-N determines it. Since the incursion of codec frame depends on the IRS-N, the generation of voice packets relies on the communication duration of IRS clients. The reserved bit of SC controls the voice state transition of the IRS clients through the knowledge of IRS-N to prevent occurrence of MAC-Overhead. In this study, in order to analyze the SC operation, a real-time client server system is employed. The IEEE 802.16 system model is constructed in the MAC layer to utilize UGS, rtPS, and E-rtPS as the on-demand resource for VoIP services similar to the studies in Refs. [10,11]. Real-time Client (RC) and Real-time Server (RS) are deployed as the real-time agents to support the voice call establishment and termination. The real-time agents use Session Initiation Protocol (SIP) message transmission using Real-time Transport Protocol (RTP)/User Datagram Protocol (UDP).

5.5 Results and Discussions

In this section, we present our promising results on the proposed vehicular-cloud system. It consists of theoretical and hardware implementation parts. The aim of the hardware implementation is to build a simple micro-scale level of the proposed system. This simple model was very helpful to obtain rough cost estimates and provide us with in-depth information about the main practical design challenges in terms of delay and system throughput. The simple model consists of three subsystems, the in-vehicle subsystem connected via cellular networks, the IP-based network (Internet), and the MCC represented by the IRS-N client. We remark here that we used 10 car-toys while obtaining the results in this section for realistic verification purposes.

The MCC is implemented via Ubuntu-PC (laptop) for experimental purposes. And the car-toys were equipped with Arduino boards attached to GPRS modules as shown in Figure 5.3.

To evaluate the performance of resource allocation framework in the central cell, 5 MHz LTE-A is considered as a wireless cellular network with 19-cells functioning at 2.0 GHz. The bandwidth of 5 MHz LTE-A is split into 25 RBs [9]. As the cell has 500 meters as an inter-site distance, the network users can be randomly distributed within the cell site [9]. For each user, the uplink power transmission is set to be 24 dBm and the threshold value η of interference avoiding mechanism is assumed to be 0.8 [9]. The selection mode of Modulation and Coding Scheme (MCS) depends on the RBs allocation for the users with minimum Signal to Noise Ratio (SINR). The mapping lookup table for SINR value to MCS mode is found in Ref. [9]. For each user, the packet arrival process follows the poison distribution with the mean arrival rate 1,000 bits/s. Table 5.1 summarizes the simulation parameter of IMS networks [9].

FIGURE 5.3 A simple prototype model of the proposed system.

TABLE 5.1 Simulation Parameters of IMS Networks

Parameters	*Values*
Frequency	2GHz
Bandwidth	5MHz
Number of cells	19
Distance (Inter-site)	500m
Number of RBs	25
Shadowing standard deviation	8 Db
Uplink device transmission power	24 dBm
Proximity distance	10m
Maximum power transmission P_m	24 dBm
Threshold η	0.8
Channel model	200Tap, Urban [9]
Path loss [distance-dependent]	$128.1+3.76\log(R)$, R in km
Modulation and coding scheme	QPSK, 16 QAM, 64 QAM (1/2, 2/3, 3/4)

In order to examine the quality of real time services, namely UGS, rtPS, and E-rtPS with RT-BMAP, an IRS core function with Proxy Call Session Control Function (P-CSCF), Interrogating Call Session Control Function (I-CSCF), Serving Call Session Control Function (S-CSCF), and Home Subscriber Station (HSS) has been installed in Ubuntu supporting LTE, as shown in Figure 5.4. This test-bed also integrates TS 23.167 [12], which has real-time service architecture not only to allocate the network resources but also to ignore the roaming restriction as referred in Ref. [13]. Since this architecture is planned for ITS systems where high safety levels are required, we assume the registered requests as Emergency-Calling (E-Calling) that contains an initial request to grant the emergency services as "Unrecognized E-Calling" [14].

To probe the voice connection as a real-time service, the packet generator is integrated with the IRS core function to generate the concurrent voice calls. These voice calls can be transmitted in either serial or parallel to analyze the effectiveness of E-rtPS with RT-BMAP in comparison with UGS and rtPS. Since the IRS core function can extend its service scalability for the device connection through the knowledge of 3GPP networks [9], we have deployed an IRS core network, which has a feature of LTE-A to enlarge the transmission region and provide better service

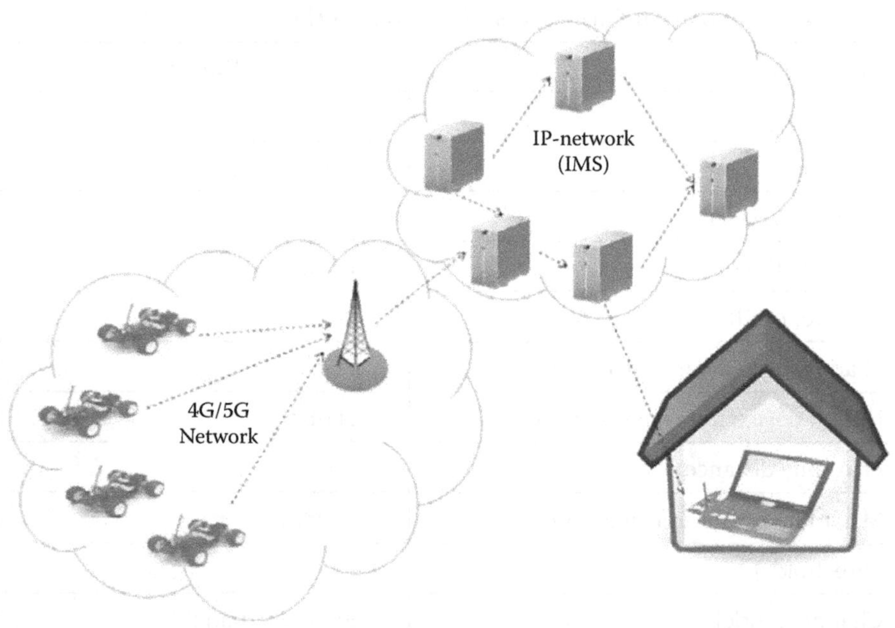

FIGURE 5.4 Experimental testbed for the vehicular-cloud system of IEEE 802.11d/e.

flexibility. In this study, the IRS core is embedded with its Call Session Control Functions (CSCF) in order to have a realistic test bed. Besides, this core network integrates the HSS to cross-examine the stability of voice service over available networks. As the service network is defined for the real-time connectivity, we increase the voice connectivity gradually for the analysis of connectivity delay and service throughput rate. An emergency voice call session is defined as the generation of anonymous voice calls to examine the session setup delay and throughput rate of the real-time systems.

For the establishment of VoIP call and service connection, we employ the IRS client (as known as UCTIMS) and packet analyzer (as known as Wireshark). This experimental setup aims not only to evaluate the SIP based VoIP performance over IRS core using wireless connectivity but also to examine voice connectivity delay and service throughput rate of the networks.

As illustrated in Figure 5.5, the throughput of the proposed algorithm is significantly higher in comparison to UGS and rtPS. In addition, as shown in Figure 5.6, the packet delay of the proposed algorithm is also superior for all the critical regions when compared with the other services. This behavior is evident even when the packet delay is set as 60 ms. Please note that the predetermined delay value is an important criterion in the system of IEEE 802.16d/e [5] for the packets with delay violation.

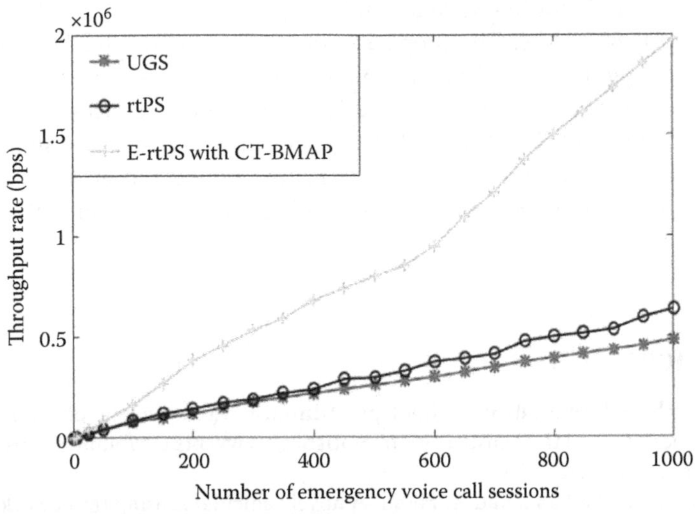

FIGURE 5.5 **Throughput versus number of voice call sessions.**

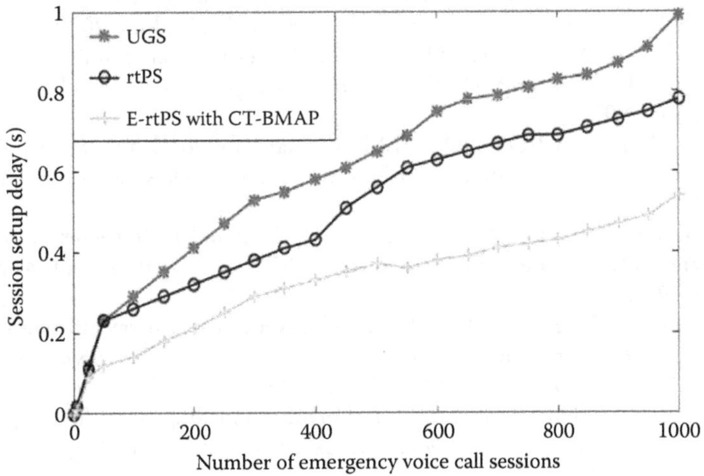

FIGURE 5.6 **Average packet delay versus number of voice connections.**

5.6 Concluding Remarks

This chapter presents an adaptive framework for dynamic speed management in smart cities. This framework utilizes the latest developments in wireless networks and use existing communication infrastructure to stream data, sound, and video in order to maximize the system adaptability and minimize the cost. A key component to the proposed system is the dynamic medium access approach for real-time communications. Accordingly, this article proposes the E-rtPS for the vehicular-cloud of IEEE 802.11d/e. The proposed approach integrates RT-BMAP to analyze the throughput rate and average packet delay. For practical assessment, the real-time system is configured with the IEEE 802.16d/e services, namely UGS, rtPS, and E-rtPS with RT-BMAP. The experimental results clearly show that the proposed approach is more efficient in terms of average packet delay, as well as throughput when compared with the results obtained from existing 802.16d/e services in the literature.

References

1. Research and Innovative Technology Administration, Bureau of Transportation Statistics, National Transportation Statistics, www.bts.gov/publications/national_transportation_statistics.
2. J. Sahoo, S. Cherkaoui and A. Hafid, A novel vehicular sensing framework for smart cities, *LCN, 2014 IEEE 39th Conference*, 8–11 Sep. 2014, Edmonton, pp. 490–493, 2014.
3. X. Baró, S. Escalera, J. Vitria, O. Pujol and P. Radeva, Traffic sign recognition using evolutionary adaboost detection and forest-ECOC classification, *IEEE Transactions on Intelligent Transportation Systems*, vol. 10, pp. 113–126, Mar. 2009.
4. J. van de Beek, M. Sandell and P. Börjesson, ML estimation of timing and frequency offset in OFDM systems, *IEEE Transactions on Signal Processing*, vol. 45, pp. 1800–1805, Jul. 1997.
5. IEEE 802.16-REVd/D5–2004, IEEE standard for local and metropolitan area networks - part 16: air interface for fixed broadband wireless access systems, May 13, 2004.
6. A. Radwan, K. Saidul Huq, S. Mumtaz, K. Fung and J. Rodriguez, Low-cost on-demand C-RAN based mobile small-cells, *IEEE Access Journal*, vol. 4, pp. 2331–2339, 2016.
7. S. Mumtaz, K.M.S. Huq, A. Radwan and J. Rodriquez, Energy efficient scheduling in LTE-A D2D communication, *IEEE Comsoc Multimedia Communications Technical Committee*, vol. 9, no. 1, pp. 282–287, Jan. 2014.
8. A. Klemm, C. Lindermann and M. Lohmann, Modelling IP traffic using the batch Markovian arrival process, *Performance Evaluation*, vol. 54, no. 2, pp. 149–173, Oct. 2003.
9. 3GPP TS 36.300, Evolved Universal Terrestrial Radio Access (EUTRA) and Evolved Universal Terrestrial Radio Access Network (E-UTRAN), Rel. 10, v10.2.0, Dec. 2010.
10. H.C. Hsieh and J.L. Chen, Distributed multi-agent scheme support for service continuity in IMS-4G-cloud networks, *Computers and Electrical Engineering*, vol. 42, pp. 49–59, Feb. 2015.
11. H. Lee, T. Kwon and D.H. Cho, Extended-rtPS algorithm for VoIP services in IEEE 802.16 systems, *Proceedings of the IEEE International Conference on Communications*, Istanbul, Turkey, pp. 2060–2065, Jun. 2006.

12. 3GPP TS 23.167, IP multimedia subsystem (IMS) emergency sessions, Release 11, Sep. 2013.
13. 3GPP TS 23.228, IP multimedia (IM) subsystem Cx and Dx interfaces, Signalling flows and message contents, Release 11, Sep. 2013.
14. M. Patel, SOS uniform resource identifier (URI) parameter for marking of session initiation protocol (SIP) requests related to emergency services, draft-patel-ecrit-sos-parameter-07.txt, Oct. 26, 2009.

Chapter 6

5G-Enabled Devices and Smart Spaces in Social-IoT*

Fadi Al-Turjman
Antalya Bilim University

Contents

* **F. Al-Turjman**, "5G-Enabled Devices and Smart-Spaces in Social-IoT: An Overview", *Elsevier Future Generation Computer Systems*, 2017. doi:10.1016/j.future.2017.11.035.

6.1 Introduction

In this information age, online activities form a significant part of our daily life. Social networks and collaborative sites such as Twitter, Facebook, and Google+ are incredibly popular online forums, offering easy and compelling ways for millions of users to post content and interact with each other. In addition to providing attractive mediums for person–person interactions, social networks also offer unprecedented opportunities for social data analysis, i.e., big-picture views of what people are saying, because they contain a deluge of opinions, viewpoints, and conversations by millions of users, at a scale that would be impossible using traditional networks such as the internet and circuit switched telephone networks. The 5th Generation (5G) of wireless technologies is the proposed next generation telecommunications standards for this kind of applications. 5G comes as the demand for better and faster wireless connection. As stated in Ref. [1], by the time 5G is fully embraced, there will be tens or hundreds of billions of devices that require the use of 5G technology not only because of personal usage, but also due to many new applications. One of the major new application that will contribute to this high number of devices is the highly anticipated Internet of Things (IoT). The term IoT has been loosely used in many scientific research areas as well as marketing and sales. According to Ref. [2], IoT can simply be defined as a dynamic global network infrastructure with self-configuring capabilities based on standard and interoperable communication protocols. There are estimates that 50 billion devices will be wirelessly connected to the Internet by 2020 [3]. Meanwhile, the slow but steadfast introduction of IPv6, in addition to the great proliferation of sensing and tagging technologies are speeding up the realization of the IoT—an Internet where everything is reachable and can communicate[4,5]. This means that a larger number of the familiar Wireless Networks will be deployed, generating great amounts of data to monitoring devices or personnel. This Ultra Large Scale communication paradigm will inevitably rise to become a dominant necessity for the various sections—both public and private. However, there are serious challenges in realizing such vision. Traditional design and deployment of wireless networks has been application-specific, that is, single-purpose networks [6]. The tradition stemmed from the increasing cost boundaries and the diminishing margins of practical feasibility when deploying practical, real-life wireless networks deployments, especially maintenance of cellular functionalities in very dense deployments or spread over large geographic areas where the small cells (femtocells) in 5G plays a key role [5,7].

As the computation and the communication process of heterogeneity hidden networks involve intelligent decision making, the human-to-machine perception on the realization of large-scale IoT environment is bridged as a key challenge for Social Internet of Things (SIoT) [8,9] in literature. In this chapter, SIoT is represented as a system of smart things, which allow people to interact with each other to share/exchange social information. Though the SIoT can improvise the interaction and the navigation between the communicated objects, the contextual data of the objects should be handled with respect to situation awareness. The SIoT classifies the contextual data as objective and subjective [10]. The former context is used to define the physical aspects of sensing objects, such as device status, time, location, service availability, etc. The latter is used to represent the short-term goals, trustworthiness, and preferences. So far, the combination of both has not been studied for SIoT in terms of security [11]. And while this view continues to thrive in the literature, the arguments for the boundaries and margins are no longer valid [12]. As well, disappointments resulting from several practical deployments based on this design view have raised concerns for the viability as a sustained design approach [13,14]. Ultra large scale networks, such as the cellular networks which depends significantly nowadays on the smartphones, shall respond to increasing activity in a recent direction of research advocating public sensing. In public sensing, systems probe resources like cell phones, laptops, and web servers to build a virtual network on the Internet—one which could be remotely queried for information collection [15]. These systems were proposed for promoting activity monitoring [16], air quality monitoring [8], social networking, and controlling mobile phones (using neural signals) [9]. These approaches can be either participatory or opportunistic, offering great flexibility in both design and adoption. This depends heavily on the smartphone utilized sensors.

Accordingly, the main contributions in this survey article can be summarized as follows. A comprehensive background about the 5G standards and their IoT-specific applications are outlined while emphasizing energy consumption and context-awareness. The recent directions in using smartphones' sensors with alternative design approaches that can functionally contribute to more scalable operations in smart social spaces are overviewed. Online versus offline mobility detection applications, and potential communication technology, namely the femtocell, which can be a strong candidate for the IoT paradigm realization in practice have been investigated. Modeling and energy aware metrics are outlined as well. Moreover, key open research issues are highlighted and discussed. In order to assist the readers, we provide in Table 6.1 a list of acronyms along with brief definitions as used throughout this article.

6.2 Smartphone Usage and Context-Awareness

Mobile phone usage statistics can be obtained in different ways. One approach is asking user to manually log their activities and surveys [17]. Second approach

TABLE 6.1 Acronyms and Definitions

Acronym	*Definition*
IoT	Internet of Things
UE	User Equipment
RFID	Radio Frequency Identification
HAN	Home Area Network
NAN	Neighborhood Area Network
ROF	Radio-Over-Fiber
ICT	Information and Communication Technologies
RAN	Radio Access Network
QoS	Quality of Service
MBS	Macrocellular Base Station
RNC	Radio Network Controller
FAP	Femtocell Access Point
LTE	Long Term Evolution
CVCs	Context-aware Vehicular Cyberphysical Systems
WBAN	Wireless Body Area Networks
MCC	Mobile Cloud Computing
EE	Energy-Efficient
BS	Base Station
PA	Power Amplifier
SE	Spectral Efficiency

is collecting data from the device with an agent application [18]. Third is combining the first and second approach [19]. However, user reported data can be prone to errors, due to personal biases and limitations, etc. Therefore, a better approach is to apply context or semantic analyzing methods over collected mobile sensor data [20,21]. Context-aware systems need to collect a variety of information about the user's current status and activities, some of which may be regarded as personal and make it user-centric system [22,23]. We have made a few attempts toward realizing context-aware networks [24]. We implemented a Smart Spaces platform, called "CAR" that integrates smart spaces with social networks through the IP multimedia subsystem, creating truly context-aware and adaptive spaces. We designed and implemented all components of the CAR including the central server, the location management system, social network interfacing components, service delivery server and user agents. Li and others presented an analysis of app usage behaviors by using a famous Android app marketplace in China, called Wandoujia [25]. They studied over 0.2 million Android apps and 0.8 million users. Their findings on usage patterns include app popularity, app category, app selection and network usage [25]. However, this work is done on a limited geographical area and does not give a global perspective.

It is elemental in the design of context-aware solutions to ensure attractiveness by ensuring interoperability to existing standards. Consider, for example, the case of user profiles in social network services such as Facebook. The authorized extraction and management of such profiles is elemental to any context-aware framework. At certain stages of a context-aware framework, it will be unavoidable to resort to capable profile extractions such as Friend of a Friend (FOAF) as it might prove useful in this specific domain. Briefly explained, FOAF is a decentralized semantic web technology, and has been designed to allow for integration of data across a variety of applications, websites, services, and software systems. However, careful investigations are required in order to verify its appropriateness to fulfill the above noted generic characterization requirements.

Application as a Service (AppaaS) described in Ref. [26] provides an overview of the most appropriate system architecture we can think of for smart spaces utilizing smartphones. The architecture involves the user device loaded with an AppaaS mobile application, in addition to a space/context management server. The AppaaS mobile application comprises of a Graphical User Interface (GUI) which is used to take inputs from users, a Space Handler, which collects different context information from users and service delivery makes sure that services are delivered in the form of applications relevant to the user's context. AppaaS mobile application provides users with an interface, which can be used to provide certain inputs. When a first time user registers with AppaaS, the user provides basic information and the information is sent to the AppaaS server. Upon successful registration, the user is taken to a login screen; here, the user provides the login information, and upon successful login, user's relevant scheduling information is fetched from AppaaS server. Obviously, the above-described context-aware setup depends greatly on facilitating state preservation, which essentially refers to applications able to save their latest status and user-specific data for future access. Current mobile platforms do not natively support state preservation at any level. However, individual applications can manage their own state at different levels depending on the objective from state preservation. For example, applications may use checkpointing techniques [20,27,28] to suspend and resume their execution or for migration purposes. Generally speaking, an application-independent state preservation of user-specific data remains challenging. To this end, new paradigms shall assume that applications would provide two proprietary APIs for the sake of state preservation. One API saves the application current state, where all user-specific data is saved in an XML file format. The other API uploads an application state from an XML file when the application launches.

In general, wearable sensors are promising technologies in the field of user behavior analysis and monitoring. In the near future, several mobile applications will be dependent on these tiny wearable devices, through which user habits, user mobility, application usage and context awareness can be further investigated. In Table 6.2, we tabulate the reviewed attempts in this regards.

TABLE 6.2 Comparison of the Context-Aware Usage in Mobile Applications

Reference	*Context Resource*	*Method*
[25]	Statistics of Apps w.r.t. management, device price, foreground vs. background app with network connection	8 M users, 260 K devices
[27]	Location (via WiFi), app duration and type	14 teenagers, 4 months, HTC Wizard, logging and interview
[20]	Stay points (GPS, WiFi) and labelling, app type and voice/sms, density	77 participants in 9 months, using Nokia N95
[18]	Location, interaction frequency, battery use, connectivity, call frequency	1277 types of devices, 16K participants, 2 years 175 countries, using Android
[29]	Number of user interaction, network traffic, app use, and energy drain	255 (33 android + 222 windows phone), using an app for logging
[28]	Battery, location, time, selected apps using app sequence histories, and connectivity	50 participants in 9 months
[30]	Stay points (GPS), app duration, <loc,app> sequences	30 students in 3 months
[19]	App frequency and type	28 participants in 6 weeks, using an app for statistics
[31]	Location: GPS, Bluetooth, WiFi, accelerometers	8 participants in 5 months, using an agent on device

6.3 Smart Space Discovery

In this section, we elaborate on the most significant components in the smartphone usage, which is the space discovery paradigms. Moreover, we categorize the existing tracking Apps in this area.

6.3.1 Space Discovery Paradigms

Facilitating location-based services is fundamental to the realization of the efficient smartphone usage. For outdoors and open areas with direct line of sight, it is possible to depend on Global Positioning System (GPS) or even terrestrial localization schemes, e.g., as offered by cellular networks. Announcing the location for indoor devices, however, becomes more problematic. In this direction, researchers investigate the viability and efficiency of a multitude of space discovery mechanisms, to facilitate rapidly identifiable location-based services. Prominent contenders in this investigation include Received Signal Strength Indicator (RSSI) models, Near Field Communication (NFC), and Wi-Fi/cellular-based models. Other investigations encompass their efficiency in indoor environments, operability under varying density constraints, privacy preservation, and distance to identification devices. For example, RFID readers for NFCs.

A prominent and novel direction of investigation encompasses employing ultrasound-based detection mechanisms. The distinct advantages of ultrasound communication include controlled transmission range, limited locale (no wall penetration), does not require additional hardware on part of the user, and has modest implementation requirements. More importantly, the behavior of ultrasound is more predictable than IR and other radio waves. Prominent disadvantages to ultrasound communication such as air speed and scattering could be addressed by careful signaling design, e.g., longer pulses. Careful considerations also need to be made when utilizing ultrasound communication in medical environments; all of which are core parameters to this research direction. To further elaborate, the intended operation of the proposed system entails sending information through beacon devices in the non-audible, ultrasound range (i.e., >20 kHz). This information can be picked up and interpreted by the user's device microphone. This setup readily indicates the wide feasibility of implementation given that any smartphone device would have a microphone. Furthermore, a few evaluations indicate the readiness of common smartphones to this. For example, Figure 6.1 shows the response of the Samsung Galaxy II for a continuous tone at 20 kHz [32]. It should be noted, however, that as the audio profiles of the device microphones vary across devices, a calibrator module might be required in order to prepare individual smartphones for this communication.

As a prototype, the beacon device can be implemented on a single-board computer can be used inside a beacon. The computer shall have enough computing power to playback an audio file with a sampling rate high enough to produce frequencies higher than 20 kHz and a wireless channel, such as WiFi, to allow for

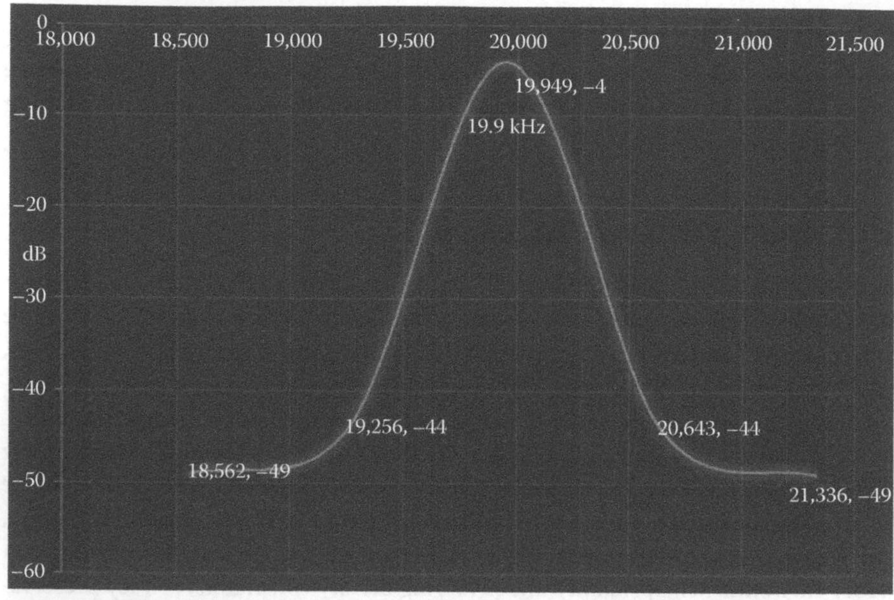

FIGURE 6.1 Frequency response of Samsung for the 20 kHz [32].

remote configuration. (Note that to be able to produce frequencies up to 21 kHz we need to generate the audio signals with a 44.1 kHz of sampling rate). Such a beacon can encode the service code on its own board without the need of external CPU processing. Meanwhile, beacon devices will be designed to support more than one "audio-tagged" device simultaneously. To achieve this, synchronization is needed between beacon devices in the same vicinity. On the client side, it is worth noting that a minimum sampling rate of 44.1 kHz is supported by most microphones, although a few experimentations indicate that ultrasound communication would work between if a sampling rate of 48 kHz was utilized. Moreover, the ultrasound system will be amended by a user configuration interface that facilitates visualization of designed locales, and aids the designer in placing the beacon devices. Figure 6.2 shows an instance of the configuration module communicating with the beacon devices for initial configuration download.

6.3.2 Online Tracking Apps

There have been a few attempts on getting user location for different purposes. Some of them include predicting application usage [28,30], friend recommendation systems [33], classifying applications or user characteristics [24]. Even though movement and location analysis are different, they share some common characteristics. First, we need to answer which applications are used when users are mobile. We need to identify user movement for this purpose. We can classify the work done

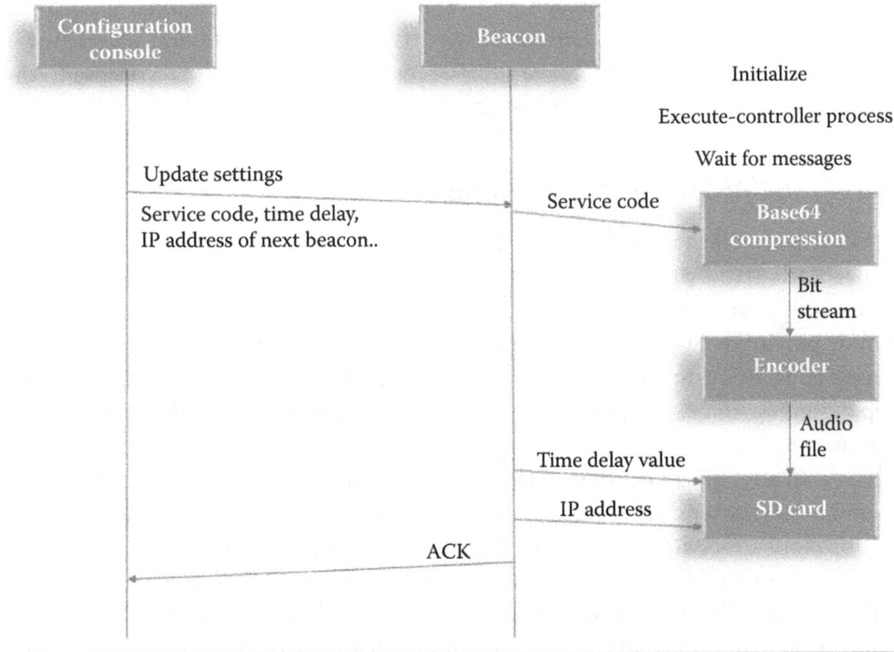

FIGURE 6.2 Initial configuration download.

in this area to online and offline tracking applications. Online applications need a stable internet connection to send collected data to a cloud-based storage [34].

Wagner et al. used mobile phone providers' Call Data Records (CDRs) and Wi-Fi-based location information in order to analyze the user's data [32]. They avoided GPS in order to reduce the battery consumption. Their application was installed on more than 16,000 contributors in 175 countries. In this work, it was found that most of the users turn off Wi-Fi connection to save battery. In addition, the majority of devices in their dataset saw three or fewer Wi-Fi access points most of time. In this work, there is no information about app usage [32].

6.3.3 Offline Tracking Apps

In this section, we look at offline tracking applications. Rahmati and Zhong have conducted an analysis in which they installed logging software on mobile phones and distributed them to users for research [27]. They collected app usage and location relation and movement information. However, this work lacks detailed analysis. Another approach is storing user and context data and then uploading them daily to a server [20]. Offline apps collect user related information and send them to the cloud or a central serve when user becomes online. Lu et al. proposed a method for predicting app usage using stay points and location information. They used semantic locations approach and created a database containing application launches and stay locations offline in the devices. Their prediction algorithm is designed to be online to predict the subsequent app needs of a user according to the previous locations and app launches [25].

6.3.4 Social Space Provisioning

A social space takes into account usually the user's identity and personal information, location and time information, relevant to the situation or activity the user is currently involved in. As well, relevant to user's space information, social spaces aim at providing the user with the most relevant services in the form of a smartphone application by customizing the behavior of the smartphone application(s) depending on the user's profile, as described above in the AppaaS context-aware system.

Several attempts in the literature have been applied to eliminate the user involvement in smart-spaces communications. Aforementioned system components adhere to this motivation, achieving a high degree of service portability as the user moves from one space to another. The ultimate motivation of the social IoT; however, is tailor service delivery to the characteristics of the wireless and mobile environment. These include intermittent connectivity, limited features (display, processing power, etc.), limited battery, in addition to heterogeneity of required resources. Moreover, while dependence on device participation in moving between spaces adds considerable flexibility, it ultimately restrains the device's lifetime and, in turn, the user's mobile experience.

To truly instill global seamlessness in social service delivery, it becomes inevitable to engage the cloud in IoT paradigms. This spans both virtualized services and resources, in addition to in-network processing. More specifically, the range of services that involve mobile devices providing data are on the rise, spanning entertainment services, such as online social gaming and networking, to crowdsourcing, such as collaborative participatory sensing as well as services that can be offered on the fly, such as video streaming of a current event. However, the rich functionalities that such applications over increasingly demand resources beyond the capabilities of inherently resource-constrained devices. Such lack of resource matching places limitations on the type of functionality and services that can be offered, restraining users from taking full advantage of their mobility passion. Cloud computing, therefore, offers the possibility to unleash the full potential of mobile devices to provide reliable data services. This expands on the notion of the aforementioned AppaaS to become Space as a Service (SpaaS). As a basis, SpaaS unfolds the full potential of service integration between the device and the space, going beyond the integrated sensing environment discussed above. SpaaS also goes beyond user migration to cater to user mobility. In such instances, the user's space expands and reduces as he moves from, say, the home to the car to the office to the mall, etc. More importantly, however, Spaas extends profile-based space customization to full space mobility, explicitly enabling a form of space virtualization by which a user can bring a distant space in his or her direct presence.

The aforementioned cloud-assisted mobile service architecture involves four key entities: a user, a mobile device, a cloud, and a data provider, as shown in Figure 6.3. The user represents the service consumer. The mobile device represents a mobile service provider and acts as the integration point where service execution plans are generated and decisions regarding offloading are made. The cloud is the supporting computing infrastructure that the mobile provider uses to offload resource-intensive tasks. Web service operations might involve third-party data processing during the execution of the service functionality, such as weather information or navigation databases. In such cases, data could be fetched from a data storage provider. In this architecture, the user sends the service request to the mobile provider. The mobile provider decides on the best execution plan and whether offloading is beneficial. The cloud offers elastic resource provisioning on demand to mobile providers. The mobile provider may collect the execution results from the cloud and generate a proper response for the user. It is also possible that the provider will delegate the cloud to forward the response directly to the user, given that no further processing is required at the mobile side.

The proposed framework encompasses the following major components, as depicted in Figure 6.4: Request/Response Handler, Context/Space Manager, Profiler, Execution Planner, Service Execution Engine, and Offloading Decision Module.

Briefly described, the profiler characterizes the offered space services operations and generates resource consumption profiles, while the execution planner investigates all possible execution plans based on locations of required data and current context information. The service execution engine evaluates these plans and selects

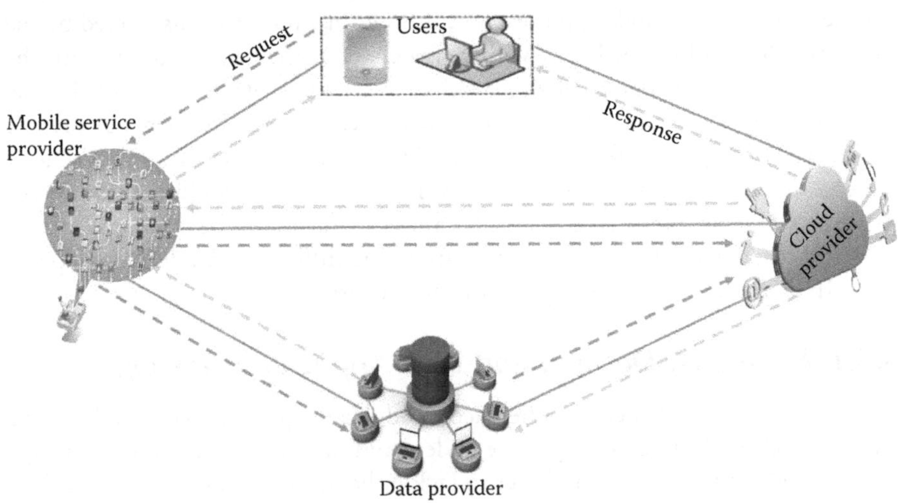

FIGURE 6.3 Architecture of Space as a Service (SpaaS) [26].

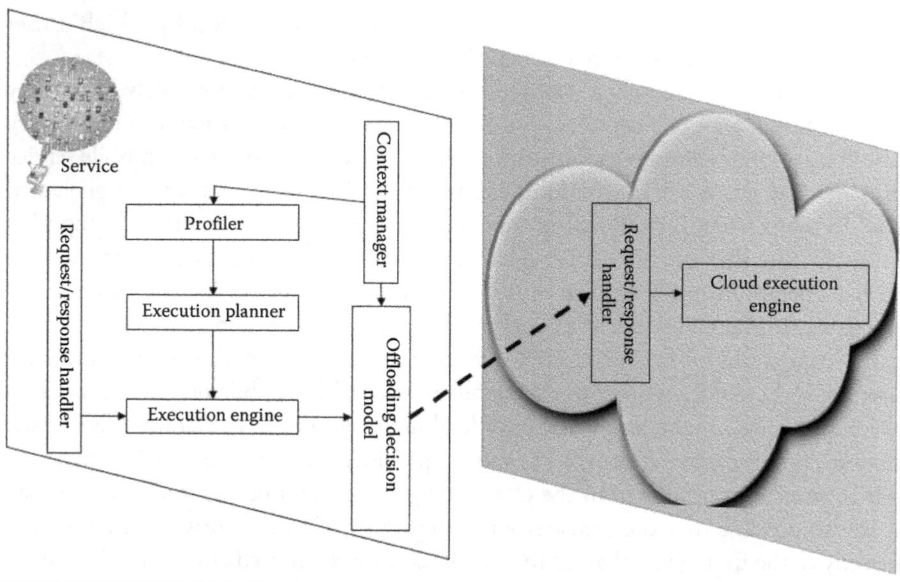

FIGURE 6.4 Details of Space as a Service (SpaaS) components [26].

the best resource efficient plan that, in addition to satisfying the resource constraints, yields better performance and lower latency.

6.4 Smartphones & Mobile Applications Testing

Mobile testing underwrites a flawless customer experience. No matter how amazing the smartphone mobile application is, if it has just one error discovered by the end-users, they will not be happy with it. Even worse, they will associate with the producer company or brand all similar kind of errors. That's why companies pay lots of attention to test all the mobile application features and behaviors before it is on air. But testing all the features and behaviors of any application is impossible or unnecessary. Instead of this, automated testing strategies for mobile applications or test cases which are prepared according to these strategies can be used to save both time and cost. In addition, mobile applications are different than desktop applications. In this section, we investigate these differences.

6.4.1 Mobile vs. Desktop and Web Application Testing

There are some fundamental differences and challenges between Mobile and Desktop Web application testing. If we understand these differences and challenges of the mobile testing, it will be easier to tackle them. Let us start with the definitions of application types. There are mainly three types of applications: desktop, Web, and mobile. First, desktop application is a native application, which executes on the user's local machine. This type of applications may have a network module to communicate with an external server. Second, the Web application is a kind of application that runs within a browser. All the information stored in an external server and browsers are used to communicate with it. Finally, mobile application called native application downloaded from an app store is intended to run on mobile devices such as the smartphones that are equipped mostly with iOS or Android operating systems. Mobile applications like desktop applications may have a network module to communicate with the external server.

Environments and testing concerns of these application types are unique by nature. Mobile applications are inherently tied to the hardware and operating system. Therefore, they have more environmental situations and concerns than other application types. First, there are variety of mobile devices. According to the Google Play Store, there are 12,402 types of mobile devices, which runs on Android operating system. These devices differ in screen sizes and hardware capabilities. In addition, there are variety of operating systems versions. There are 12 main versions of Android OS and 10 main versions of Apple iOS. Accordingly, almost every year a new version of these operating systems is introduced. Furthermore, there are variety of mobile network operators. There are over 400 mobile network operators in the world [35]. Each mobile operator supports different network technologies including Long-Term Evolution (LTE), Code Division Multiple Access (CDMA), Global System for Mobile Communication (GSM), and some other local networking standards. In addition, mobile applications have some unique test cases, such as interruptions, battery consumption, and GPS. However, while testing a mobile application, testers need to simulate an environment, which considers these cases, this is the hardest part of mobile application testing. Accordingly, there are two ways to test these cases. The first way is Emulator testing. Testers need an Android or iOS emulator to do that, but simulating these features for all possible cases systematically is not easy because of the differences between real device and emulator. On the other hand, testers can use real devices to test the mobile application. This option provides more reality, but you cannot create an environment specifically for any situation.

6.4.2 Challenges of Mobile Applications Testing

Nowadays, there are nearly 13,000 types of mobile devices running 100 operating systems and versions serviced by 400 carriers worldwide [35]. If the goal is to test every permutation possible, then every test case will be run 9,100,000,000 times.

6.4.2.1 Device Fragmentation

The most complicated aspect of mobile application testing is device fragmentation. The fragmentation is coming from mostly screen sizes and hardware capabilities. As a fact iOS device fragmentation is not a huge problem, but device fragmentation is an issue for the Android operating system. In 2012, there were 4000 separate

Android devices available. In 2016, the number is exploded to 12,000. Accordingly, it is not observed yet the total number of devices an app needs to be tested before releasing it in the market.

6.4.2.2 Operating Systems and Fragmentation

Not just newer or smarter mobile devices but also new versions of operating systems are being launched almost every year. The problem is compatibility issues that mobile applications face while being deployed across devices having different operating systems like Android, iOS, Windows, BlackBerry etc. or versions of an operating systems such as iOS 9.x and iOS 10.x.

6.4.2.3 Simulation Environment

Mobile emulators and simulators are important and main method as a testing tool to verify the general functionality and perform regression testing. The testing is conducted in a simulated environment that is not real. Testers can set their environmental conditions while debugging the application manually or automatically.

A mobile application while functioning may face several interruptions like incoming text messages, incoming calls, incoming notifications, network coverage outage and recovery, battery removal, or power cable insertion and removal. A well-tested and developed application should be able to handle these interruptions by going into a suspended state and resuming afterwards. To do this, while setting simulation environment, testers needs to add these interruptions into the test cases.

The innovation in the battery usage duration field has not been quick as in the application consumption. End-users are running and using lots of applications during the day and some of these applications keep running in the background without even being noticed. These applications keep consuming power till the death of the battery. While testing these applications, testers need to set not only possible battery percentages but also battery usage. If application consumes too much power, this may cause less or non-usage of the application.

6.5 5G/Femtocells in IoT Smart Spaces

Given that IoT is strongly connected to the 5G future-networking paradigm, in this chapter, we are spotting the light on the 5G network where the femtocell is a main player. Femtocell is a cellular network Base Station (BS) that connect standard mobile devices to a mobile operator's network using residential Digital Subscriber Line (DSL), cable broadband connections, optical fiber or wireless last-mile technologies [36]. It is an inexpensive compact BS providing equal radio access interface as a common Macrocell Base Station (MBS) toward User Equipment (UE) [37]. It is a solution to offload from overloaded macrocells and increase the coverage area [36].

They are generally used for increasing indoor coverage and designed for use in a home or small business where there is a lack of cellular network or increasing the Quality of Service (QoS). It has advantages for both cellular operator and the smartphone user. For cellular operator, the main advantages stem from the increased coverage and capacity. Coverage area is widening due to eliminated loss of signals through buildings and capacity is increased by a reduction in total number of UEs that uses the macrocellular network. They use Internet instead of using cellular operator network. For customers, they have better service and improved coverage and signal strength since they are closer to the BS. Moreover, using femtocells leads to prolonged UE battery lifetime due to the close distance to the femtocell [36].

Typical femtocell structure consists of five main parts: femtocell device, DSL router, ISP (Internet link), mobile operator network, and cellular tower (macrocell) [38] (as shown in Figure 6.5). Femtocell does not require a cellular core network since it contains Radio Network Controller (RNC) and all other network elements. It acts like a Wi-Fi access point and it needs data connection to the DSL or Internet connected to a cellular operator core network [36]. Although it does not need to be under the macrocell coverage area, there are many examples for deployments under the macrocell in order to increase the capacity and QoS in presence of huge user demands or significant shortage in coverage. Apart from this, it can be used in rural areas in order to provide cellular coverage where there is no macro coverage.

Although femtocell technology first designed to use indoor, there are many outdoor applications of femtocell technology. To illustrate, it can be deployed in transit systems such as bus, train, etc. In this application, mobile users connect to femtocell instead of macrocells or satellites. There is a transceiver connected to Femtocell Base Station (FBS) through wired connection and to macrocell or satellite through

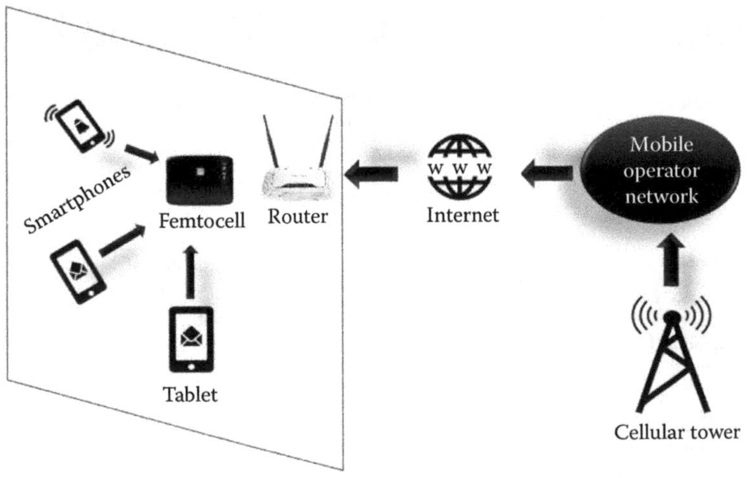

FIGURE 6.5 Typical femtocell structure [36].

a wireless link [39]. Moreover, femtocells can be good solution to increase coverage and capacity in public outdoor areas, especially in crowded areas. The key point behind the femtocell is to bring cellular network closer to user and with this approach it manages to be low-power and low-cost technology [38]. It is usually difficult for a macrocell to provide indoor service since there is a signal loss. Moreover, 50% of voice calls and 70% of data calls comes from indoor [38]. There is an estimation that 10% of active femtocell household deployment can offload 50% of the overall macrocell load [40]. Thus, it increases revenues of cellular operators, and thus it is expected that there will be about 28 million unit of femtocell by 2019 [41].

IoT touches every facet in our life and our smart applications that can improve quality of our lives in different regions and domains such as home, highways, hospitals, etc. [42]. IoT describes communication capabilities of these objects with each other and elaborates information perceived. IoT includes very wide range of applications (see Figure 6.6) and these can be grouped as the following: (i) transportation and logistics, (ii) healthcare, (iii) smart environments, (iv) personal and social, and (v) futuristic applications.

Femtocell can be used as a communication mechanism for the IoT paradigm, especially in the smart grid [43]. In Refs. [44,45], the authors proposed the use of femtocells in a home area network (HAN) as a cost-effective technology. Health-care IoT applications are another example for using a femtocell in smart spaces. In Ref. [45], the authors proposed an IoT-oriented healthcare monitoring system where sensors gather the particle measurements of an Android application and then they used LTE-based femtocell network in order to send the data with new scheduling techniques. In Ref. [46], the authors underline that femtocell is susceptible to man-in-the-middle attacks when it is used to fix shadow area problem. Moreover, they propose

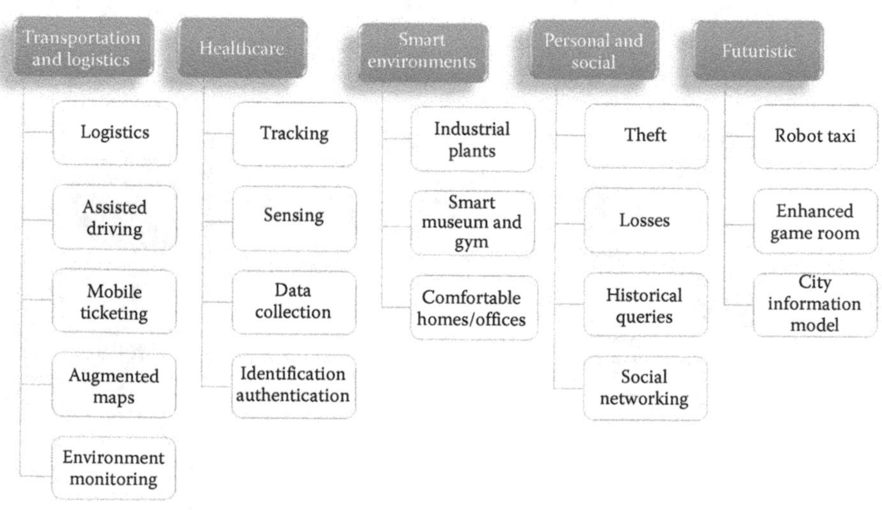

FIGURE 6.6 Application domain of IoT.

the interlock protocol to protect the confidential information. In Ref. [47], the benefits of using 5G femtocells for supporting indoor generated IoT traffic is highlighted. Authors underline the fact that supporting the traffic produced from IoT is a major challenge for 5G and significant percentage of the traffic will be generated indoor.

Thus, it can be said that femtocells can be used in most of the IoT applications since there is a desperate need for a capable communication mechanism/tool. Indeed, it is expected that small cells such as the femtocell will be central to 5G network architectures both for human users and for IoT embedded systems.

6.6 Energy Aspects

Energy is a key design factor in all enabling technologies, nowadays, which realizes the smart-space vision in IoT. Measuring energy in the targeted enabling technologies in this study including the smartphones and femtocells is of utmost importance toward realizing the smart spaces in practice. There are different Energy-Efficient (EE) metrics in the literature and they are applied at three different levels: the UE/component level, the BS level, and the network level. In component level, millions of floating-point operations per second per watt (MFLOPS/W) and Millions of Instructions Per Second per Watt (MIPS/W) can be used to calculate processor related energy consumption and the ratio of output to input (ROI) power can be used to calculate energy efficiency of power amplifier component. In BS level, EE metrics can be evaluated under two main categories. Bits per second per hertz per watt is for trade-off between energy consumption and Spectral Efficiency (SE) and bits times meters per second per hertz per watt is an energy efficiency metric when energy consumption, spectral efficiency and the transmission range of the base station is taken into consideration. In network level, obtained service relative to the consumed energy is evaluated by EE metrics which is power per area unit (watts per square meter) in order to evaluate the coverage energy efficiency. Table 6.3 gives the summary of the reviewed EE metrics.

Since the structure of femtocells is similar to that of the macrocell, the macrocell component level energy metrics are also suitable for femtocells. However, for the BS and network level, the difference between the femtocell-supported and macrocell provided services and the interference between femtocells and macrocell should also be considered; as femtocells are overlapped by a macrocell and use the same spectrum as the macrocell. In Ref. [48], the authors proposed an EE metric for the femtocell and macrocell heterogeneous network that considers the service rate and power consumption in both FBS and MBS.

6.6.1 Energy Modeling

Generally, energy consumption of a wireless network [49] is evaluated at two different levels as described in Table 6.4. The first one, embodied energy consumption,

TABLE 6.3 Energy Efficiency (EE) Metrics

Energy Efficiency Metrics	*Levels*	*Descriptions*
ROI	Component level	Used to evaluate energy efficiency of power amplifiers
MIPS/W or MFLOPS/W	Component level	Used to calculate processing associated energy consumption
bits/s/Hz/W	BS level	For trade-off between energy consumption and spectral efficiency (SE)
(b*m)/s/Hz/W	BS level	Taking into consideration energy consumption, SE, and the transmission range of BSs
W/m^2	Network level	Used to evaluate the coverage EE

TABLE 6.4 Energy Consumption of a Femtocell BS in Different Levels

Energy Consumption Level	*Definition*	*Average Power*
Embodied energy	The total primary energy that is consumed for making the product	162 MJ (1 W/s)
Operational energy	The amount of the energy spent during system's lifetime	6 W/s

calculates the total primary energy that is consumed for making the product. Embodied energy of for a femtocell for example is assumed as 162 MJ, same as a mobile terminal and the average lifetime of a femtocell can be assumed as 5 years. In other words, embodied energy of the femtocell per second is calculated as 1 W [50]. The second level is operational energy consumption, which can be defined as the amount of energy spent during a system's lifetime and it changes depending on different configurations such as age of the facility and load of the femtocell. In Ref. [50], average operational power of a femtocell is assumed as 6 W. The total power consumption of a femtocell mainly depends on radio frequency power amplifier and the power amplifier of power supply.

In the literature, there are few models about the power consumption of a femtocell. On–off model is the most basic one, which can be used for theoretical analysis where FBS is assumed to consume unit power in active mode and zero power when it is off. However, it does not reflect actual power consumption. In Ref. [51], they proposed a linear power model that considers the traffic load. This model is used for analysis and more accurate than on–off model since the traffic load is defined. In Ref. [52], the authors proposed a more detailed model and they argued that the traffic load has no effect on power consumption so it can be omitted.

In Ref. [53], the authors proposed a simple analytic model to predict the FBS power consumption based on offered load and datagram size. They tried to fit their model to real experiment that they measured power consumption of a FBS is idle, as the offered load is varied and as the datagram size is varied. In the experiment, they used one femtocell that supports up to four simultaneous end-user devices and it is connected to campus network. They also predicted the energy consumption of voice and File Transfer Protocol (FTP). In their prediction, they neglected that radio energy must be expended by the BS when making downlink transmission. Since in voice call, the downlink is active for a small period and active for the majority of the time during FTP download, they expected that radio power consumption should be higher than voice calls.

In Ref. [54], the authors proposed power consumption model, it is based on a femtocell that consists of three interacting blocks: microprocessor, Field Programmable Gate Array (FPGA) and radio frequency transmitter and power amplifier. They used energy efficiency, which is defined as power consumption needed to cover a certain area, in order to compare different technologies. They used ITU-R P.1238 propagation model and office scenario is assumed. They considered frequency, the floor penetration loss factor, the number of floors between BS and terminal and the distance power loss coefficient to calculate the range. Based on this model, they compared energy consumption for different bit rates and different technologies. Moreover, they used this model in a deployment tool that allows to design energy efficient femtocell networks by using genetic algorithm. Table 6.5 summarizes the aforementioned energy consumption models.

6.6.2 Energy-Efficient (EE) Deployments

Deployment strategy of 5G network is a very important design factor that has caught the attention of many researchers. Even though 5G technology is engineered to operate in different fields e.g., IOT, it's the cellular phone applications that will be the first to truly embrace the full capabilities of 5G technology. The authors in Ref. [55] claim that massive Multiple-Input Multiple-Output (MIMO) has emerged as one leading technology enabler for the next generation mobile communications i.e., 5G. Most of the investigated MIMO scenarios consider preferably wide area outdoor deployments [56]. However, indoor users [57], generate most of the mobile traffic. In 5G, the main traffic volume of mobile services is occupied by high data rate services at indoors and at hotspots [58]. In this order to be more specific, we focus in the following on 5G femtocells deployments.

Femtocells can be deployed in heterogeneous networks with different combinations of macrocells, microcells, picocells, and femtocells and the reduction in total power consumption can be obtained from these schemes. The amount of energy

TABLE 6.5 Comparison of Different Energy Consumption Models

Reference	*Model Type*	*Input*
[53]	Analytical	Datagram size (byte), offered load (Mbps), baseline power consumption when the femtocell is idle (Watts)
[54]	Experimental	Power consumption (Watts) of the microprocessor, the FPGA and the power amplifier (input power of antenna and the efficiency of the power amplifier which is the ratio of radio frequency output power to electrical input power)
[51]	Experimental	Static power consumption (idle mode power consumption of power amplifier, base band transceiver units, feeder network and cooling system) and dynamic power consumption (load on the base station (BS) and backhaul power consumption of BS)
[52]	Analytical (Log-normal distribution)	Radio frequency output power at maximum load, minimum load and in sleep mode and the dependency of the required input power on the traffic load

efficiency depends on different variables. However, how to create this combination is also a challenging issue and there are some criteria to choose which cell to deploy. Although energy efficiency is not the first goal in some of the combinations modeled, energy efficiency is achieved. In general, cells are deployed depending on mobile user density, traffic and coverage in the network models. It is also important to provide same or better QoS when energy efficiency is subjected. In majority of the aforementioned studies, the network criteria mainly depend on the mobile user density, traffic and coverage. Mobile user density and traffic is considered to choose the correct cell size. Urban and rural areas are also important criteria in order to choose the cell. In the areas that have very weak signals can be supplemented with smaller cells such as femtocell so that coverage can be increased and there will be reduction in total power consumption.

There are different models in the literature to analysis energy efficiency [59] using heterogeneous network includes femtocell and macrocell. In Ref. [60], authors considered a network in which cells of different sizes have been deployed depending on mobile user density, traffic and coverage such that power consumption can be minimized without compromising with the QoS. They developed analytical models of power consumption in the proposed five different schemes and obtained the reduction in power consumption compared to macrocell network. In the first one, they used femtocell-based network instead of macrocells in the area fully covered with femtocells and they obtained between 82.72% and 88.37% reduction in power consumption. In the second schema, they divided the area into three part as urban, suburban and rural. They considered mobile user density, mobile user traffic and required coverage and they covered urban areas by femtocells, suburban areas with macrocells and rural areas with portable femtocells. Because of this simulation, they succeed between 78.53% and 80.19% reduction in power consumption. In the third schema, they allocated femtocells to densely populated urban area, picocells to sparsely populated urban areas, microcells to suburban areas and portable femtocells to rural areas and the reduction in power consumption rate was obtained as between 9.19% and 9.79%. In the fourth schema, they allocated microcells, picocells and femtocells to the border region and macrocell to remaining region. The reduction in power consumption is between 5.52% and 5.98% for this combination. The last one, femtocells are allocated between the boundaries of macrocell, where signal is not enough for a call. The result of the last model was between 1.94% and 2.66% reduction in power consumption and macrocell coverage.

In the analysis conducted by Bell Labs, scientists analyzed the efficiency of a hybrid network of both femtocell and macrocell, with open access femtocell, where all subscribers can connect to these femtocells like any BS [61]. The implementation area was 10 km by 10 km urban area of New Zealand and the population was about 200,000 people that means 65,000 homes and the 95% of population uses mobile equipment. They deployed varying number of femtocells that can serve up to eight users in 100 m by 100 m area with 15 W energy consumption. The reason of higher consumption rate compared to other analysis is that it is an open access model. They used continuously operated macrocells with 2.7 kW energy consumption and femtocells deployed randomly in the area. The result of this analysis was depending on the network used: voice call and data connections. When femtocells mainly used for voice call, there was no big saving on total energy. On the other hand, when femtocells used for data connections, femtocells were able to reduce total energy consumption up to 60%. In this analysis, macrocells were used to ensure coverage and femtocell to offload capacity.

Another one was from Ofcom (the UK telecoms company) and Plextek as consultant [53]. Researchers there have analyzed two approaches; first, they deployed femtocells to 8 million households, which is nearly 25% of the UK. Each femtocell consumed 7 W every day and total annual energy consumption was about 490 GWh. In the second approach, they modeled macrocell network in order to provide indoor coverage, same with first approach and they would need 30,000 BSs in order to provide coverage. The model estimated that it takes 40 times more energy to deliver signal to indoor from macrocell, compared to femtocell. Because of this analysis, in order to provide same coverage provided by femtocells, the total annual energy consumption of macrocell was 700 GWh per operator and 3500 Gwh for five operators. Thus, the ratio of energy consumption in order to provide same indoor coverage in the UK was 7:1 over using macrocells.

6.6.3 Power Consumption at Femtocells

The Femtocell access typesets the rules about who can connect to FBS. It can be categorized into three types: open, closed, and hybrid. In the literature, most of the studies about access type examine interference, QoS, and handover issues. On the other hand, although there is no work directly compares the energy efficiency related the access type, energy consumption can be compared with examining different models in the literature. In open access, all available sources are shared between users and everyone can connect the network. They are mainly deployed in public places and there is no restriction to connect the network. Closed access, unlike open access, only closed subscriber group can connect the network but there can be different service levels between users and they are mainly used in small buildings. Hybrid access is combination of open and closed so that it allows particular outside users to access a femtocell. However, the conditions to connect the network outside the user group are defined by the operator and new entries to the system are requested by owner. These users (outside) can get only limited service, depends on the operator management. The comparison of the access types is given in Table 6.6.

Although there is no work that compares the power consumption based on different access mode, it can be resulted as open access consumes more energy. In Ref. [61], power consumption of the femtocell is assumed as 15 W although it is assumed as 6 W in Ref. [50]. The reason of the high power consumption in Ref. [61] is the open access. Bit rate is another important factor that affects energy consumption of a femtocell. It can be defined as number of bits that are conveyed or processed per unit of time. As it reaches higher rates, the speed of connection becomes better since it is speed-based measurement. Cellular operators try to increase the bitrates since it is important in market. On the other hand, bit rate has impact on energy consumption. In general, as bit rate increases, energy consumption also increases. Moreover, energy consumption for different bit rates is not same in different wireless technologies.

TABLE 6.6 Comparison of Access Types

	Open Access	*Closed Access*	*Hybrid Access*
Deployment	Public places	Residential deployment	Enterprise deployment
Number of handovers	High	Small	Medium
Owner preference	No	Yes	Yes
High user densities	Yes	No	No
QoS	Low	High	High
Interference between femtocell and macrocell	Increase	Decrease	Decrease

As aforementioned, energy consumption of femtocell is not stable for different wireless technology standards such as Worldwide Interoperability for Microwave Access (WiMAX), High Speed Packet Access (HSPA), and LTE, which are rivals in the sector. WiMAX can be used for transferring data across an Internet service provider (ISP) network, as a fixed wireless broadband Internet access, replacing satellite Internet service or as a mobile Internet access. HSPA is another wireless technology standard, which is enhanced version of 3G. LTE is considered as 4G and provides better capacity and speed. They provide different bitrates and energy consumption of a femtocell is different for these technology standards. On the other hand, it is hard to say which one provides better energy efficiency since it changes with different bitrate ranges.

In Ref. [54], authors investigated energy efficiency of a FBS and compared bit rates and for different wireless technologies, WiMAX, HSPA, and LTE. Based on this model, they found that femtocell consumes nearly 10 W for a range between 9 and 130 m and WiMAX is the most energy efficient technology for bit rates more than 5 Mbps and LTE is the most energy efficient technology for bit rates between 2.8 and 5 Mbps. They used this model in a deployment tool that allows to design energy efficient femtocell networks by using genetic algorithm and they concluded WiMAX is the most energy efficient one for this scenario.

Network type also affects the energy consumption of femtocell. To illustrate, its power consumption is different for voice call and data transfer. Moreover, there are different data transfer protocols such as FTP and User Datagram Protocol (UDP). Datagram size and offered load have also impacts on the consumption. In Ref. [53], the authors studied the effects of network type, datagram size and offered load. As the result of the analysis conducted by Bell Labs [61], the efficiency was depending on the network used: voice call and data connections. When femtocells mainly used for voice call, there was no big saving on total energy consumption. On the other hand, when femtocells used for data connections, femtocells were able to reduce total energy consumption up to 60%.

The last factor that has impact on femtocell power consumption is sleep mode. Since femtocell provides very small coverage compared to macrocell and very few users connect it, especially when it is deployed indoor, at most of the times it becomes idle. However, it consumes energy even if it serves no user. In this case, it is better to switch off the BS so this can be implemented with sleep mode. Sleep mode is energy efficient and feasible but the decision is also important. There are few works in the literature that analyzes the impact of sleep mode. In Ref. [54], the authors examined sleep mode to reduce power consumption and it reduces the power consumption supporting up to eight users and it led to 24% of power consumption in the network. Table 6.7 gives the summary of the section.

6.6.4 Smartphone/UE Power Consumption

One of the main idea of femtocell is to become closer to the UE which is a way to increase the system capacity and reduce energy consumption of both cellular

TABLE 6.7 Factors Affect the Power Consumption of a Femtocell BS

Factor	*Energy Consumption*
Access mode	Higher in open access, less in close access
Bit rate	Increases with higher bit rate
Wireless technology standards	Depends on bit rate range
Network type	Higher in data transfer
Sleep mode	Less energy consumed if sleep mode is an option

network and UE [53]. However, since the coverage area of femtocell is not wide, the number of handover is very high in femtocell network. Moreover, UE uses most of its energy for handover process [62]. Another issue about handovers is that they decrease QoS and network capacity. Thus, handover decision algorithm is an important issue in femtocells. Although there are many studies about handover decision, only few of them depends on energy efficiency.

In Ref. [62], the authors studied the energy efficiency of a femtocell per UE battery. They worked on the fact that there is a reduction in power consumption of UE under femtocell coverage, compared to macrocell connection and they proposed a handover decision algorithm that aims to reduce UE power consumption while maintaining QoS. The suggested algorithm enhances the strongest cell handout policy using an adaptive handout hysteresis margin. Although there is a need for increased LTE network signaling in the proposed algorithm, they derived power consumption and also interference. They compared the results with different algorithms in the literature and compared to a strongest cell based handout decision algorithm, the proposed algorithm reduced the UE energy consumption per bit by up to 85% with respect to femtocell deployment in LTE network.

6.7 Open Issues

Our view of the core design problem in smartphones usage depends strongly on a clear understanding about their functionalities, and on required representations of the following: (i) the environment they operate in; (ii) the available resources (nodes, components, their costs, and their accuracies); (iii) the lifetime expectancy; and (iv) the development in their operations and maintenance. The design process should further overcome the complications resulting from coupling application requirements and interfaces, and the underlying components and topology required for each. A significant challenge also lies in adequately representing the resources available in a given region of interest: those belonging to sensing nodes, such as transceivers, sensors, and processors, and those that exist under other systems. The most prominent sources of those that exist under other systems fall under the umbrella of industrial and municipal resources. They are deployed in abundance to collect information and control different operations and machinery.

Since user privacy is a concern for many users, researches design applications accordingly [27, 32]. Many users do not install apps that collect unnecessary information thinking of privacy. Games, entertainment apps and shopping apps necessarily does not need permissions that can frustrate users. On the other hand, some application categories really need to get some private information from users. For example, navigation apps stop working when they cannot get location permission. Friend finding apps, health apps also needs some private information. Moreover, some app need to encrypt network communications in order to secure user related information [32]. Application categories that take encryption as a primary design factor can be banking apps, social media and messaging. On the other hand, some application categories can consider encryption as a secondary design factor like alarm apps, translation apps, etc. Further research shall pitfalls of the aforementioned security/privacy issues in addition to the single application-specific design, which stands as a major hindrance in realizing public sensing in smartphone usage. It shall allow for great flexibility in design and an expandable and/or upgradeable set of functionalities. It shall allow for the reutilization of existing sensing resources and infrastructures across applications, resulting in extended functional operations and increased return on investment.

Moreover, this survey shows that there are variety of issues on the energy efficiency of femtocell networks needing to be investigated in the future, including the following:

- Energy metrics and energy consumption models considering not only femtocells but also femtocell and macrocell heterogeneous networks.
- Energy consumption models, considering more details in order to make estimation that is more accurate.
- Comparison of energy efficiency of femtocells when they are deployed with different access types.
- The optimized deployment of cellular networks in huge area, including one macrocell and many overlapping femtocells.
- Femtocell handover decision while considering the network energy efficiency.
- The effect of femtocell on UE energy consumption, considering different parameters such as the access type.
- Evaluating the efficiency of femtocell in terms of energy consumption as a communication technology in IoT applications and comparison with other communication technologies.
- The effect of interference problem on femtocell energy consumption.

All the aforementioned issues are still open research problems that necessitate further investigations in order to realize the vision of 5G enabled devices and smart spaces.

6.8 Concluding Remarks

In this research, we believe that the idea of monitoring certain types of smartphone activities will minimize the users concerns about accessibility and tracking in smart environments. The main drawback of the existing methods is their intrusive and disruptive nature, due to the fact that users may need to be interrupted from what they are doing in order to provide a piece of information such as a password. Although some methods are less disruptive (such as face recognition), they are still intrusive as the use of a camera or other sensor types means highly personalized information are collected and stored by the system to be used in the verification process of any context aware system [11]. This also raises a question concerning users' perception and tolerance to such methods as people are concerned how, where else, and by whom their information is being used. Similarly, context aware systems need to collect a variety of information about the user's current status and activities, some of which may be regarded as personal. New proposals are required to address this issue by only collecting information related to the user's access to resources, which has to be collected in any case. Collecting such information does not require the users to perform additional disruptive activities in the process of verifying their identity. As the information needed by the framework is collected anyway and no additional information is required (collected), the authentication decisions can be made more quickly. Furthermore, energy efficiency of femtocell networks becomes more important with the increasing deployment of huge numbers of femtocells. Moreover, it can be used in order to manage energy efficiency when it is used in smart grids. In this article, we mentioned energy efficiency of femtocell networks in IoT, considering energy metrics, energy consumption models, energy efficiency schemes for deployment, factors that affects energy consumption of femtocell networks and energy efficiency of UE under femtocell coverage. Energy efficiency of femtocell in IoT still needs to be investigated.

References

1. A. Gupta and R. K. Jha, A survey of 5G network: architecture and emerging technologies. *IEEE Access*, vol. 3, pp. 1206–1232, 2015.
2. D. Uckelmann, M. Harrison, and F. Michahelles, *An Architectural Approach Towards the Future Internet of Things. Architecting the Internet of Things*. Springer, Berlin Heidelberg, 2011, pp. 1–24.
3. L. M Ericsson, More than 50 billion connected devices, Feb. 2011, www.ericsson.com/res/docs/whitepapers/wp-50-billions.pdf.
4. A. P. Castellani, N. Bui, P. Casari, M. Rossi, Z. Shelby, and M. Zorzi, Architecture and protocols for the internet of things: a case study, IEEE *Pervasive Computing and Communications Workshops (PERCOM Workshops), 2010 8th IEEE International Conference on PERCOM*, Mannheim, Germany, pp. 678–683, 2010.
5. L. Atzori et al., The internet of things: a survey. *Computer Networks*, vol. 54, no. 15, pp. 2787–2805, 2010.

6. G. Barrenetxea, F. Ingelrest, G. Schaefer, and M. Vetterli, The hitchhiker's guide to successful wireless sensor network deployments, *Proceedings of the 6th ACM Conference on Embedded Network Sensor Systems (SenSys). ACM*, New York, pp. 43–56, Nov. 2008.
7. F. Al-Turjman, H. Hassanein, W. Alsalih, and M. Ibnkahla, Optimized relay placement for wireless sensor networks federation in environmental applications. *Wireless Communications & Mobile Computing Journal*, vol. 11, no. 12, pp. 1677–1688, Dec. 2011.
8. F. Al-Turjman, H. Hassanein, and M. Ibnkahla, Optimized relay repositioning for wireless sensor networks applied in environmental applications, *Proceedings of the International Wireless Communications and Mobile Computing, Conference*, Istanbul, pp. 1860–1864, June, 2011.
9. F. Al-Turjman and H. Hassanein, Enhanced data delivery framework for dynamic Information-Centric Networks, *Proceedings of the Local Computer Networks (LCN)*, Sydney, pp. 831–838, Oct. 2013.
10. G. Anastasi, M. Conti, M. Francesco, and A. Passarella, Energy conservation in wireless sensor networks: a survey. *Ad Hoc Networks*, vol. 7, no. 3, pp. 537–568, May 2009.
11. F. Al-Turjman, Cognition in information-centric sensor networks for IoT applications: an overview. *Annals of Telecommunications*, vol. 72, no. 1, pp. 3–18, 2017.
12. A. Salhieh, J. Weinmann, M. Kochhal, and L. Schwiebert, Power efficient topologies for wireless sensor networks, *International Conference on Parallel Processing (ICPP)*, Valencia, Sept. 3–7, 2001.
13. M. Biglarbegian and F. Al-Turjman, Path planning for data collectors in precision agriculture WSNs, *Proceedings of the International Wireless Communications and Mobile Computing Conference (IWCMC)*, Nicosia, pp. 483–487, 2014.
14. R. Kuntz, A. Gallais, and T. Noel, Medium access control facing the reality of WSN deployments. *SIGCOMM Computer Communication Review*, vol. 39, no. 3, pp. 22–27, Jun. 2009.
15. N. Lane, E. Miluzzo, L. Hong, D. Peebles, T. Choudhury, and A. Campbell, A survey of mobile phone sensing. *IEEE Communications Magazine*, vol. 48, no. 9, pp. 140–150, Sep. 2010.
16. F. M. Al-Turjman, H. S. Hassanein, and M. Ibnkahla, Efficient deployment of wireless sensor networks targeting environment monitoring applications. *Computer Communications*, vol. 36, no. 2, pp. 135–148, 2013.
17. A. Rahmati and L. Zhong, Human–battery interaction on mobile phones. *Pervasive and Mobile Computing*, vol. 5, no. 5, pp. 465–477, 2009.
18. D. T. Wagner, A. Rice, and A. R. Beresford, Device analyzer: understanding smartphone usage, *Mobile and Ubiquitous Systems: Computing, Networking, and Services, Springer*, pp. 195–208, ISBN: 978-3-319-11568-9, 2014.
19. R. Ferdous, V. Osmani, and O. Mayora, Smartphone app usage as a predictor of perceived stress levels at workplace, *Proceedings of the 9th International Conference on Pervasive Computing Technologies for Healthcare*, Istanbul, Turkey, pp. 225–228, May, 2015.
20. T. M. T. Do, J. Blom, and D. Gatica-Perez, Smartphone usage in the wild: a large-scale analysis of applications and context, *Proceedings of the 13th ACM International Conference on Multimodal Interfaces*, Alicante, Spain, pp. 353–360, Nov. 2011.

21. F. Al-Turjman, Mobile couriers' selection for the smart-grid in smart cities' pervasive sensing. *Elsevier Future Generation Computer Systems,* 2017, doi:10.1016/j.future.2017.09.033.
22. J. Altmann and R. Sampath, UNIQuE: a user-centric framework for network identity management, *Network Operations and Management Symposium, NOMS, 10th IEEE/IFIP,* Vancouver, pp. 495–506, Apr. 3–7, 2006, doi:10.1109/NOMS.2006.1687578.
23. H. Koshutanski, M. Ion, and L. Telesca, Distributed identity management model for digital ecosystems, *International Conference on Emerging Security Information, Systems, and Technologies,* Athens/Glyfada, Greece, pp. 132–138, Oct. 2007.
24. F. Al-Turjman and M. Gunay, CAR approach for the internet of things (IoT). *IEEE Canadian Journal of Electrical and Computer Engineering,* vol. 39, no. 1 (Winter), pp. 11–18, 2016.
25. H. Li, X. Lu, X. Liu, T. Xie, K. Bian, F. X. Lin, and F. Feng, Characterizing smartphone usage patterns from millions of android users, *Proceedings of the* 2015 *ACM Conference on Internet Measurement Conference, ACM,* pp. 459–472, Oct. 2015.
26. K. Elgazzar, A. Ejaz, and H. Hassanein, AppaaS: offering mobile applications as a cloud service. *Journal of Internet Services and Applications,* vol. 4, p. 17, 2013.
27. A. Rahmati and L. Zhong, Studying smartphone usage: lessons from a four-month field study. *IEEE Transactions on Mobile Computing,* vol. 12, no. 7, pp. 1417–1427.
28. Z. X. Liao, S. C. Li, W. C. Peng, P. S. Yu, and T. C. Liu, On the feature discovery for app usage prediction in smartphones, *IEEE 13th International Conference on Data Mining (ICDM),* pp. 1127–1132, Dec. 2013.
29. H. Falaki, R. Mahajan, R. Mahajan, D. Lymberopoulos, R. Govindan, and D. Estrin, Diversity in smartphone usage, *Proceedings of the 8th International Conference on Mobile Systems, Applications, and Services,* San Francisco, CA, pp. 179–194 June 15–18, 2010.
30. E. H. C. Lu, Y. W. Lin, and J. B. Ciou, Mining mobile application sequential patterns for usage prediction, *IEEE International Conference on Granular Computing (GrC),* pp. 185–190, 2014.
31. R. Montoliu and D. Gatica-Perez, Discovering human places of interest from multimodal mobile phone data, *Proceedings of the 9th International Conference on Mobile and Ubiquitous Multimedia,* p. 12, Dec. 2010.
32. Samsung S II [Available online]: https://www.anandtech.com/show/4686/samsung-galaxy-s-2-international-review-the-best-redefined/13.
33. Y. Zheng, L. Zhang, Z. Ma, X. Xie, and W. Y. Ma, Recommending friends and locations based on individual location history. *ACM Transactions on the Web (TWEB),* vol. 5, no. 1, pp. 2–44, 2011.
34. F. M. Al-Turjman, H. S. Hassanein, and M. Ibnkahla, Quantifying connectivity in wireless sensor networks with grid-based deployments. *Journal of Network and Computer Applications,* vol. 36, no. 1, pp. 368–377, 2013.
35. F. M. Al-Turjman, H. S. Hassanein, and M. Ibnkahla, Testing strategies and tactics for mobile applications, keynote white paper (PDF), Keynote.com. Retrieved 2012-05-02.ASD.
36. R. A. Saeed, (Ed.), *Femtocell Communications and Technologies: Business Opportunities and Deployment Challenges,* IGI Global, London, 2012.
37. Smart grids and meters - Energy - European Commission. (n.d.). Retrieved November 21, 2016, from http://ec.europa.eu/energy/en/topics/markets-and-consumers/smart-grids-and-meters.

38. V. Chandrasekhar, J. G. Andrews, and A. Gatherer, Femtocell networks: a survey. *IEEE Communications Magazine*, vol. 46, no. 9, pp. 59–67, 2008.
39. National Intelligence Council, Disruptive civil technologies—six technologies with potential impacts on us interests out to 2025, *Conference Report CR 2008-07*, 2008.
40. M. Z. Chowdhury, S. Q. Lee, B. H. Ru, N. Park, and Y. M. Jang, Service quality improvement of mobile users in vehicular environment by mobile femtocell network deployment, *ICTC 2011, IEEE*, Seoul, pp. 194–198, Sep. 28–30, 2011.
41. S. R. Hall, A. W. Jeffries, S. E. Avis, and D. D. N. Bevan, Performance of open access femtocells in 4G macrocellular networks, *Wireless World Research Forum 20 (WWRF 20)*, Ottawa, Apr. 22, 2008.
42. M. Domingues, and A. Radwan, Optical fiber sensors for loT and smart devices. *Springer Briefs in Electrical and Computer Engineering*, Mar. 2017, doi:10.1007/978-3-319-47349-9.
43. A. Radwan, K. Huq, S. Mumtaz, K. F. Tsang, and J. Rodriguez, Low-cost on-demand C-RAN based mobile small-cells. *IEEE Access*, vol. 4, pp. 2331–2339, May 2016, doi:10.1109/ACCESS.2016.2563518.
44. A. D. Domenico, R. Gupta, and E. Calvanese, Dynamic traffic management for green open access femtocell networks, *Proceedings of IEEE VTC-Spring*, 2012.
45. Z. Fan, P. Kulkarni, S. Gormus, C. Efthymiou, G. Kalogridis, M. Sooriyabandara, and W. H. Chin, Smart grid communications: overview of research challenges, solutions, and standardization activities. *IEEE Communications Surveys & Tutorials*, vol. 15, no. 1, pp. 21–38, 2013.
46. M. N. Hindia, T. A. Rahman, H. Ojukwu, E. B. Hanafi, and A. Fattouh, Enabling remote health-caring utilizing IoT concept over LTE-femtocell networks. *PloS*, vol. 11, no. 5, 2016.
47. T. H. Cho and G. M. Jeon, A method for detecting man-in-the-middle attacks using time synchronization one-time password in interlock protocol based internet of things. *Journal of Applied and Physical Sciences*, vol. 2, no. 2, pp. 37–41, 2016.
48. ABI Research, High inventory and low burn rate stalls femtocell market in 2012, (July 5, 2012 [online] http://www.teletopix.org/news-telecom/high-inventory-and-low-burn-rate-stalls-femtocell-market-ending-2012/).
49. F. Al-Turjman, H. Hassanein, S. Oteafy, and W. Alsalih, Towards augmenting federated wireless sensor networks in forestry applications. *Springer: Personal and Ubiquitous Computing Journal*, vol. 17, no. 5, pp. 1025–1034, June, 2013.
50. J. Zhang et al., A novel power control scheme for femtocell in heterogeneous networks, *Proceedings of IEEE CCNC*, Las Vegas, NV, pp. 86–95, Jan. 2012.
51. Y. Hou, and D. I. Laurenson, Energy efficiency of high QoS heterogeneous wireless communication network, *Vehicular Technology Conference Fall (VTC 2010-Fall), 2010 IEEE 72nd*, pp. Sept. 1–5, 2010.
52. M. W. Arshad, A. Vastberg, and T. Edler, Energy efficiency gains through traffic offloading and traffic expansion in joint macro pico deployment, *Proceedings of IEEE WCNC*, pp. 2230–2235, 2012.
53. Z. H. Hashmi, Adaptive and efficient resource management for emerging wireless networks, *Electronic Theses and Dissertations (ETDs) 2008*, May 4, 2013, doi:10.14288/1.0073687.
54. R. Riggio and D. J. Leith, A measurement-based model of energy consumption in femtocells, *2012 IFIP Wireless Days*, pp. Nov. 1–5, 2012, doi:10.1109/wd.2012.6402872.

55. B. Panzner et al., Deployment and implementation strategies for massive MIMO in 5G, *Globecom Workshops (GC Wkshps)*, pp. 346–351, IEEE, Austin, Dec. 8–12, 2014.
56. E. G. Larsson, O. Edfors, F. Tufvesson, and T. L. Marzetta, Massive MIMO for next generation wireless systems. *IEEE Communications Magazine*, vol. 52, no. 2, pp. 186–195, Feb. 2014.
57. J. Zhang and G. de la Roche (Eds.), *Femtocells: Technologies and Deployment*, Wiley, West Sussex, 2013.
58. S. Chen, and J. Zhao, The requirements, challenges, and technologies for 5G of terrestrial mobile telecommunication. *IEEE Communications Magazine,* vol. 52, no. 5, pp. 36–43, 2014.
59. F. Al-Turjman, Energy–aware data delivery framework for safety-oriented mobile IoT. *IEEE Sensors Journal*, vol. 18, no. 1, pp. 470–478, 2017.
60. M. Deruyck, D. D. Vulder, W. Joseph, and L. Martens, Modelling the power consumption in femtocell networks, *IEEE Wireless Communications and Networking Conference Workshops (WCNCW)*, pp. 31–35, Paris, Apr. 1, 2012.
61. Z. Feng and Z. Yuexia, Study on smart grid communications system based on new generation wireless technology, *International Conference on Electronics, Communications and Control (ICECC)*, Ningbo, Sept. 9–11, 2011.
62. A. Mukherjee, S. Bhattacherjee, S. Pal, and D. De, (2013). Femtocell based green power consumption methods for mobile network. *Computer Networks*, vol. 57, no. 1, pp. 162–178, doi:10.1016/j.comnet.2012.09.007.

Chapter 7

SWARM-Based Data Delivery in Social Internet of Things

Mohammed Zaki Hasan and Fadi Al-Turjman

Antalya Bilim University

Contents

7.1 Introduction

Wireless Sensor Networks (WSNs) form the basis of the future networks and several applications of extreme importance in our present and future life in various aspects. "Social Internet of Thing (SIoT)," as shown in Figure 7.1, is an exceptionally complex network model where varieties of components are deployed as consumer electronic devices that interact with one another in a complex way [1]. However, these devices operate under strict energy constraints, thereby making the dedicated energy budget for fault tolerant routing very limited [2]. The fault tolerant routing problem has received significant attention in the literature [3]. We believe that the emerging needs for SIoT applications, such as in smart homes, smart cities, and healthcare, will further increase the importance of fault tolerance in various aspects, due to its requirement constant mode of operation. Therefore, special effort has been made to develop fault tolerance in routing [3].

WSNs often operate in an autonomous mode without human supervision in the loop [4]. Moreover, sensor nodes are often deployed in uncontrolled and sometimes even hostile environments [5]. Therefore, accurately predicting an optimal way to treat fault tolerance within a particular WSN routing approach is difficult because both the technology and envisioned applications for WSNs and SIoTs are changing at a rapid pace [6]. Given the limitations in power supply, the energy consumption in WSN is dominated by radio communication. The energy consumption of radio communication mainly depends on the number of bits of data to be transmitted with sensor network. In order to avoid the loss of significant data from the sensor node, the available communication energy is significantly lower than the computation energy [7]. For example, the energy cost of transmitting one bit is typically around 500–1,000 times greater than that of a single 32-bit computation [8]. Therefore, to recover from path failure, fault tolerance routing algorithms that require only a limited amount of communication energy regardless of any additional computational energy must be developed. Otherwise, any unpredictable events may cause the devices to fail, partition the network, and disrupt the network functions.

These problems necessitate the development of fault tolerant routing approaches that require minimal additional computation regardless of any additional communication requirements, hence as to construct or recover the selected path [9]. Basically, multipath routing protocols provide tolerance to failures and increase the network reliability [9]. The fault tolerant routing problem is often formulated as Multi-objective Optimization Problem (MOP) to establish k-disjoint paths that guarantee connectivity even after the failure of up to

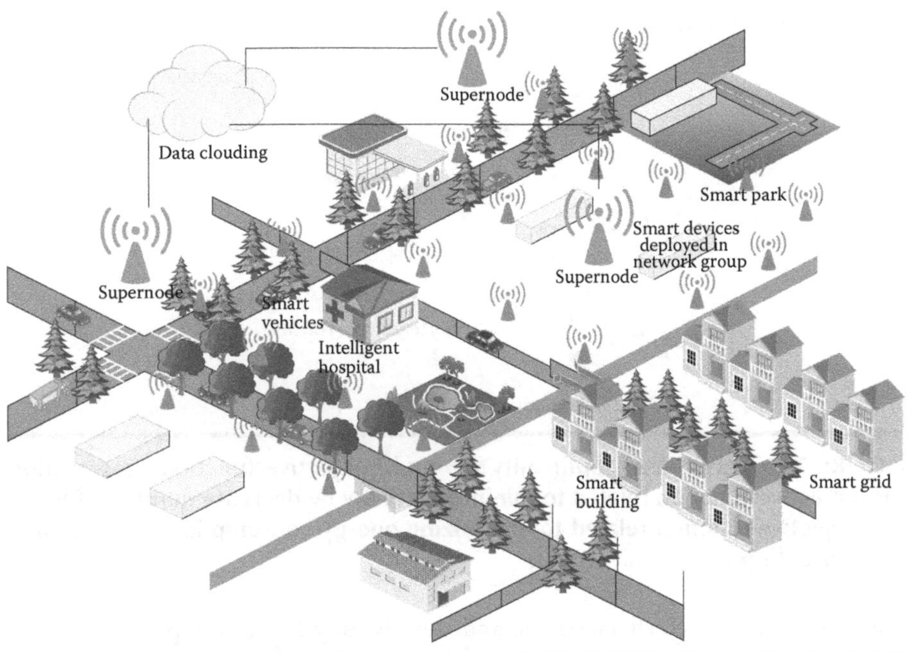

FIGURE 7.1 The devices deployed in SIoT

$k-1$ paths. To have a more realistic analysis of our model, we formulate the strong fault tolerant routing problem as a MOP that is are treated simultaneously while being subjected to a set of constraints. Multiple objective may or may not be conflicting; therefore, multiple objectives cannot achieve their respective optimal values at the same time contrary to the problems of Single-Objective Optimization (SOP) as illustrated in Figure 7.2. Furthermore, a single globally optimal solution that is considered the best with respect to all objective functions may be non-existent. Therefore, MOP means shall be identified in order to formulate a SOP and achieve Pareto-optimal solutions. The decision maker has to choose the best solution depending on the priorities of the Quality of Service (QoS) objectives to achieve.

Strong fault tolerance requires enormous computational efforts, which induce large control message overhead and lack of scalability as the problem size increases [10]. Solving these problems on an individual sensor node may require extreme memory and computational resources and yet produce average results [11]. We develop a bio-inspired Particle Swarm Optimization (PSO) routing algorithm to achieve fast recovery from path failure. Owing to its simple concept and high efficiency, PSO has been actively utilized in these problems with promising results [12]. Despite its competitive learning performance, precisely solving the fault tolerant routing problem remains a challenging task for PSO due to premature convergence. Unfortunately, most premature convergence traps are caused by

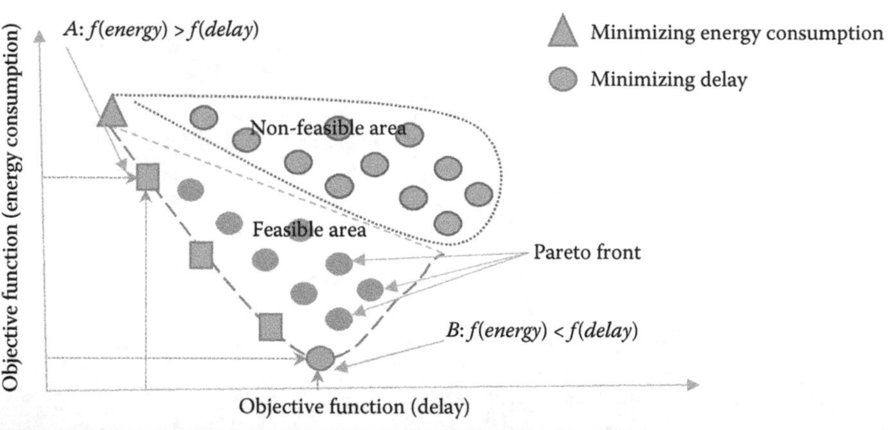

FIGURE 7.2 Pareto-front optimality for multi-objective function presentation (A): objective function related to minimizing delay by decrease number of hops. (B): objective function related to minimizing energy consumption by increasing number of hops.

the rapid convergence characteristic and the diversity loss of the particle swarm, thereby leading to undesirable quality solutions [13]. We face another challenge in the ability of PSO to balance exploration and exploitation searches. Neither exploration nor exploitation must be overemphasized as exploration inhibits swarm convergence, while the exploitation tends to cause the particle swarm to hastily congregate without the feasible region that leads to premature convergence [14]. Motivated by these challenges, especially the actual connectivity model inside the WSN that has been integrated into SIoT, we propose two new routing algorithms, namely, Fully Particle Multi-Swarm (FPMS) and Canonical Particle Multi-Swarm (CPMS), which are based on multi-swarm computationally efficient alternatives to analytical methods that tolerate the failure of multipath with reconstructed multipath and satisfy the QoS parameters in terms of energy consumption, delay, and throughput. Our contributions can be listed as follows:

- We develop a bio-inspired PSO routing algorithm to achieve fast recovery from path failure by attempting to extend an existing approach for finding an optimal solution in SOP for MOP. We define the objective functions and then optimize the effective values of these objective functions, which are computed at each sensor node that is selected to construct a *k*-disjoint multipath.
- We investigate the performance of the proposed multipath routing algorithms by comparing them with Canonical Particle Swarm Optimization (CPSO) [15,16] to provide an alternative learning strategy for particles.

These algorithms are very similar to one another and only differ in a few details of their learning strategies in different situations with respect to convergence, exploitation, exploration, and jumping out of the basins of attraction of optimal solutions. Additionally, increasing the number of paths requires the exchange of more messages and additional communication overhead [17]. Therefore, by looking at the similarity and differences among these algorithms, we employ complex network connectivity to represent the population of swarm topology. We also adopt the multipath routing algorithm to balance the trade-off between fault tolerance and communication overhead by taking advantage of the combination of proactive and reactive routing mechanisms to exchange demanding information of calculation and maintain on every particle and records the objective function value for the selected paths. Afterward, the particles are adaptively increased or decreased and connected with their matching velocity to make a proper selection by considering the optimized objective function.

The rest of the paper is organized as followed. Section 7.2 introduces the related works. Section 7.3 introduces the concepts of the system model. Section 7.4 presents the problem solving mechanism. Section 7.5 discusses the simulation results. Section 7.6 concludes the paper.

7.2 Related Work

SIoT is attracting considerable attention from governments, universities, and industries for its role in assisting the development of new applications for healthcare, environmental monitoring, and smart cities. These applications have a great impact on the quality of life of people and also lead to several health and economic benefits. Fault-tolerant routing is often used to guarantee the availability, reliability, and dependability of the network [3]. Therefore, fault-tolerant routing is important for SIoT applications because WSN devices are attached to the environment to perform advanced mentioning or detecting functions that assure data transmission with low delay, latency, loss rate, and minimum energy consumption for various IoT applications. Fault tolerance in WSNs can be achieved through several approaches, of which the most popular is multipath routing; this approach is employed at high levels as one of the possible solutions with limited network resources [18]. Multipath routing protocols can be divided into two categories, namely alternative path routing or retransmission and concurrent path routing or replication. These categories depend on two mechanisms to construct, select, and maintain multipaths, namely, disjoint multipath and braided multipath. However, many multipath routing protocols have been proposed in the literature and they can be classified according to their utilized mechanisms. Previous studies have considered different types of optimization, such as meta-heuristic strategies, to employ multipath routing [19]. The authors in Ref. [12] solve the fault-tolerant routing problem by constructing multipath routing to provide fast recovery from path failure. The authors in Ref. [20] proposed an energy-aware multipath routing scheme based on PSO to find an optimal loop-free path and to solve the disjoint multipath problem in a Mobile Ad-hoc Network (MANET). The authors in Ref. [21] used PSO to construct an appropriate path routing scheduling to distribute routing load over selected paths. The authors in Ref. [22] proposed an Enhanced PSO-based Clustering Energy Optimization (EPSO-CEO) scheme for minimizing the power consumption of each node by constructing clusters in a centralized manner and optimizing the selection of the cluster head. However, none of these studies discusses how to formulate an objective function with constraints that can lead to the convergence of the solution within the search scope as proposed in our routing algorithm. The difference between these studies and our work is that we try to find an optimal multipath to route the data in two-tiered heterogeneous WSNs with the multi-objective of minimizing the total transmission, average delay, and maximize the throughput of the nodes, while other works focus only on a flat homogenous sensor node topology. Additionally, we focus on the connectivity between sensor nodes and other components in the network model as well as on the k-disjoint multipath to route from the sensor nodes to the sink node that can tolerate at least $k-1$ network failure. Meanwhile, previous studies do not employ k-disjoint multipath between sensor nodes and thus cannot guarantee fault-tolerance in case of $k-1$ network failure.

The meta-heuristic able sensor nodes in WSNs has to configure their own network topology and must be able to localize, synchronize, and calibrate or learn themselves, to coordinate internode communication, and to determine other important operating parameters. Therefore, we define a neighborhood relationship among sensor nodes according to their information exchange mechanism. Our motivation for using this mechanism is to combine the network topology information for exploitation and to converge toward finding the optimal configuration or route, and the sensor nodes can help maintain diversity. Several studies [23,24] proposed the concept of diversity to avoid the premature convergence of swarms by setting a lower and upper bound of search space to ensure that the swarm has a good search ability in finding optimal solutions for various real-world applications. The authors in Ref. [25] identified the performance of load the distribution model as an optimization problem in order to facilitate the selection of an optimal network. The authors take into consideration both the network bandwidth and errors to determine the optimal load distribution among heterogeneous networks with minimal system cost. The authors in Ref. [26] aimed to improve the fault tolerance and cluster lifetime of the hybrid energy-efficient distributed clustering protocol by introducing non-probability waiting time, residue energy, and centers. Meanwhile, some researchers aim to change the parameters of PSO, such as inertia weight and acceleration coefficient of updating velocity, in order to improve the performance of the Comprehensive Learning Particle Swarm Optimizer (CLPSO) for multimode problems. The authors in Ref. [27] proposed an adaptive learning PSO with historical learning in which the particles can adjust their learning adaptively. Although, CLPSO suffers from a slow resolution, the authors adopt this algorithm and improve its search behavior to optimize the continuous solutions. The authors in Ref. [28] improved the CLPSO algorithm by introducing a new adaptive parameter strategy. The algorithm evaluates the objective functions of individual particles and the whole swarm based on which values of inertia weight and acceleration coefficient are dynamically adjusted to achieve a more effective search. However, none of these studies discusses how to formulate MOP with constraints, which can lead to a converged solution within the search scope as proposed in our routing algorithm. Based on CLPSO, some researchers use the concept of pareto dominance to solve MOP problems. The diversity of the search space can be promoted by using the Strength Pareto Evolutionary Algorithm (SPEA) to find the optimal solutions by coordinating the relationship of the objectives in a fitness function. We distinguish several techniques, such as the adaptive grid adopted in Pareto Archived Evolution Strategy (PAES) [29], PAES2 [30], clustering in SPEA, and multi-objective immune algorithm [31–33] which has improved version of SPEA. We differ between these techniques that we improved algorithm based on CPSO to coordinating this relationship by introducing multi-swarm method to share the information fairness among nearest neighbor density estimation to construct and select multipath routing. However, the effect of sharing all information fairness (i.e., personal best fitness), each sensor node can update its velocity and position the distribution of personal-best fitness. None of these studies uses this strategy, which full share of personal-best information converting MOP to SOP for fault-tolerant routing.

7.3 System Model

The proposed routing model employs fault-tolerant topology control in two-tiered heterogeneous WSNs consisting of resource-rich supernodes and simple sensor nodes with batteries of limited capacity and unmitigated QoS constraints. This mixed deployment of two-tiered heterogeneous WSN can balance of performance and cost of WSN [34]. However, to obtain a strongly fault-tolerant network topology, we consider a topology by constructing a k-disjoint multipath that poses a major challenge to the deployment of sensor nodes. Moreover, we consider a many-to-one traffic pattern where the supernodes and simple sensor nodes are able to ensure the required connectivity degree. Before introducing the model, we give some necessary definitions as follows:

Definition 7.1. n paths are said to be energy-node-disjoint $\leftrightarrow$ they have no common nodes.

Every node-disjointness has to be used in building the network topology with k-disjoint multipath routing in order to increase the number of alternative paths. Therefore, the network becomes fault tolerant. Our model is based on the observation that a node can connect and/or disconnect the links with neighbors that are not on one of the k-disjoint multipaths from the node to one of the supernodes. Therefore, the model needs to determine which neighbors are on one of such multipaths and which are not.

Definition 7.2. n paths are said to be node-disjoint $\leftrightarrow$ they have no common nodes.

Definition 7.3. A WSN is k-vertex supernode connected if the removal of any $k-1$ sensor nodes does not result in network partition.

Definition 7.4. Given a node ν_i, that ν_j next hops toward the destination node, and the forwarding selection rule F is associates node ν_i with another node ν_j in $|V| \varsigma \{v_i\}$, in such a way that the path $P(v_{source}, v_i, v_j), \ldots, v_{destination}$ obtained by applying the rule from the source to the destination.

7.3.1 Network Model

According to Definitions 7.1–7.4 the node-disjointness relations are modeled as a directed graph $G(V, E)$, where $|V| = \{v_1, v_2, \ldots, v_N, v_{N+1}, \ldots, v_{N+M}\}$ is a finite number of nodes (i.e., particles). We hereby note that both the particle and sensor node will be interchangeably used throughout the whole paper to refer to a "sensor node." Therefore, N denotes the sensor node and M denotes the supernodes. ε is the matrix of a defined set of k-disjoint multipaths from the source–destination nodes bypassing to one of the supernodes. The relation between a pair of nodes is represented by the number of edges E in G, which is a set of paths. Meanwhile, $E = \{(v_i, v_j) \mid Hop(v_i, v_j) \leq \tau\}$, where τ is the transmission power range of the node, and $Hop(v_i, v_j)$ depicts the distance between ν_i and ν_j. ν_j is indicated by the forwarding selection rule F that selects the next hop of node ν_i toward the ν_d with τ according to mechanism for selecting F. Therefore, ε induces multipaths among any possible connection between source–destination pair in the network. Thus,

$$F : \varepsilon \rightarrow k^{sd}_{v(i,j)}(\varepsilon), \tag{7.1}$$

where $k^{sd}_{v(i,j)}(\varepsilon) = 1$ iff the connection between node i and j is part of the path between the source and destination node. Moreover, the path $P(v_i, v_j)$ from vertex ν_i to vertex ν_j in graph G is a sequence of edges that are traversed when going from ν_i to ν_j, where $i \neq j = 1, 2, \ldots, N + M$. Therefore, $p(v_i, v_j)$ represents a set of alternative paths. $E(v_i, v_j \in p(v_i, v_j))$ represents a node-disjoint between $p(v_i, v_j)$ and (v_N, v_{N+M}), and $e(e \in p(v_i, v_j), (v_N, v_{N+M}))$ represents a direct link between any two nodes. Thus, we can get the k-disjoint path in G by considering the QoS parameters that affect the selection mechanism of the optimal multipaths, including energy consumption, delay, and throughput. These parameters are used to evaluate the objective function of the selected multipaths, and we use the derivation in Refs. [35,36] to solve the objective functions of minimizing energy consumption, average delay, and maximizing throughput. Table 7.1 lists all notations, which have been used throughout the paper.

TABLE 7.1 Notation

Symbol	*Quantity*
N	Sensor nodes or particle
M	Supernodes
T	transmission range of node
$Hop(v_i, v_j)$	Distance between indicating two nodes ν_i and ν_j
P_v^o	Energy dissipation for running the transmitter and receiver circuitry
SNR	Signal to noise ratio
$Energy_{v_{sd}}$	Total energy consumption of the given path
P^t	Power transmit
P^r	Power receive
p	Adjustment parameter
λ_s	Failure rate
t_s	Specific time period
ε	The definition matrix of k-disjoint multipaths
F	Forwarding rule to select next hop of node
k	Connectivity indicator between node i and node j
$P(v_i, v_j)$	The path in network
$\aleph_{i,j}$	The set of nodes disjoint of the k-disjoint path
$d_{v(i,j)}^{\alpha}$	Distance between node i and node j
L_p	Bits of data frame
λ_v^a, λ_v^b, and λ_v^{μ}	The time duration of data acquisition, processing and data packet transmission time
$P_{\min}$ and $P_{\max}$	Lower and upper energy constraints value bound of selected path
$\wp(\xi_i, \xi_j)$	Delay of routing data packets from source to destination
$L_e^{\wp}$	Hop delay requirement along the path from the source to the sink

(*Continued*)

TABLE 7.1 (*Continued*) Notation

Symbol	*Quantity*
$\Delta\wp$	Bounded delay
f	Objective function
Z	Set of feasible solutions
B	Bandwidth
p_{best}	Personal-best position
g_{best}	Global-best position
ϕ_1	Personal best coefficient
ϕ_2	Neighbor best coefficient
χ	Constriction coefficient
ν	Velocity toward selecting optimal solution

7.3.2 Problem Formulation

Using the network model, we aim to construct a k-disjoint multipath routing in the fault-tolerant network topology to route the data collected by sensor nodes to the supernodes for two-tiered WSNs. By combining these components into the network model, the sensor nodes become multi-functional depending on their electronic mechanical and communication limitations as well as their application-specific requirements. However, how we can use such resources-constrained sensor nodes to meet certain application requirements, including power transmission, end-to-end delay, and throughput? WSNs may face several challenges by integrating them within other co-exiting wireless heterogeneous wireless systems. This coexistence and integration may substantially affect the performance of sensor nodes that rely on diverse performance metrics to optimize QoS. Given that these metrics often conflict with one another, the trade-off among sensor nodes must be balanced to optimize the overall performance of WSNs. Network connectivity is closely related to energy efficiency; thus, we need to define the relationship between the number of sensor nodes that remain active and the connectivity with acceptable achieving QoS. Therefore, we model topology control as a transmission-range assignment problem for each sensor node in the network. Our objective is to minimize the assigned QoS parameters in terms of transmission power range and average delay for all sensors while maintaining k-disjoint multipaths from each sensor to the set of supernodes to determine the optimal multipaths. In this topology, each sensor node in the network must be connected to at least one supernode with k-disjoint multipaths to exchange information with one another. We can define the problem as follows: a k-disjoint multipath constructed by connecting a group of supernode and energy-constrained sensor nodes that can adjust their transmission range up to a predefined optimal value. The transmission range of each sensor is minimized, and the resulting topology is still k-disjoint multipath to ensure that all QoS application requirements are satisfied.

7.3.3 Energy Model

To obtain the appropriate constraints value, we depend on the determination of two variables, namely, the number of hops and the intermediate distance between two sensor nodes along the selected path, where $\tau_{v(i,j)}$ is the distance from one node to the next hop node. Neighborhood topology is used to define a neighborhood relationship among the nodes as mentioned in Section 7.3. The next hop is determined by exploiting the nearest neighbor; therefore, each sensor node has a transmission range that can communicate with the neighbors within that range. Suppose that the transmission range is given by $\tau > 0$, then its next neighborhood is given by:

$$\aleph_{i,j} = \{i, j \neq \imath \,\|\, \| v_i - v_j \| \leq \tau_v\}. \tag{7.2}$$

Here, $\aleph_{s,d}$ which denotes the set of nodes disjoint of the k-disjoint path. The transmission range can vary during transmission and reception of information and may lead to the disconnection of several neighbors and the partition of multipaths, unless the constraints are satisfied. Therefore, these constraints can be considered a dynamic point of the objective function that may be the in the feasible region. Therefore, these constraints can extremely change the topology connectivity degree, which can be used to solve the objective function of energy consumption as the optimization process evolves. The lower and upper bound values of the number of hops and transmission range are determined by [35] according to the method for determining the cut-off method values. The cut-off method for the integer programming (IP) problem as discussed in Ref. [35] is closely related to a certain global optimization problem. This method has been modified and combined with other methods, such as lines searches, and quadratic approximation [37]. However, cut-off methods have always been challenged by constraint dropping strategies. Given that a new constraint is added to the existing set of constraints and that no constraint is ever deleted; in each iteration, the size of the problem to be solved increases from one iteration to another. Thus, we depend on determining the lower and upper bounded for constraints similar to [35], whereas the number of constraints in each iteration is bounded. Furthermore, the energy dissipation for running the transmitter and receiver circuitry is denoted as the operational power P_v^o. We assume that

$$P_v^{trans} = P_v^{rec} = P_v^o. \tag{7.3}$$

Meanwhile, the transmit power level that must assigned to sensor node i to connect sensor node j with an acceptable Signal to Noise Ratio (SNR) is denoted as is $P_v^t = \varepsilon_{mp} d_{v(i,j)}^{\alpha}$, where α denotes the energy loss due to channel transmission under the assumption that the WSN is relatively free of obstacles, $d_{v(i,j)}^{\alpha}$ is distance between sensor node i and sensor node j and $\tau_{v(i,j)}$ is transmission range. Therefore, the overall expression for power consumption can be simplifier to

$$P_v^t = 2P_v^o + \varepsilon_{mp} d_{v(i,j)}^{\alpha}, \tag{7.4}$$

where $d_{v(i,j)}^{\alpha} = \tau_{v(i,j)}$ is sensor i's transmission range, s.t. $\tau_{\max} \geq \tau_{v(i,j)} \geq \tau_{\min}$, where $\tau_{\max}$ is a fixed maximum communication distance that is considered by the maximum power that the sensor nodes can transmit $P_{\max}$ while $\tau_{\min}$ is a minimum communication distance that is considered by minimum the power that the sensor node can transmit $P_{\min}$. Therefore, the optimal theoretical hop number is obtained as an integer number from

$$HOP = \sqrt[\alpha]{\tau_{v(i,j)} \left(\frac{3\varepsilon_{mp} d_{v(i,j)}^{\alpha}}{2P_{v(i,j)}^{o}} \right)}. \tag{7.5}$$

The total energy consumption of the given path at specific interval period λ is given as:

$$Energy_{v_{sd}}(\lambda) = L_p \left\{ \sum_{v(i,j)=0}^{N+M} (\lambda_v^a + \lambda_v^b) P_v^o + \lambda^{\mu} P_v^t \right\}, \tag{7.6}$$

where λ_v^a, λ_v^b, and λ_v^{μ} indicate the time duration of data acquisition, processing, and data packet transmission at sensor node i, respectively. We can assume the energy cost for each path $P(v_i, v_j)$ can be expressed as:

$$Energy_{(\aleph)}(\lambda) = \sum_{(s,d)\in \Pi_{P(v_i,v_j)}(\aleph)} L_p \left\{ \left(\lambda_v^a + \lambda_v^b \right) P_v^o + \lambda^{\mu} P_v^t \right\} \tag{7.7}$$

with

$$\Pi_{P(v_i,v_j)}(\aleph) = \left\{ (s,d) \ s.t.\ k_{v(i,j)}^{sd}(\varepsilon) = 1 \text{ for at least one } j \right\}. \tag{7.8}$$

The set $\Pi_{P(v_i,v_j)}(\aleph)$ defines the binary constraints that ensure all links exist between the source–destination pair as well as between the two sensor nodes i and j. Suppose, transmission by sensor node can be received by all sensor nodes within its transmission range, that is $\Leftrightarrow$ and that there is a connection between i and j, and if $d_{v(i,j)}^{\alpha} \geq \tau_{v(i,j)}$, then the connectivity constraint is computed as:

$$d^{\alpha}_{v(i,j)} = \tau_{v(i,j)} = \begin{cases} 1 & \text{if a connection exist between a node } i \text{ and } j \\ 0 & \text{otherwise} \end{cases}. \quad (7.9)$$

The objective function for minimizing the energy spent by a node to transmit a data packet of length L_p bits at distance τ is given by:

$$\min Energy_{v_{sd}}(\lambda), \quad (7.10)$$

subject to

$$d^{\alpha}_{v(i,j)} = \tau_{v(i,j)} \quad (7.11a)$$

$$\tau_{v(i,j)} \leq HOP \quad (7.11b)$$

$$\tau_{\min} \leq \tau_{v(i,j)} \leq \tau_{\max} \quad (7.11c)$$

$$2 \leq \alpha \leq 4 \quad (7.11d)$$

$$\sum_{v(i,j)} d^{\alpha}_{v(i,j)} \leq HOP. \quad (7.11e)$$

The energy of the selected path can adjust its transmitting power within a closed interval of lower and upper bound, such as $[P_{\min}, P_{\max}]$, where $P_{\min}$ and $P_{\max}$ determine the minimum and maximum constraint values of the path.

7.3.4 Delay Model

The definition of delay depends on the optimal hop number that can have different delay guarantees denoted as $\wp(\xi_i, \xi_j)$. The increased number of nodes results in more paths becoming available for simultaneously routing packets to their destinations, which benefits the reduction of delay. The number of paths may also increase proportionally to the number of nodes on the invoked path. Therefore, determine the optimal number of hops using Eq. (7.5), which minimizes the delay of the successful transmission of a packet, and then jointly optimize the hops and estimate the delay constraint to derive a scaling for minimizing the delay. Therefore, to solve the optimization problem, all source nodes and intermediate nodes periodically calculate the delay when generated from the one-hop neighborhood of each node because the one-hop is easier to acquire [35]. Suppose that one QoS requirement is satisfied at each hop, then the end-to-end QoS requirement is also met [35]. Specifically, a node can satisfy the hop requirement by selecting the next hop, which allows the bounded delay to be evenly divided at each hop. The end-to-end delay between any two sensor nodes ξ_{Source} and $\xi_{Destination}$ over the set of paths P is given by:

$$\wp_{sourcessink}(L_p) = \min\left\{\sum_{\xi_i} \wp(\xi_i, \xi_j)\right\}, \quad (7.12)$$

where $\wp_{SourcesDestination}$ is the minimum achievable delay when the generated data are routed along the set of paths between ξ_{Source} and $\xi_{Destination}$. The delay $\wp(\xi_i, \xi_j)$ between two nodes is the time required to successfully transmit a packet after being received by the first node. This time may include queuing, contention, transmission, retransmission, idle, propagation, and processing. The mean delay of each sensor node can be computed as:

$$\xi = D_{quening} + D_{propagation} + D_{processing} \\ + D_{transmission} + D_{retransmission} + D_{idle} \tag{7.13}$$

subject to

$$\sum_{v=1}^{N+M} \wp(\xi_i, \xi_j) \le \Delta\wp, \tag{7.14}$$

where $\Delta\wp$ is the bounded delay, which depends on two factors, namely, the number of hops taken and the delay of a node, both of which are expressed in additive form and denoted as η_{ij} and $\wp^e$, respectively. Therefore,

$$\Delta\wp = \wp_0^{Source} + \wp_{\eta_1}^{\xi+1} + \wp_{\eta_2}^{\xi+2} + \cdots + \wp_{\eta_{N+M}}^{Destination}. \tag{7.15}$$

$L_e^{\wp}$ is the hop delay requirement along the path from the source to the sink. This requirement composed of η_i and depends on the partition requirements at sensor node ξ_i. The hop delay requirement is equal to

$$L_e^{\wp} = \frac{\Delta\wp - \wp^e}{\eta_i}. \tag{7.16}$$

Then, rewrite the constraint

$$\sum_{v=1}^{N+M} \wp(\xi_i, \xi_j) \le L_e^{\wp}. \tag{7.17}$$

7.3.5 Throughput Model

Each sensor has a maximum transmitting power P_{max} and a maximum bandwidth. The bandwidth is denoted as B is defined as the sum of total of all transmitted and received loads. This sum should not exceed the bandwidth capacity of the node. We use the throughput, which is defined in Ref. [36], as the amount of data packets successfully transmitted. Therefore, the total amount of data packet successfully transmitted the optimal number of hops is calculated as:

$$Throughput = \left(\frac{D_{transmission}}{\xi}\right) Tx_{datarate}. \tag{7.18}$$

The objective function for maximizing the throughput for all outgoing packets at the selected path is given by:

$$\max Throughput_{\upsilon_{sd}}(\lambda), \tag{7.19}$$

subject to

$$\sum_{(s,d)} \aleph_{s,d} \le B. \tag{7.20}$$

7.4 Convergence Behavior of Different Particle Swarm Optimization Algorithms

The k-disjoint multipath algorithm assigns each sensor particle/node with a transmission range level according to the hop-distance as referred in Eq. (7.5) for each neighbor to search on diversity characteristics of the particle swarm. Each node has the ability to improve cooperative learning behavior by exchanging messages with its neighbors. Upon receiving these messages, each node computes the disjoint paths and updates the local path information according to the constraints as referred in Eqs. (7.11), (7.17) and (7.20). According to the calculation of the objective functions in Eqs. (7.10) and (7.4) a new potential path will be generated, constructed, selected, and maintained. Therefore, the nodes that select adaptively with a velocity $v_{(i,j)}$ are updated at each iteration to satisfy the right direction of selected paths.

Assume that $Z_t = (z_1, z_2, z_3, \ldots, z_m)_t$ is the feasible solution, and $Z = (z_1, z_2, z_3, \ldots, z_w)$ is the solution space of the multi-objective problem with k-disjoint multipath. The position and velocity of particle ν are represented by m dimensional vectors $|V| = \{v_1, v_2, \ldots, v_N, v_{N+1}, \ldots, v_{N+M}\}$ and $\mathrm{v} = \{\mathrm{v}_1, \mathrm{v}_2, \ldots, \mathrm{v}_N, \mathrm{v}_{N+1}, \ldots, \mathrm{v}_{N+M}\}$. Two positions, namely, personal-best position p_{best} and global-best g_{best} are defined in proposed algorithm.

During solving the multi-objective functions Eqs. (7.10), (7.13) and (7.21) represent energy consumption, average delay, and throughput, respectively. The nodes will be connected in the whole searching range at each iteration to generate arbitrary feasible solutions (i.e., paths of each objective function). However, solution Z_1 may dominate Z_2 and/or Z_3 if and only if $\{\forall_t = 1, 2, \ldots, n, f_t(Z_1) \le f_t(Z_2)\} \wedge \{\exists j = 1, 2, \ldots, n, f_j(Z_1) \le f_j(Z_2)\}$; this observation is denoted by $Z_1 \succ Z_2$. These nodes are influenced by individual exchange message, which is denoted as an extreme value point and global extreme value point, thereby leading the nodes to select the next hop toward the extreme value point within the scope of the search space at each iteration process. We indicate with $v_N = d_v^F(v_{N+1}), (p_{best(v_N)})$ or $v_N = d_v^F(v_{N+1}), (g_{best(v_{N+M})})$ is the next hop of node ν_i toward ν_j with an extreme value of either p_{best} or g_{best} within $\tau_{(i,j)}$ according to the F rule of updating the velocity as referred in Eqs. (7.21) and (7.23). Therefore, the nodes will deviate from the constraints domain and hard to converge to the extreme value point of constraint domain. The whole personal-best positions of the swarm imply the distribution of good objective functions as referred Eqs. (7.10) and (7.13), related to exchange messages under satisfying constraints in Eqs. (7.11), (7.18), and (7.20) that use to update the values of velocities and then select the optimal multipath route. In each path, the personal-best position of particle $v_{(i,j)}$ is denoted as $p_{best(v_{(i,j)})} = \left(p_{best(v_1),best(v_2),\ldots,best(v_N),best(v_{N+M})}\right)$ and global-best position is denoted as: $g_{best(v_{(i,j)})} = \left(g_{best(v_1),best(v_2),\ldots,best(v_N),best(v_{N+M})}\right)$.

The degree of influence of the personal-best position p_{best}. Thus, the velocity update function, which drives the CPSO, is defined mathematically.

$$\overrightarrow{v_v} := \chi \cdot v_v \tag{7.21a}$$

$$+\vec{Z}(0,\phi_1) \otimes \left(\overrightarrow{p_v} - \overrightarrow{v_{N+M}}\right) \tag{7.21b}$$

$$+\vec{Z}(0,\phi_2) \otimes \left(\overrightarrow{g_v} - \overrightarrow{v_v}\right), \tag{7.21c}$$

where $\vec{Z}$ is a distribution of objective functions sampled under satisfying constraints in Eqs. (7.11), (7.17), and (7.20). χ is a constriction coefficient that helps to balance global exploration and local exploitation. It is defined as:

$$\chi = \frac{2}{\phi + \sqrt{\phi^2 - 4\phi}}, \quad \text{with } \phi = \phi_1 + \phi_2 > 4. \tag{7.22}$$

During the evolution process, the velocity update function, Eq. (7.21a), is referred as momentum that represents the current selection direction of the node, Eq. (7.21b) is referred as social component, that is being attracted toward the best solution so far evaluated by the neighbors, and Eq. (7.21c) is referred as the cognitive component that is being attracted toward the previous solutions known by the node. The difference between the CPSO and Fully Particle Multi-Swarm Optimization (FPMSO) algorithms is the velocity update function, which describes that not only the best position node is taken into account but also all of its neighbors to increase the diversity of the search space. Additionally, the multi-swarm algorithm generates multipaths when the velocity update function from the main path does not show any changes in its objective function. The non-change of the objective function for the selected path is detected when the path is not dominated by any path among the set feasible paths. However, when a new multipath is formed from the main path, the sensor nodes that have triggered the construction and selection are merged into a new multipath and are removed from the main path. Algorithm 2 shows the process construction process of a new multipath. Therefore, the velocity update function is defined as:

$$\overrightarrow{v_v} = \chi\left(\overrightarrow{v_v} + \frac{1}{N}\sum_{v=1}^{N+M} \vec{Z}(0,\phi_1) \otimes \left(\left(\overrightarrow{p_v} - \overrightarrow{v_v}\right)\right)\right). \tag{7.23}$$

The details of the proposed Algorithm 1 are detailed as follows:

Algorithm 1: Main Multi-Swarm Algorithm

Calculate p_{best}'s objective function as referred in Eqs. (7.2) and (7.4) in terms of energy consumption and average delay. Then figure out the minimal value of objective function among these p_{best}'s objective function for k-disjoint multipath.
Calculate a constriction coefficient χ as referred in Eq. (7.10) in order to help in preventing velocity explosion.
Update the velocity value.
The node will select the optimal solution to improve the fault-tolerant multipath routing.

We devise a CPMS optimization as shown in Algorithm 3 in order to solve the objective function. In sum, we apply multi-swarm after the construction and then select the paths according to Algorithm 2. Multi-swarm employs a biologically inspired population-based optimization algorithm to solve the tolerance problem. However, multi-swarm is encouraged to explore the search space to maintain the swarm diversity and to learn from the global best particle to refine a promising optimal solution. Taking full advantage of the exchanged information of all personal-best messages will contribute to ignoring the fault tolerance error messages of several nodes that are trapped in the local optimal solution, thereby strengthening ability of nodes to learn from the experience of other nodes and to guide its selection direction. Therefore, the performance of each algorithm depends on the way the nodes are influenced, select in the search space for the analysis, and achieve the goal objective functions.

Algorithm 2: Construct and Selection Mechanism

input: The network topology;
for $i \in \aleph$, $j \in \aleph$ where $i \neq j$;
Calculate distance range $d^{\alpha}_{v(i,j)} = \tau_{v(i,j)}$ measure between i and j;
if $d^{\alpha}_{v(i,j)} \leq \tau$ then;
Initialize candidate particle to be solution for construct and select the multipaths
Add the candidate to a candidate set
set the connectivity $k^{\alpha}_{v(i,j)} = 1$ in G
end if
end for
Assign the fitness function for each path in term of
Applying the CPMS to compute a minimal energy consumption, average delay and maximize throughput
return selected the candidate set of construction and selection paths

The pseudo-code of the proposed CPMSO algorithm is shown in Algorithm 3

Algorithm 3: Canonical Particle Multi-Swarm Optimization

---

input:Objective functions $\vec{Z}$;
$V := \left\{\vec{v}_1,\ldots,\overrightarrow{v_{N+M}}\right\} := \mathit{InitNode}\left(\overrightarrow{lowerbound},\overrightarrow{upperbound}\right) \rightarrow \forall_v \in \{1,\ldots,N+M\} : \overline{N+M_v} := \vec{U}\left(\overrightarrow{lowerbound},\overrightarrow{upperbound}\right)$
$\vec{V} := \left\{\vec{v}_1,\ldots,\vec{v_v}\right\} := \mathit{InitParticleVelocities}\left(\overrightarrow{lowerbound},\overrightarrow{upperbound}\right) \rightarrow \forall_v \in \{1,\ldots,N+M\} : \overline{v_1,\ldots,N+M_v} := \left(\overrightarrow{upperbound} - \overrightarrow{lowerbound}\right) \otimes \vec{U}(0,1) - \frac{1}{2}\left(\overrightarrow{upperbound} - \overrightarrow{lowerbound}\right)$
$\Upsilon = \left\{\vec{\Upsilon}_1,\ldots,\overrightarrow{\Upsilon_{N+M}}\right\} := \mathit{EvaluateObjectfunction}\left(\vec{Z}\right) \rightarrow \forall_v \in \{1,\ldots,N+M\} : \Upsilon_v := f\left(\overrightarrow{Z_v}\right)$
$P = \left\{\vec{p}_1,\ldots,\overrightarrow{p_{N+M}}\right\} := \mathit{Initlocallyoptimal}(Z) \rightarrow Z$
$P = \left\{p_1^Z,\ldots,p_v^Z\right\} := \mathit{InitObjeectivefunction}(\Upsilon) \rightarrow \Upsilon$
$G = \left\{\vec{g}_1,\ldots,\overrightarrow{g_{N+M}}\right\} := \mathit{Initgloballyoptimal}(P,T) \rightarrow P$
$G = \left\{g_1^Z,\ldots,g_1^Z\right\} := \mathit{Initgloballyoptimal}\left(P^Z,T\right) \rightarrow P^Z$
while termination condition nor met **do**
  for each particle/node v of $N+M$ **do**
    $\vec{v}_v = \chi.\left(\vec{v}_v + \vec{Z}(0,\phi_1) \otimes \left(\overrightarrow{p_v} - \vec{v}_{N+M}\right) + \vec{Z}(0,\phi_2) \times \left(\overrightarrow{g_v} - \vec{v}_v\right)\right)$
    $\vec{v}_p := \vec{Z}_p + \vec{v}_p$
  end for
  $\Upsilon := \mathit{EvaluateObjectivefunction}\left(Hop, Z\right)$
  $P, P^Z := \mathit{Updatelocallyoptimal}(Hop, Z) \rightarrow \forall_p \in \{1,\ldots,N+M\} : \vec{p}_p, p_p^Z := \begin{cases} \overrightarrow{Z_p}, y_i & \text{if } \Upsilon_v \text{ better than } p_p^f \\ \vec{p}_p, p_p^f & \text{otherwise} \end{cases}$
  $G, G^f := \mathit{Updategloballyoptimal}\left(P, P^Z, T\right) \rightarrow \forall_p \in \{1,\ldots,N+M\} : \overrightarrow{g_p}, g_p^Z := \mathit{best}\left(P_{T_p}, P_{T_p}^Z\right)$, where T_p are the neighbors of p
end while
best solution found

7.5 Performance Evaluation

In order to assess the performance of the proposed algorithm, we have performed extensive simulations. We have implemented our algorithms using MatLab [38] to develop and generate network topology, evaluate the objective functions and visualize the outputs of the evolution. We employ 30, 40, and 50 sensor nodes that are distributed uniformly over the area of 1,000 m × 1,000 m as seen in Figure 7.3. The supernodes are also distributed uniformly in this area. The path loss exponent for the wireless channel α ranges between 2 and 4. The transmission range of the nodes is initially set to 12.00 m in order to guarantee that the connection among the supernodes and nodes will satisfy the constraints of problems. Three main test experiments were performed to investigate the usability of the algorithms (CPMSO, FPMSO, and CPSO) in finding and selecting a robust optimal multipath according the objective functions.

The first test experiment investigates the performance of the three algorithms in finding a robust optimal multipath that is generally usable with PSO algorithms. The second experiment investigates the performance of the three algorithms when the number of sensor nodes is increased. The third test experiment investigates the behavior of sensor nodes (i.e., the type of topology and constraints processing) in selecting the optimal path. The robust objective function in terms of energy consumption, average delay, and throughput is estimated by approximating the effective or expected fitness function with different k-disjoint multipath for the

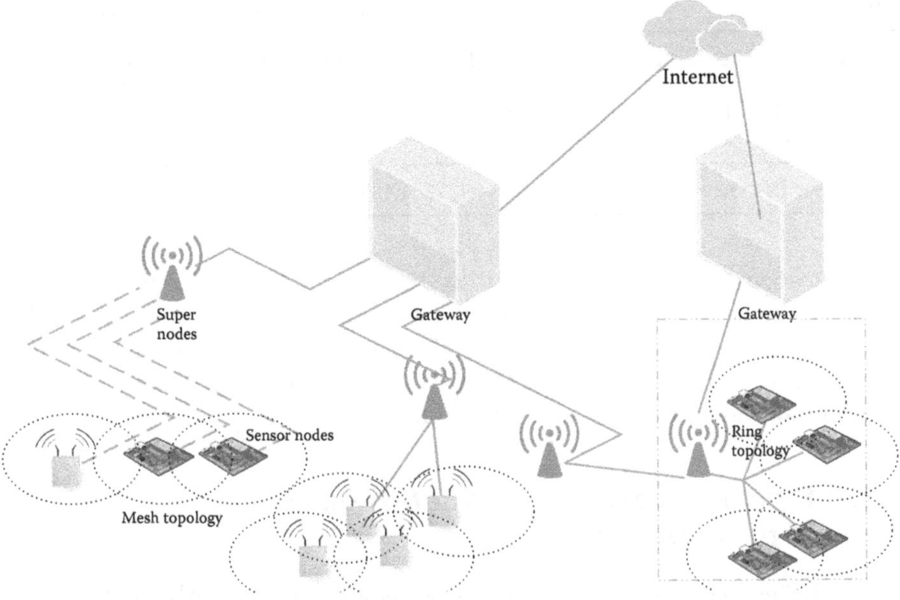

FIGURE 7.3 The devices deployed in SIoT.

TABLE 7.2 Definition of Parameters

Parameter	*Value*
E_{elec}	50 nJ/bit
ε_{fs}	10 pJ/bit m^2
ε_{mp}	0.0013 pJ/bit m^2
Topology structure	Square (1,000 m × 1,000 m), sensor node distributed uniformly
Total number of sensor nodes	30, 40, 50 sensor nodes
Message payload	64 bytes
Data length p	2,000 bits
Transmission range	12.00 m
Tx data rate	250 kbps
Personal best coefficient ϕ_1	2.8
Neighbor best coefficient ϕ_2	1.3
Robust fitness estimation	f_{eff}
Initialization	Energy consumption, delay, and throughput
Boundary processing	Cut-off method and ignore cut-off method
Degree of Disjoint connectivity k	2,3,4
Number of hops	*hops* = 1, 2, 3, 4, 5

approximation. Given that the coefficients $\phi_1=2.8$ and $\phi_2=1.3$ are set $\phi_1+\phi_2 \geq 4$. Therefore, to ensure that these algorithms are optimized with bounds of feasible search space, a cut-off boundary constraints processing is performed, where lb and ub are derived from [35]. All variables, parameters (e.g., positions and velocities), and constant values are summarized in Table 7.2.

7.5.1 Experiment Set (1) General Investigation of the Performance

The first experiment examines the performance of CPMSO, FPMSO, and CPSO with the maximum number of hops (hop=5) in finding the robust optimal multipath. The findings are summarized as follows:

7.5.1.1 Total Energy Consumption

Figure 7.4 presents the total energy consumption resulting of the three algorithms. The total energy consumption in the *k*-disjoint generated by CPMSO and FPMSO is better than that generated by CPSO because solving the objective function utilized by CPSO presents some challenges in discovering the *k*-disjoint multipath after recovering the fault tolerance error messages. Moreover, given its length, the search space cannot be substituted with *k*-disjoint multipath and requires a high total transmission energy. Another important observation related to the exchange of messages between supernodes and nodes is that the CPSO performs significantly worse than FPMSO and CPMSO because CPSO requires significantly more control messages to exchange between the neighborhoods. Therefore, CPSO needs to find a *k*-disjoint multipath in its reachable neighborhoods, while FPMSO and CPMSO can directly search for paths using a smaller number of control messages between the reachable neighborhoods. Thus, the *k*-disjoint multipath for FPMSO and CPMSO can achieve a lower total energy consumption compared with CPSO.

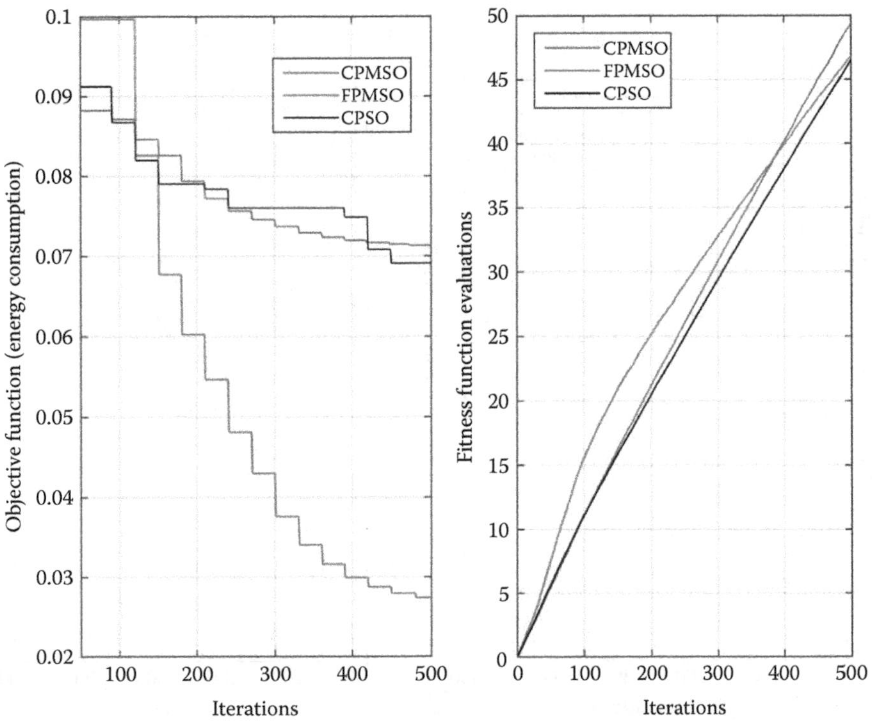

FIGURE 7.4 Multi-swarm optimization routing algorithm for optimizing energy consumption.

7.5.1.2 Total Average Delay

Figure 7.5 shows the average delay of the selected optimal multipath from the source to the sink. FPMSO and CPMSO have demonstrated a minimize delay per hop compared with CPSO. Therefore, can be returned to the selection and maintenance of the *k*-disjoint multipath for fault tolerance, which can satisfy the hop requirement by selecting the next hop in the neighborhood of each node. These algorithms also require significantly less control messages for fault tolerance compared with CPSO for selecting and maintaining a 1-hop neighborhood. Therefore, both FPMSO and CPMSO are more feasible for the *k*-disjoint multipath than CPSO.

7.5.1.3 Total Throughput

Throughput may be frequent due to high-bit error rate or other conditions, such as the environmental ones. Therefore, we present in Figure 7.6 the effect of solving the objective function on throughput as referred in Eq. (7.4). As the delay minimizes along with the increasing the optimal number of hops, throughput degrades significantly. This result is expected because minimizing the delay under

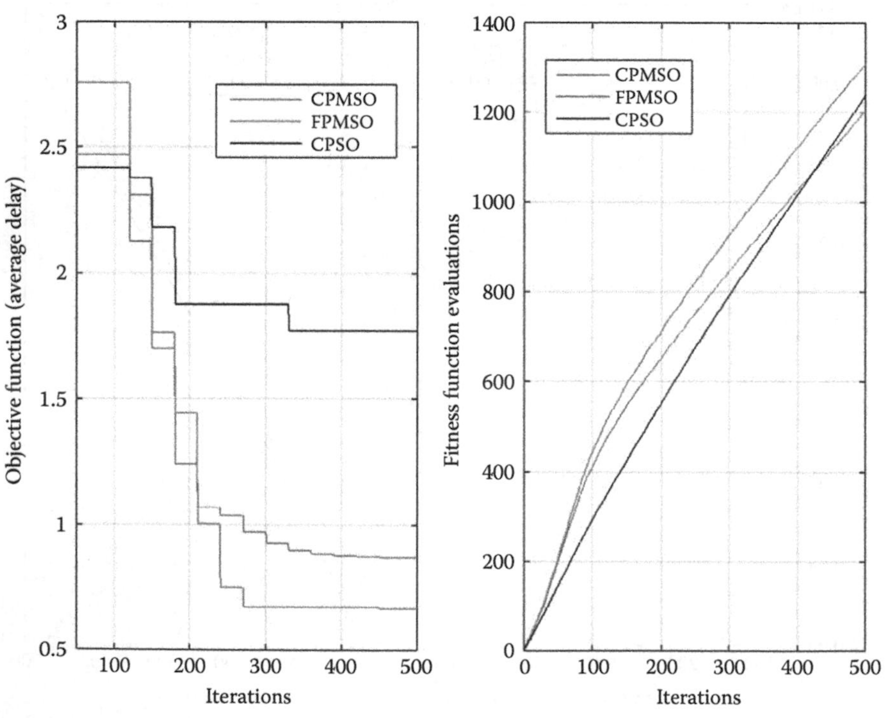

FIGURE 7.5 Multi-swarm optimization routing algorithm for optimizing average delay.

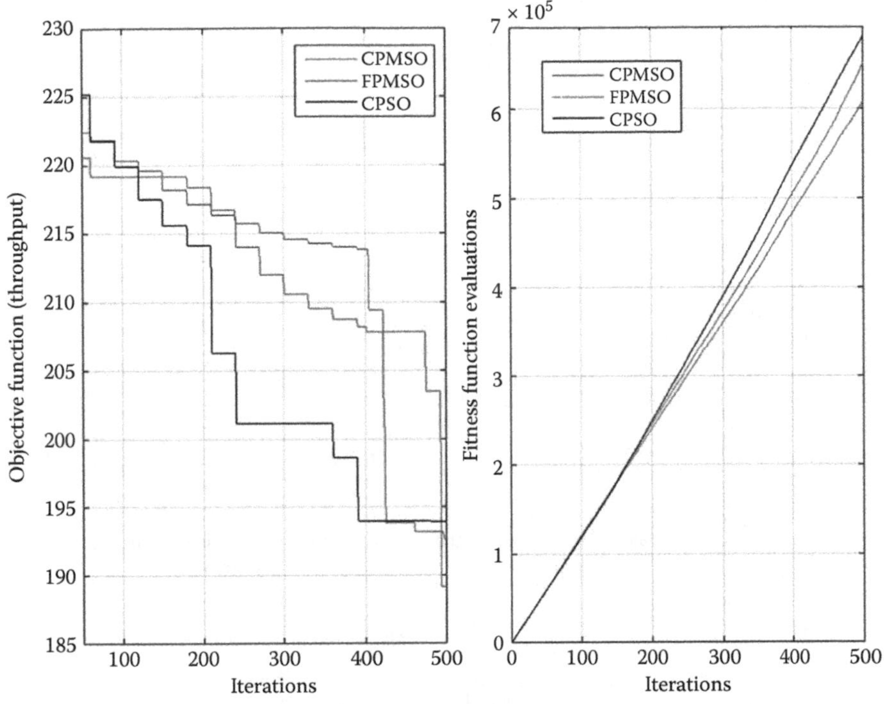

FIGURE 7.6 Multi-particle swarm optimization routing algorithm for optimizing throughput.

the aforementioned constraints with the optimal number of hops leads to a lower number of exchanged control messages for fault tolerance and thereby allowing to can be obtained completely. The observed performance depends on the network topology construction. Although CPSO achieves a fully connected topology (i.e., when each node is connected to all nodes in the swarm as neighbors), this algorithm has exhibited a particularly poor performance compared with CPMSO and FPMSO. The simultaneous attraction of the *k*-disjoint multipath provokes a random behavior from each node, as shown and this behavior can support an optimal performance in FPMSO and CPMSO with full connectivity.

7.5.2 Experiment Set (2) General Investigation of the Performance

In this experiment, the objective function is conducted with varying number of sensor nodes in the network topology. The results for each algorithm in Figures 7.7–7.9, which show the performance of each algorithm relative to increasing the number of sensor nodes. Interestingly, for the multi-objective functions, the algorithms show better

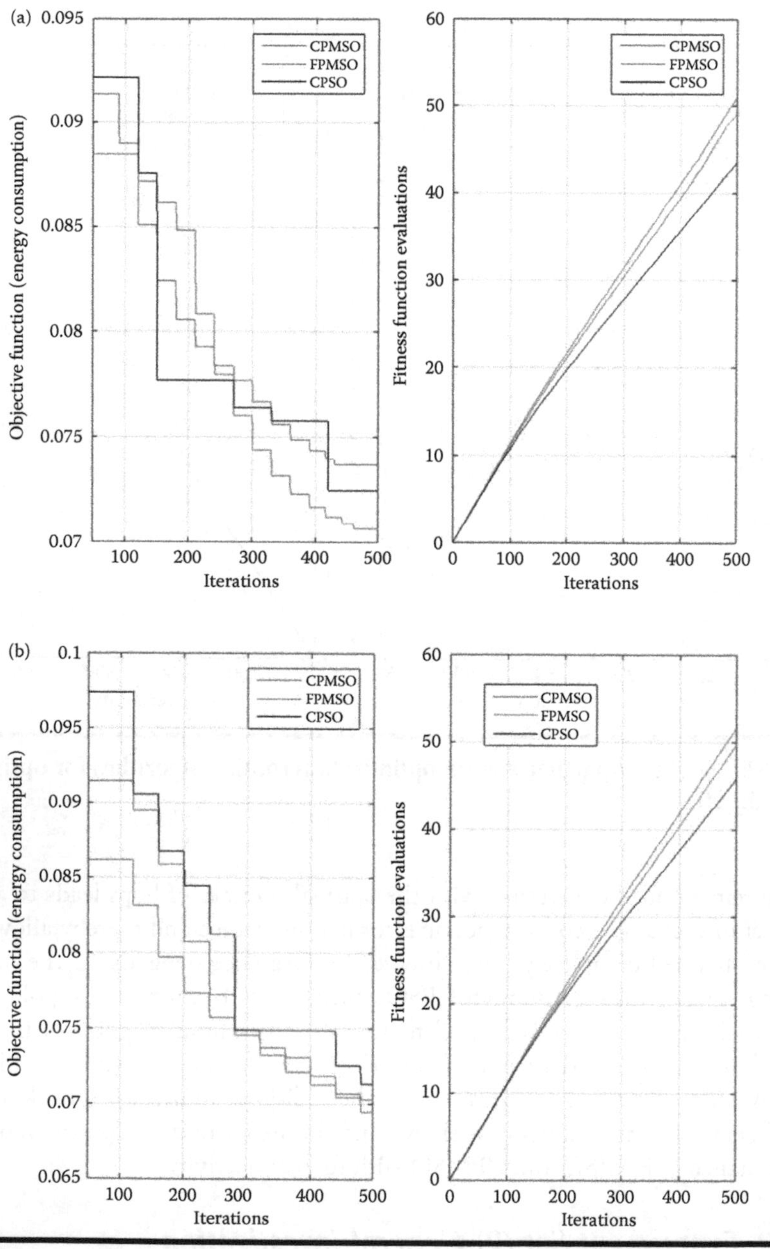

FIGURE 7.7 Total power consumption with varying number of nodes deployed in the IoT with different transmission range. (a) Energy consumption of 30 nodes deployed in the IoT. (b) Energy consumption of 40 nodes deployed in the IoT.

(Continued)

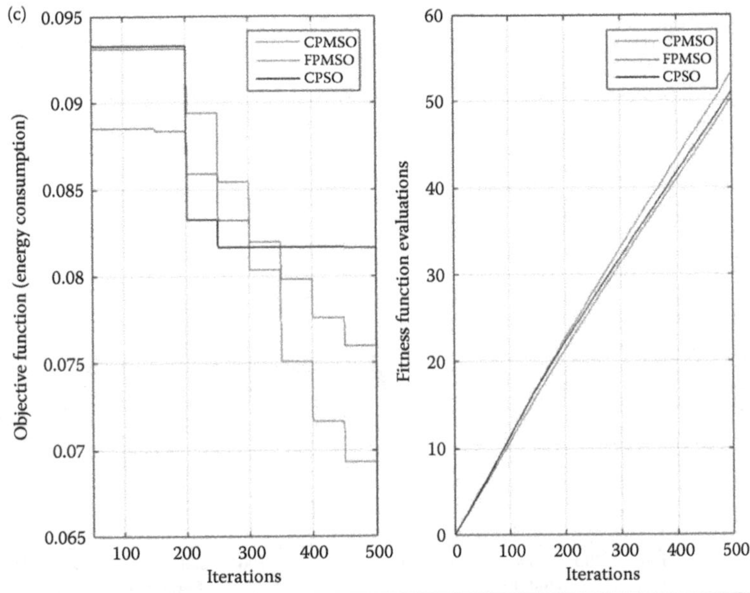

FIGURE 7.7 (CONTINUED) (c) Energy consumption of 50 nodes deployed in the IoT.

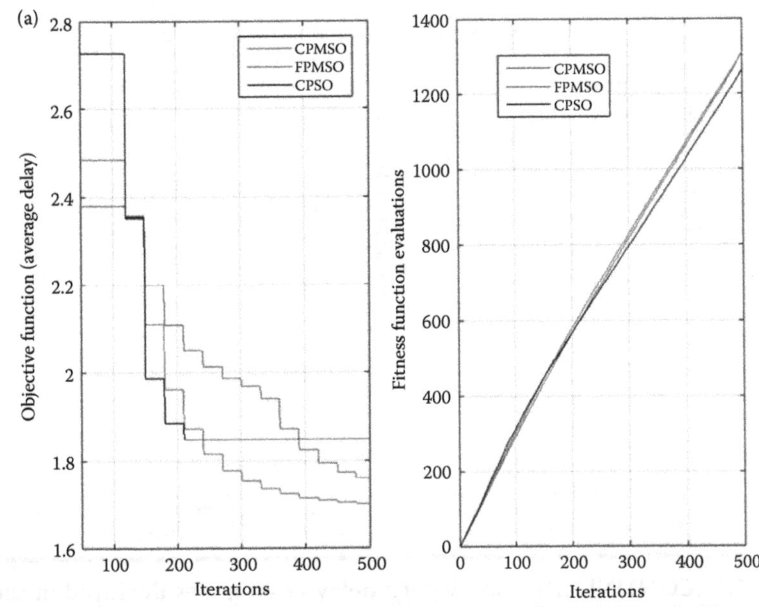

FIGURE 7.8 Total average delay with varying number of nodes deployed in the IoT with different transmission range. (a) Average delay of 30 nodes deployed in the IoT.

(Continued)

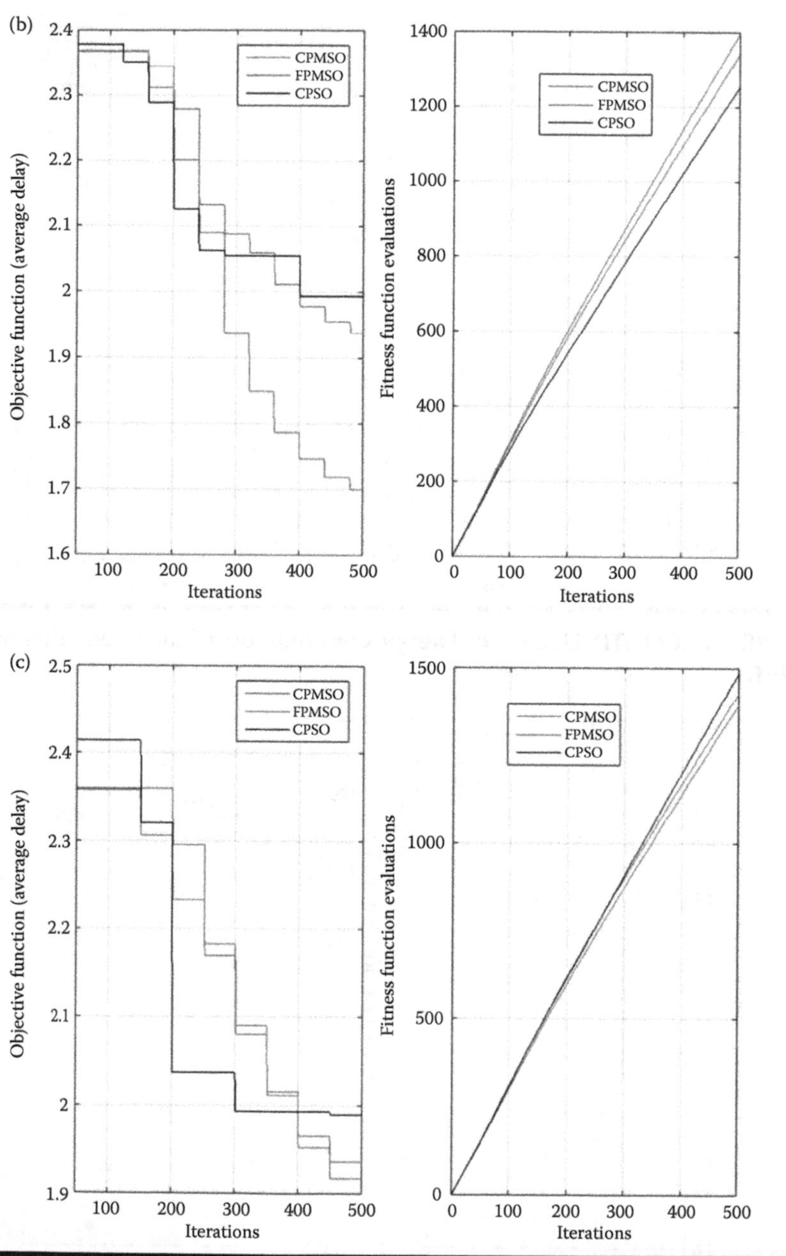

FIGURE 7.8 (CONTINUED) (b) Average delay of 40 nodes deployed in the IoT. (c) Average delay of 50 nodes deployed in the IoT.

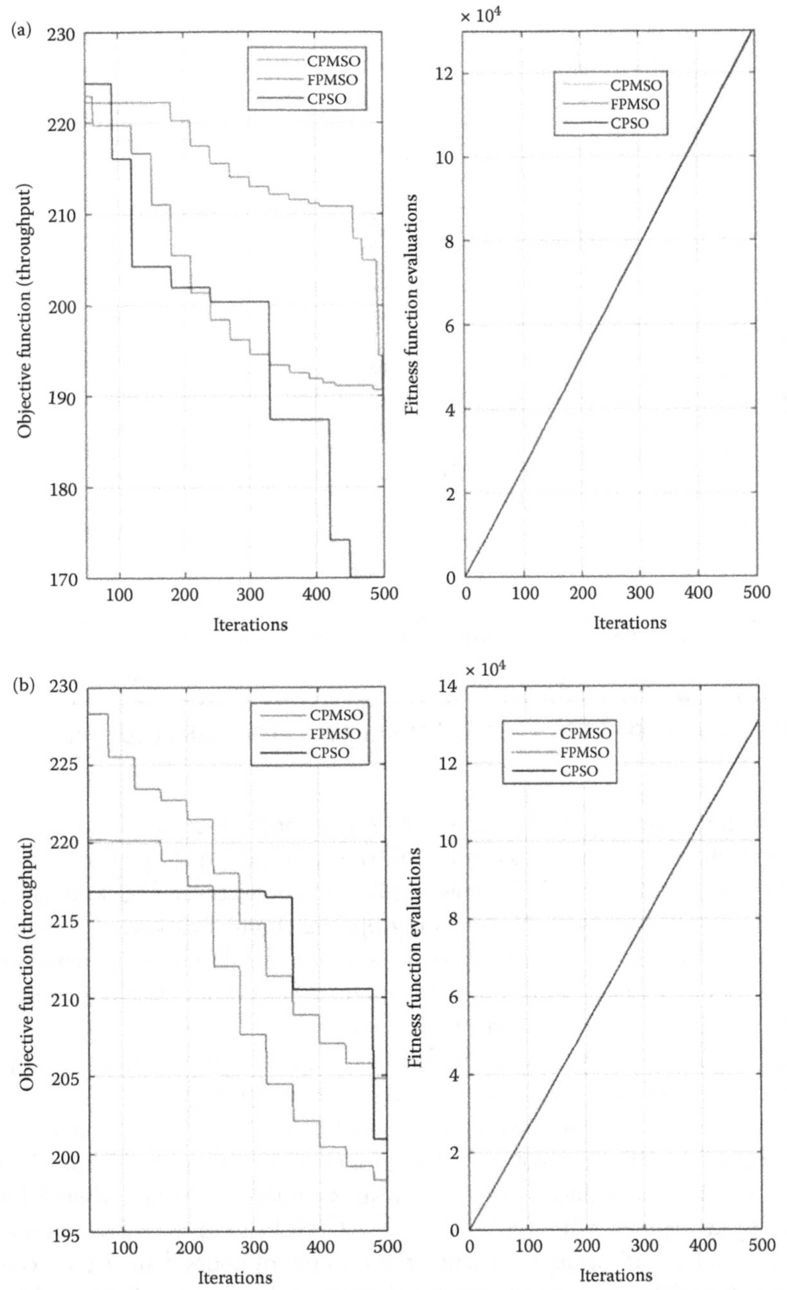

FIGURE 7.9 Total throughput with varying number of nodes deployed in the IoT. (a) Total throughput of 30 nodes deployed in the IoT. (b) Total throughput of 40 nodes deployed in the IoT.

(Continued)

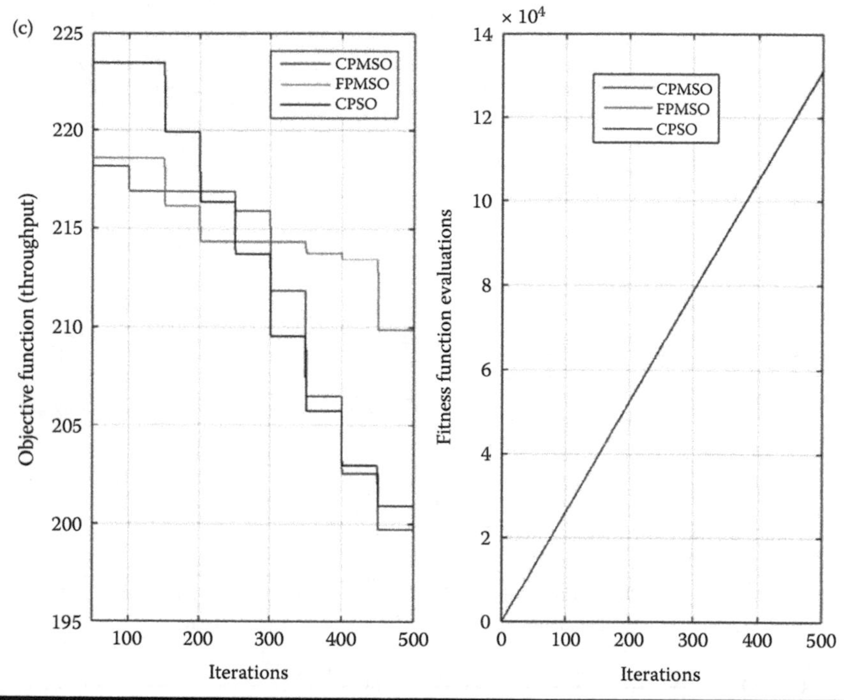

FIGURE 7.9 (CONTINUED) **(c) Total Throughput of 50 nodes deployed in the IoT.**

results in the beginning, and these results sometimes worsened over time. Having a lower number of sensor nodes can benefit the creation of quality optimal solutions by using the cut-off method. The low number of connectivity $k=3$ also causes the algorithm to make mistakes in generating optimal approximations of the objective functions. However, the behavior of particles under a small number of sensors in the network topology can be used for approximating the robust objective function that can the algorithm to move toward a more favorable area in the search space.

Each algorithm shows a better performance in the beginning and that improves in the ending when 40 nodes are deployed as shown in Figures 7.7b–7.9b as well as when 50 nodes are deployed as shown in Figures 7.7c–7.9c. The sensor nodes deployment experiment involves the deployment of 30 nodes. Thus, FPMSO and CPMSO have produced very good results under all settings when 30 nodes are deployed. Figures 7.7a–7.9a indicate that CPSO shows the worst energy consumption, delay, and throughput when the number of nodes deployed exceeds 30. However, CPSO has managed to produce better results over time because this algorithm can construct and select optimal paths from an unfavorable area in the search space. Specifically, the low number of generated for the paths for the objective functions may explain why the convergence of CPSO is slightly off the global optimal solution when 30 nodes are deployed compared with the other algorithms.

7.5.3 Experiment Set (3) Investigating the Standard Topologies

The experiment reports the performance of each algorithm. Figures 7.10 and 7.11 show the differences in the quality of each algorithm, which can be attributed to the amount of information that is available among the sensor nodes and that can be used to construct the paths in mesh and ring topologies. The information in mesh and ring topologies is capsuled locally in their multipaths. However, for the ring topology, the information is shared only between two neighboring sensor nodes; while for the mesh topology, the information is shared among the sensor nodes of a per-defined size of the multipaths. Therefore, the mesh topology does not have all information shared, but the sensor nodes tend to explore further, thereby leading to construction and selection more multipaths, which can be beneficial for finding the optimal paths. Generally, all algorithms show their best performance in both mesh and ring topologies. In CPSO, the mesh topology as shown in Figure 7.10a–c presents a better performance than the ring topology as shown in Figure 7.11a–c. Such slight little bit difference in quality can be attributed to the amount of exchanged information between the nodes, which is defined by the connectivity among the sensor nodes and supernodes. However, in construction of all topologies the information between nodes is available, while in the ring topology,

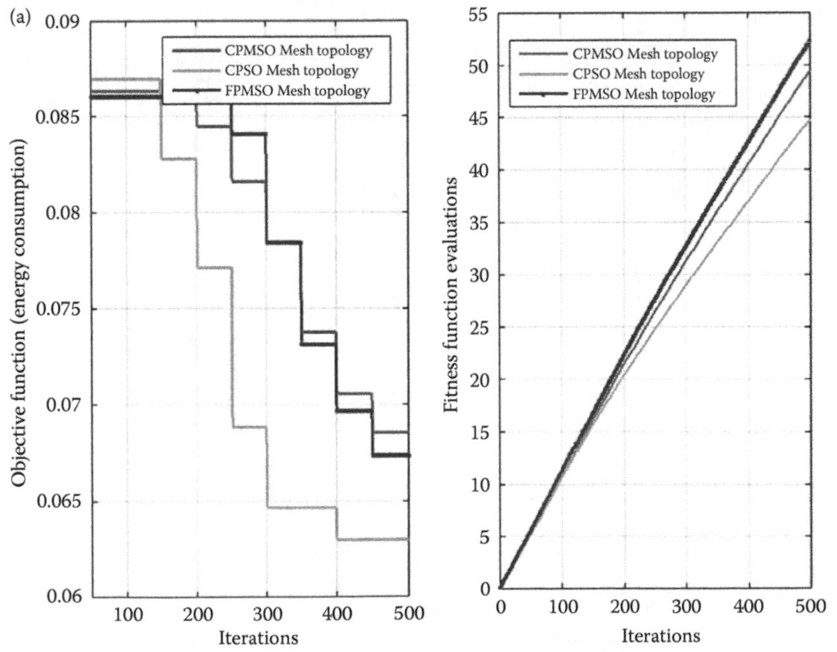

FIGURE 7.10 Performance comparisons of each algorithms for the mesh topology. (a) Energy consumption comparisons for the mesh topology.

(Continued)

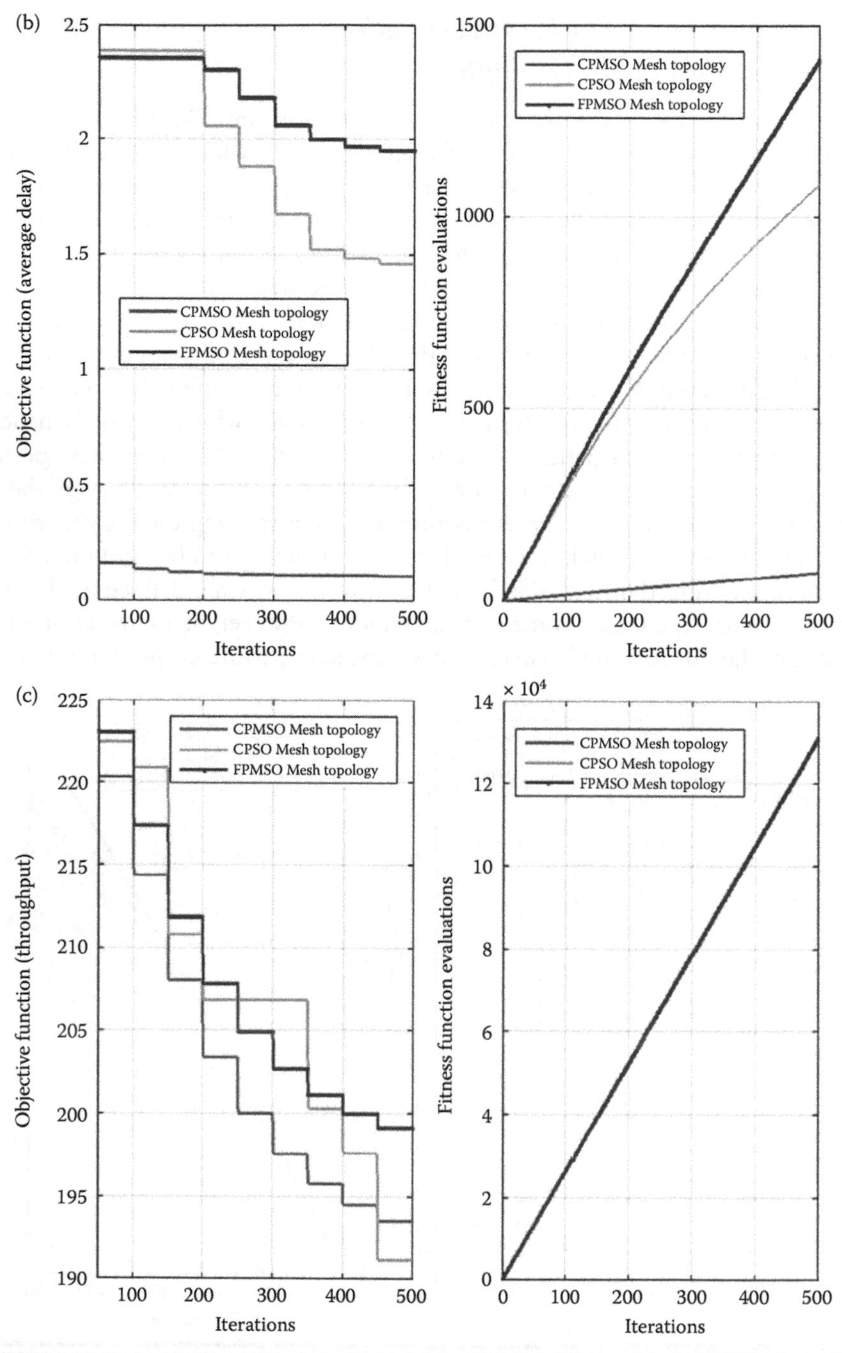

FIGURE 7.10 (CONTINUED) (b) Average delay comparisons for the mesh topology. (c) Throughput comparisons for the mesh topology.

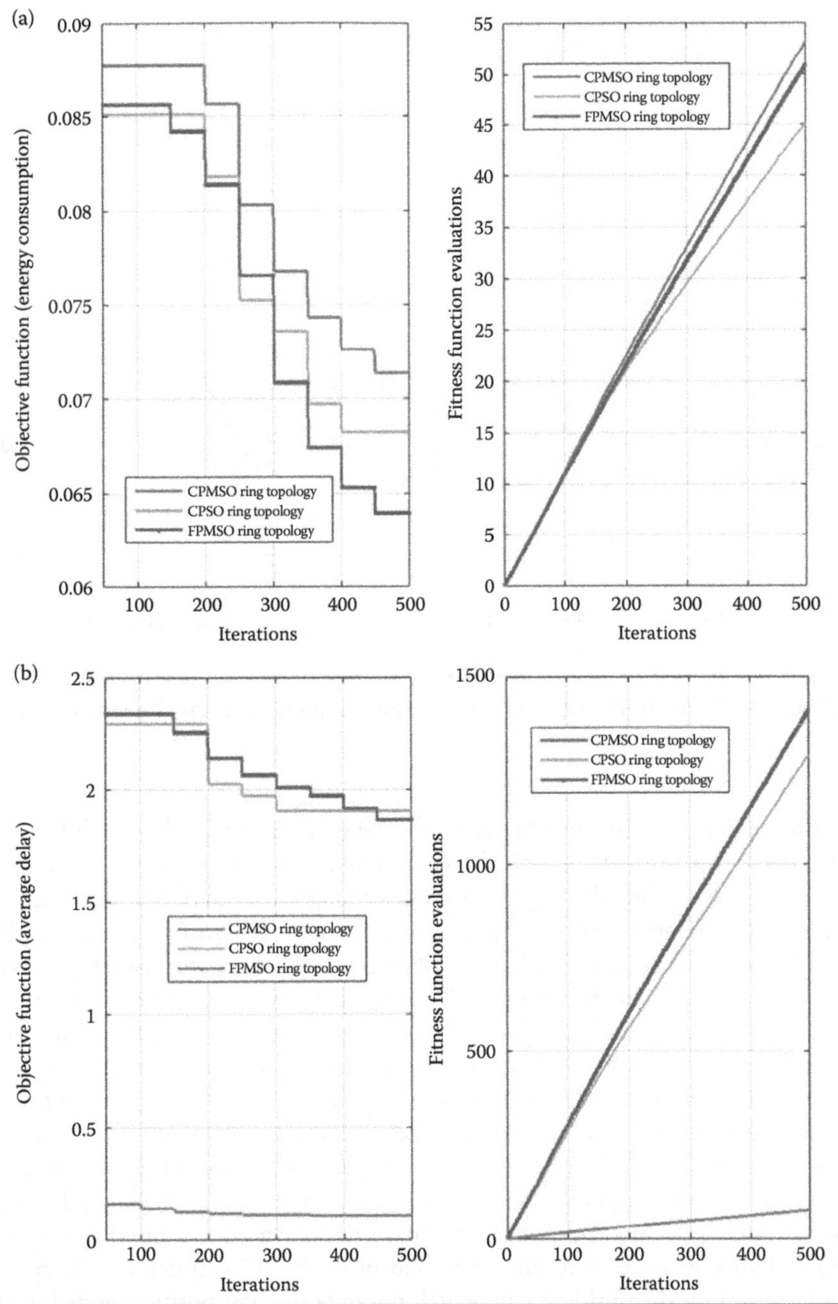

FIGURE 7.11 Performance comparisons of each algorithms for the ring topology. (a) Energy consumption comparisons for the ring topology. (b) Average delay comparisons for the ring topology.

(Continued)

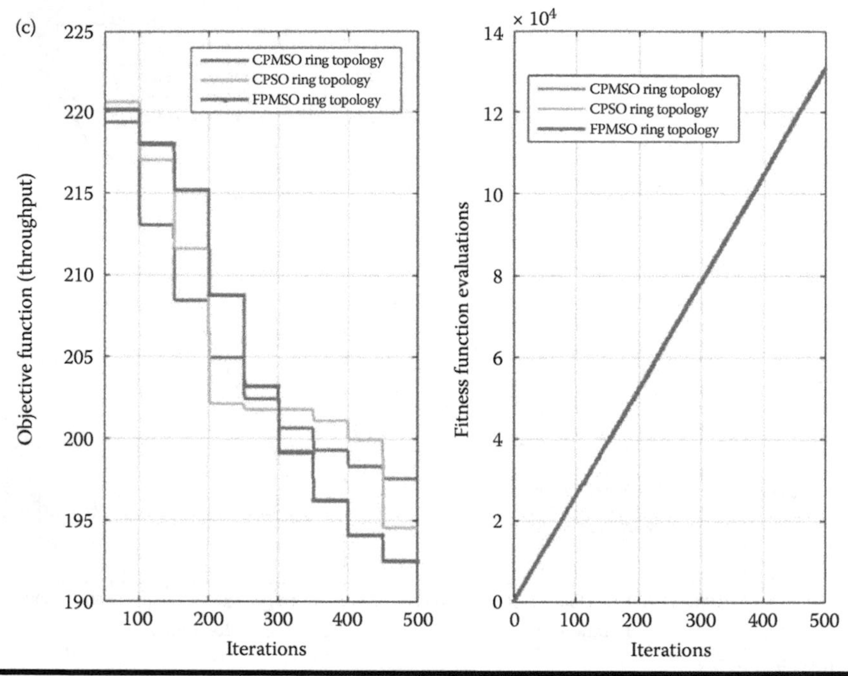

FIGURE 7.11 (CONTINUED) (c) Throughput comparisons for the ring topology.

the information is shared only between two neighboring nodes. Meanwhile, in the mesh topology, the information is shared among nodes with a pre-defined connectivity degree. Therefore, the ring topology has enough information to be shared and tends to explore more neighboring nodes in order to create more paths. FPMSO and CPMSO use the same settings as CPSO to make a fair comparison between these algorithms as shown in Figures 7.10a–c and 7.11a–c. Several conclusions can be obtained. First, both FPMSO and CPMSO present a better performance than CPSO because of the mesh topology in the multi-swarm algorithm doing local searches. Having all information from the neighbors is certainly helpful in finding the optimal paths. Second, as referred in Eq. (7.21) that is divided into two phases, the first phase Eq. (7.21b) depicts the convergence of the current node selection, which only uses the cognitive component in the normal swarm. Meanwhile, in the multi-swarm, the main swarm uses the cognitive and momentum components of the standard velocity update. In the second phase Eq. (7.21c) depicts the convergence behavior of the multi-swarm, which not only the best position node but also all the neighbors. Therefore, we assume that the nodes need a finite time to stabilize their objective function values and positions, which in turn suggest that setting the optimal transmission range of the node that uses the full PSO model will increase the diversity of search space and generate much better results than CPSO.

7.5.4 Experiment Set (4) Investigating the Standard Topologies

In our last experiment, we address another factor that can affect the performance of each algorithm in finding the optimal paths. We apply cut-off method for processing the constraints and ignoring processing the constraints. Figures 7.12–7.14 show the performance of CPMSO, CPSO, and FPMSO in solving the objective functions. The energy consumption of these algorithms is shown in Figures 7.12a–7.14a, their delay is shown in Figures 7.12b–7.14b, and their throughput is shown in Figures 7.12c–7.14c, respectively. We consider the cut-off method in which cuts will be admitted for bounded constraints and parts of the set of the feasible area will never be cut. By applying processing on the constraints using the cut-off method, we obtain the most satisfactory results for solving the optimization problem, whereas the number of constraints to be considered in each iteration is bounded. However, these results concerning the stability of constraints can be used to find the global solution because the cut-off method is strongly related to duality and partitioning approach for solving optimization problem as perform in Ref. [39]. On the contrary, by ignoring the processing of constraints, a certain set of constraints will not be reduced in such a way that does include the optimal solution in a feasible area. Thus, the cut-off method will reduce a certain set in such a way that does not exclude a feasible solution but include the optimal solutions.

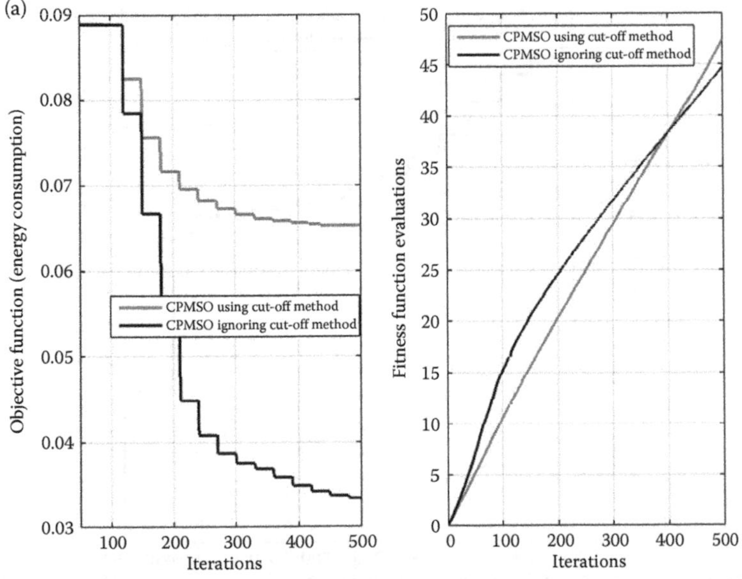

FIGURE 7.12 Constraints processing using cut-off and ignoring cutting-off method for the particle multi-swarm optimization algorithm. (a) Energy consumption comparisons for processing and ignoring processing the constraints.

(Continued)

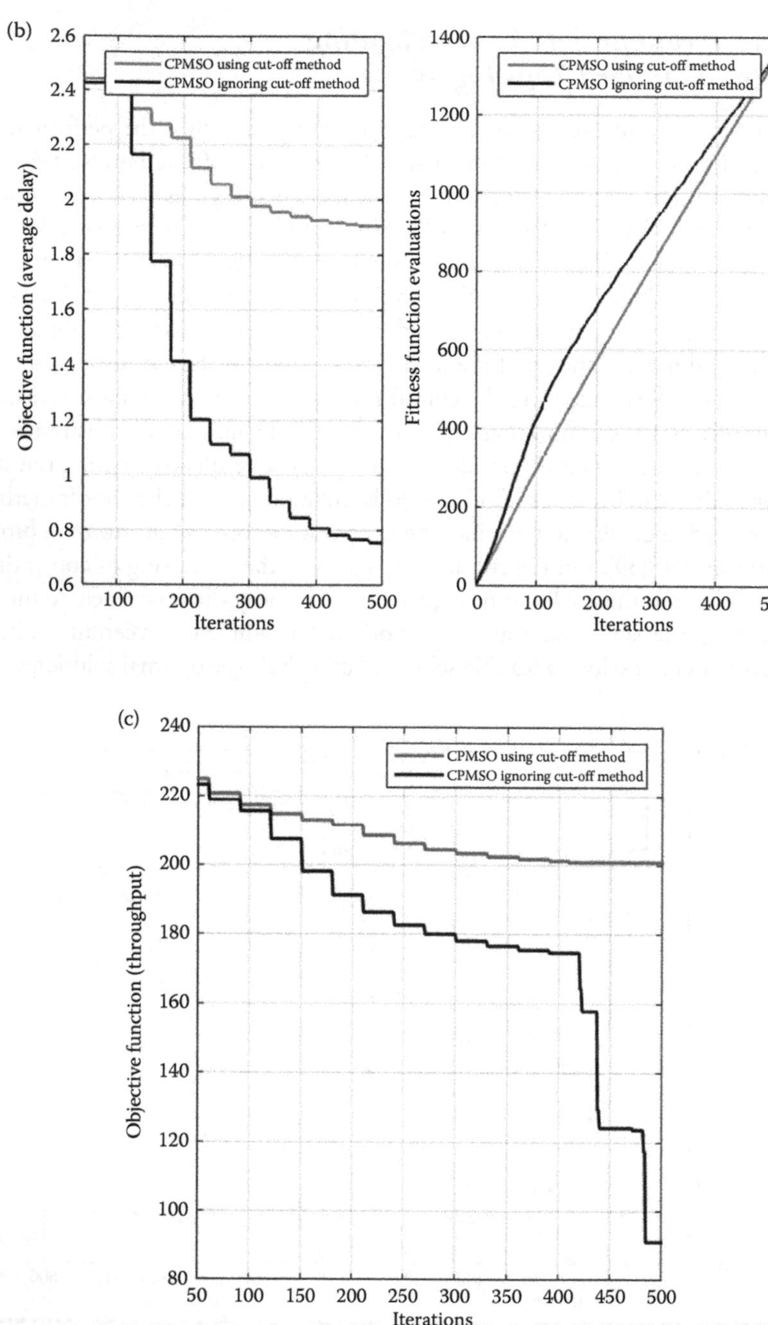

FIGURE 7.12 (CONTINUED) (b) Average delay comparisons for processing and ignoring processing the constraints. (c) Throughput comparisons for processing and ignoring processing the constraints.

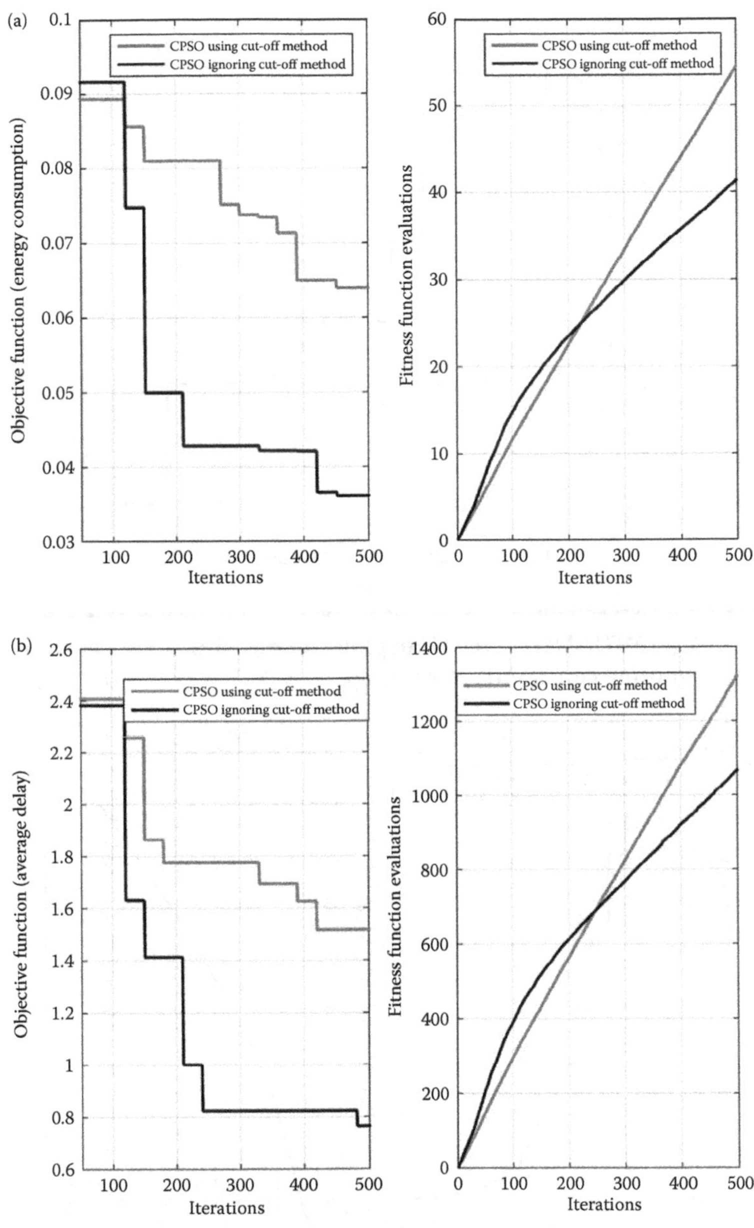

FIGURE 7.13 Constraints processing using cut-off and ignoring cutting-off method for the canonical particle swarm optimization algorithm. (a) Energy consumption comparisons for constraints processing and ignoring constraints processing. (b) Average delay comparisons for constraints processing and ignoring constraints processing.

(Continued)

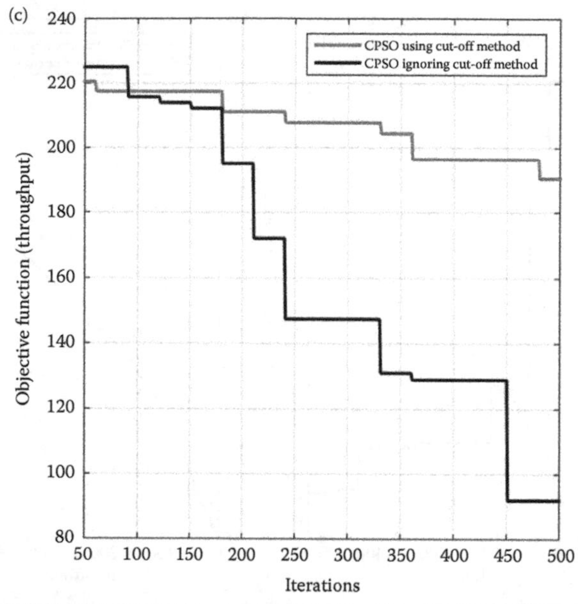

FIGURE 7.13 (CONTINUED) **(c) Throughput comparisons for constraints processing and ignoring constraints processing.**

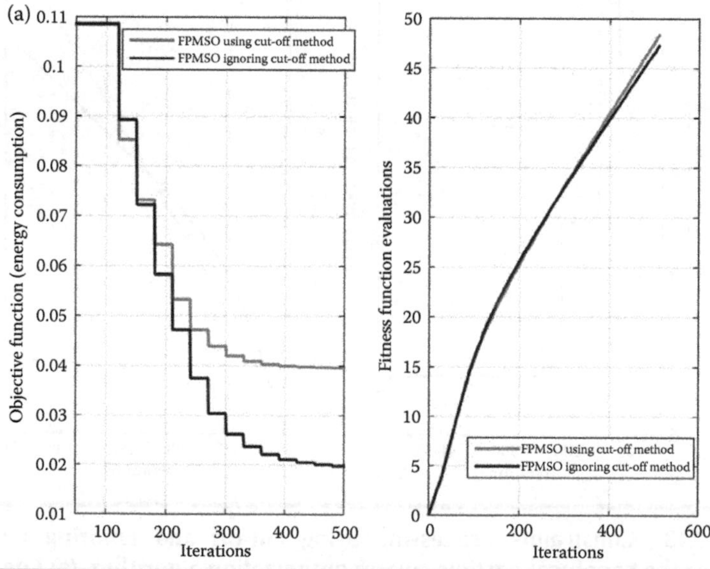

FIGURE 7.14 **Constraints processing using cut-off and ignoring cutting-off method for the fully particle multi-swarm optimization algorithm. (a) Energy consumption comparisons for processing and ignoring processing the constraints.**

(Continued)

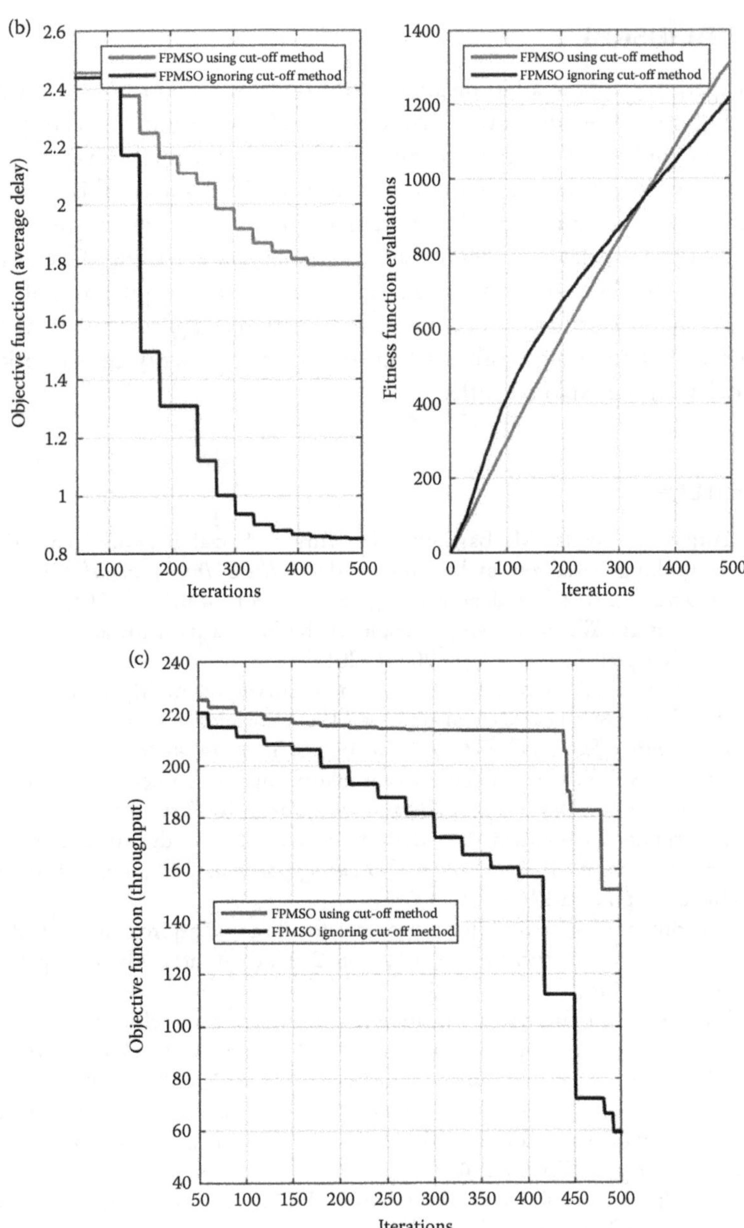

FIGURE 7.14 (CONTINUED) (b) Average delay comparisons for processing and ignoring processing the constraints. (c) Throughput comparisons for processing and ignoring processing the constraints.

7.6 Conclusion

In this paper, we propose a bio-inspired PMSO strategy to construct, recover and select *k*-disjoint multipath routes. Two position-information in terms of personal-best position, and the global position are introduced in the form of velocity update to enhance the performance of routing algorithm. To validate this strategy, we assessed objective function, which considers the average energy consumption and average in-network delay. Our results show that the strategy using the characteristics of all personal-best information is a valid strategy for the purposes of improving the PMSO performance. Moreover, the proposed algorithm has also been compared with similar algorithms, which optimize the energy consumption and average delay on the explored paths.

References

1. M. Akbas, R. Avula, M. Bassiouni, D. Turgut, Social network generation and friend ranking based on mobile phone data, *IEEE International Conference on Communications (ICC)*, Budapest, Hungary, pp. 1444–1448, June 2013.
2. V. Petrov et al., When IoT keeps people in the loop: a path towards a new global utility. arXiv preprint arXiv:1703.00541, 2017.
3. M.Z. Hasan, H. Al-Rizzo, F. Al-Turjman, A survey on multipath routing protocols for QoS assurances in real-time wireless multimedia sensor networks. *IEEE Communications Surveys & Tutorials*. vol. 19, no. 3, pp. 1424–1456, 2017 , 2017.
4. Y. Zeng, L. Xu, Z. Chen, Fault-tolerant algorithms for connectivity restoration in wireless sensor networks. *Sensors (Basel, Switzerland)*. 16(1):3, 2016.
5. F. Al-Turjman, Price-based data delivery framework for dynamic and pervasive IoT. *Elsevier Pervasive and Mobile Computing Journal*. 2017, DOI: http://dx.doi.org/10.1016/j.pmcj.2017.05.001.
6. F. Al-Turjman, Cognitive routing protocol for disaster-inspired internet of things. *Elsevier Future Generation Computer Systems*. 2017, DOI: http://dx.doi.org/10.1016/j.future.2017.03.014.
7. M.Z. Hasan, F. Al-Turjman, Evaluation of a duty-cycled asynchronous X-MAC protocol for vehicular sensor networks. *EURASIP Journal on Wireless Communications and Networking*. 2017, DOI: http://dx.doi.org/10.1186/s13638-017-0882-7.
8. M.Z. Hasan, H. Al-Rizzo, M. Günay, Lifetime maximization by partitioning approach in wireless sensor networks. *EURASIP Journal on Wireless Communications and Networking*. 2017(1):15, 2017.
9. A. Hadjidj, A. Bouabdallah, Y. Challal, HDMRP: an efficient fault-tolerant multipath routing protocol for heterogeneous wireless sensor networks. *7th International Conference on Heterogeneous Networking for Quality, Reliability, Security and Robustness*, Houston, TX, pp. 469–482, Nov. 2010.
10. M. Akbas, M. Brust, D. Turgut, C. Ribeiro, A preferential attachment model for primate social networks. *Computer Networks*. 76:207–226, 2015.
11. S. Jiang et al., Linear decision fusion under the control of constrained PSO for WSNs. *International Journal of Distributed Sensor Networks*. 8(1):871596, 2012.

12. Y. Hu, Y. Ding, K. Hao, An immune cooperative particle swarm optimization algorithm for fault-tolerant routing optimization in heterogeneous wireless sensor networks. *Mathematical Problems in Engineering.* 2012:19, 2012.
13. W.H. Lim, N.A. Mat Isa, Particle swarm optimization with adaptive time-varying topology connectivity. *Applied Soft Computing.* 24:623–642, 2014.
14. C.-H. Wu, Y.-C. Chung, Heterogeneous wireless sensor network deployment and topology control based on irregular sensor model, *Proceedings in Advances in Grid and Pervasive Computing,* C. Cérin and K.-C. Li, Editors, Springer, Berlin, Heidelberg, pp. 78–88, 2007.
15. R. Sauermann, D. Boja, F. Kirschbaum, O. Nelles, Particle Swarm Optimization for Automotive Model-Based Calibration. *6th IFAC Symposium Advances in Automotive Control Munich,* Germany, pp. 733–738, July 12–14, 2010 .
16. M.A. Montes de Oca et al., Convergence behavior of the fully informed particle swarm optimization algorithm, *Proceedings of the 10th Annual Conference on Genetic and Evolutionary Computation,* ACM: Atlanta, GA. pp. 71–78, 2008.
17. H. Bagci, I. Korpeoglu, A. Yazici, A distributed fault-tolerant topology control algorithm for heterogeneous wireless sensor networks. *IEEE Transactions on Parallel & Distributed Systems.* 26(4):914–923, 2015.
18. M. Akbas, M. Brust, C. Ribeiro, D. Turgut, Deployment and mobility for animal social life monitoring based on preferential attachment, *IEEE 36th Conference on Local Computer Networks (LCN),* Bonn, Germany, pp. 484–491, 2011.
19. M.A. Adnan, M.A. Razzaque, I. Ahmed, I.F. Isnin, Bio-mimic optimization strategies in wireless sensor networks: a survey. *Sensors.* 14(1):299–345, 2014.
20. Y.H. Robinson, M. Rajaram, Energy-aware multipath routing scheme based on particle swarm optimization in mobile ad hoc networks. *The Scientific World Journal.* 2015:9, 2015.
21. M. Azharuddin, P.K. Jana, A PSO based fault tolerant routing algorithm for wireless sensor networks, *Information Systems Design and Intelligent Applications: Proceedings of Second International Conference India 2015,* vol. 1, J.K. Mandal et al., Editors, Springer, New Delhi, India. pp. 329–336, 2015.
22. C. Vimalarani, R. Subramanian, S.N. Sivanandam, An enhanced PSO-based clustering energy optimization algorithm for wireless sensor network. *Scientific World Journal.* 2016:11, 2016.
23. M. Pant, T. Radha, V.P. Singh. A simple diversity guided particle swarm optimization, *in 2007 IEEE Congress on Evolutionary Computation.* 2007.
24. H.-L. Shieh, C.-C. Kuo, C.-M. Chiang, Modified particle swarm optimization algorithm with simulated annealing behavior and its numerical verification. *Applied Mathematics and Computation.* 218(8):4365–4383, 2011.
25. Y.C. Yee et al., Application of particle swarm optimizer on load distribution for hybrid network selection scheme in heterogeneous wireless networks. *ISRN Communications and Networking.* 2012:7, 2012.
26. Y. Zhou et al., Fault-tolerant multi-path routing protocol for WSN based on HEED. *International Journal of Distributed Sensor Networks.* 20(1):37–45, 2016.
27. J.J. Liang, P.N. Suganthan, Adaptive comprehensive learning particle swarm optimizer with history learning, *Simulated Evolution and Learning: 6th International Conference,* SEAL 2006, Hefei, China, October 15–18, 2006. Proceedings, T.-D. Wang, et al., Editors, Springer, Berlin, Heidelberg, pp. 213–220, 2006.

28. Y.-J. Zheng, H.-F. Ling, Q. Guan, Adaptive parameters for a modified comprehensive learning particle swarm optimizer. *Mathematical Problems in Engineering.* 2012:11, 2012.
29. E. Zitzler, L. Thiele, Multi-objective evolutionary algorithms: a comparative case study and the strength Pareto approach. *IEEE Transactions on Evolutionary Computation.* 3:257–271, 1998.
30. E. Zitzler, M. Laumanns, L. Thiele, SPEA2: improving the strength Pareto evolutionary algorithm, TIK-Report 103, Computer Engineering and Networks Laboratory (TIK), Department of Electrical Engineering, Swiss Federal Institute of Technology (ETH), Zurich ETH Zentrum, Zurich, 2001.
31. A. Wahid, X. Gao, P. Andreae. Multi-objective clustering ensemble for high-dimensional data based on strength Pareto evolutionary algorithm (SPEA-II), *2015 IEEE International Conference on Data Science and Advanced Analytics (DSAA)*, Paris, France, Oct. 2015.
32. H. Meng, S. Liu. ISPEA: improvement for the strength Pareto evolutionary algorithm for multiobjective optimization with immunity, *Proceedings Fifth International Conference on Computational Intelligence and Multimedia Applications*, Xi'an, China. Oct. 2003.
33. Y. Qi et al., An immune multi-objective optimization algorithm with differential evolution inspired recombination. *Applied Soft Computing.* 29(C):395–410, 2015.
34. X. Han et al., Fault-tolerant relay node placement in heterogeneous wireless sensor networks. *IEEE Transactions on Mobile Computing.* 9(5):643–656, 2010.
35. M.Z. Hasan, F. Al-Turjman, H. Al-Rizzo, Optimized multi-constrained quality-of-service multipath routing approach for multimedia sensor networks. *IEEE Sensors Journal.* 17(7):2298–2309, 2017.
36. M.Z. Hasan, F. Al-Turjman, H. Al-Rizzo. Evaluation of a duty-cycled protocol for TDMA-based Wireless Sensor Networks. *International Conference in Wireless Communications and Mobile Computing (IWCMC),* pp. 964–969, 2016.
37. B. Grimstad, A. Sandnes, Global optimization with spline constraints: a new branch-and-bound method based on B-splines. *Journal of Global Optimization.* 65(3): 401–439, 2016.
38. M.U.s.G. MathWorks, *Fundamental Coordinate System Concepts.* The Mathworks. Inc., Natick, MA, 1992.
39. M.Z. Hasan, T.-C. Wan, Optimized quality of service for real-time wireless sensor networks using a partitioning multipath routing approach. *Journal of Computer Networks and Communications.* 2013: 18, 2013.Article ID 497157, doi:10.1155/2013/497157.

Chapter 8

Mobile Couriers' Selection in Smart Cities' IoT*

Fadi Al-Turjman
Antalya Bilim University

Contents

* F. Al-Turjman "Mobile Couriers' Selection for the Smart-Grid in Smart Cities' Pervasive Sensing", *Elsevier Future Generation Computer Systems*, 2017. doi: 10.1016/j.future.2017.09.033.

8.1 Introduction

Wireless Sensor Network (WSN) has come a long way, from their support in area specific deployments such as irrigation systems, healthcare, and supply chains, to supporting multiuser systems that enable simultaneous access of application that operate in large scale Internet of Things (IoT) paradigm [1,2]. Smart cities are examples of such applications that support multiuser access on a multi-application platform. These users may want to access information such as the availability of parking space in the city, electric grid information from the smart meters around the city such as electricity consumption levels, peak hours, etc., and/or major road accidents and any other reported emergencies. Smart cities are expected to have a grid of WSNs to provide access to large-scale information [3]. In smart cities, different challenges related to type/nature of surrounding buildings and obstacles' need to be taken into account. In this study, cities with high-rise buildings are assumed, which can severely attenuate the signal strength/quality and accordingly users may experience poor service. In addition, varying distances between the base station and users due to mobility is assumed. This can also be a key challenge in the assumed smart city setup where fixed transmission range from the base station can lead to significant waste in energy in case it is set to the maximum unnecessary coverage distance. Additionally, in dense deployment of wireless user equipment, interference may happen and degrades Quality of Service (QoS) and bandwidth utilization in homogenous networks functioning over unique frequencies.

Pervasive Sensing (PS) is one great example that uses low cost sensory devices in mobile devices to create a large-scale network for transferring data among users for the greater good of the public in smart cities [2,4]. The proliferation of wireless sensors has given rise to PS as a vibrant data-sharing model. This vision can be extended under the umbrella of the IoT to include versatile data sources within smart cities such as cell phones, radio frequency identification tags, and sensors on roads, beaches, and living spaces. The facilitation of such a vision faces many challenges under the outdoor harsh operational conditions in terms of inter-operability, resource management, and pricing. With PS incorporated into IoT, it will be able to extend to data generating/sharing systems including Wireless Networks (WNs), IoT operating systems, database centers, and personal and environmental monitoring devices deployed in both cities and urban regions. Smart city projects in the urban regions introduce innovation in the provision of infrastructure services making applications accessible directly and in large-scale. Smart grids initiatives for utilities represent part of this big scenario of smart cities. The innovation of the smart grid can affect almost all services necessary for the economic development and the well-fare state of human beings in smart cities. Smart grid has been evolved recently in managing our vast electricity demands in a sustainable, smart, and economic manner, while utilizing already existing Heterogeneous Networks' (HetNets) infrastructures. The smart grid is simply an energy network that can automatically monitor the flow of electricity in a city and adjust to changes in users' demands accordingly. It comes with smart meters, which are connected to the Internet to provide consumers/suppliers with smart-decisions on their on-going energy usage/production. For example, a number of smart home appliances viz., dishwashers, washing machines, and air conditioners can communicate with the grid using these smart meters and automatically manage their electricity usage to avoid peak times and make more profit. Moreover, these smart meters can be used in "smart parking lots" projects where arrival and departure times of various vehicles are traced and measured all over the city. Therefore, these parking lots have to be planned in a way that takes the average number of cars in every region into account. This service is applicable based on sensors deployed on the roads of the city and intelligent displays which inform drivers about the best place for parking in the nearby region while being synchronized with these smart meters. In this way, drivers can find a cost-effective place for parking faster, and reduce CO_2 emissions and traffic congestions in the city.

Within a smart and green grid, mobility has to be intended as the way in which customers can access and explore the grid resources using advanced and eco-friendly electricity modes. This implies being aware of the available energy resources and their real values (in term of cost, time, and carbon emissions), as well as a simple and unified access to mobile carriers in a PS paradigm. These mobile carriers can connect the different system entities, including mobile user devices, power lines, parking lots, and appliances at home by simply establishing a dynamic infrastructure between the distributed smart meters and their base stations in a PS paradigm. PS in smart cities can significantly affect citizens' satisfaction levels and can be realized by combining private and public networks, including the IoT as the main platform.

In order to support smart grid applications, utility companies have widely adopted wireless Mesh topologies that aim to be open and interoperable HetNets. However, these HetNets solutions have yet lots of work to be done. For IoT services including smart grids there are genetic-based meta-heuristic solutions including the Effective Fitness (EF), and Hybrid Search (HS) algorithms [5,6]. EF considers the motion direction of the mobile carrier with some environmental effects, such as surrounding obstacles and available paths. Whereas HS is a modified effective method that takes into consideration speed of the motion to compute the fitness value. On the other hand, a non-genetic up-to-date approach, called Long Range Wide Area Network (LoRaWAN) [7], has been recently proposed to address data delivery problem between the smart grid meters. LoRaWAN represents a convergent alternative, which also adopts open and interoperable standards, but at a lower cost if compared to existing mesh HetNets including Worldwide Interoperability for Microwave Access (WiMAX), Long-Term Evolution (LTE), Wireless Fidelity (WiFi), Light Fidelity (LiFi), and Point-to-Multipoint Radio Systems.

In this chapter, we change the meaning of the term "sensor" to include any data source device that is moving or stationary. Hence, this huge network created by these sensors provides a multitude of services that improve the standards and quality of living in smart cities. There are abundant sensor devices in such a setting, onboard private or public vehicles, and\or deployed on roads or buildings. Such a comprehensive PS system introduces challenging implications in practice regarding the sensory system's limitations in terms of energy consumption, available bandwidth capacity, and delay [8,9]. Moreover, management is also put to test, as the data transferred across the system is of wide variety. Such a system with multiple applications and multiple sensors on the same platform is depicted in Figure 8.1a. In this model of a smart city, there are a number of Access Points (APs), which communicate with Sensor Nodes (SNs) (or smart meters), users and each other, to collect or gather requested information and send it to data sink (or Base Station [BS]) based on service requirement.

In a smart city, it is common to sometimes have more than one BS collaborating. We assume this setting in this research in contrast with other related works in the literature [10,11]. In smart cities, we exercise the ability to handle multiple users with different attributes such as latency, reliability, and throughput simultaneously. This is one important area that has not received sufficient research attention yet. The complexity involved in handling the heterogeneous traffic flows in the underlying sensor networks, comes from simultaneous user request with diverse requirement. To curb the complexity problem, we propose a method that utilizes mobile smart devices called Data Couriers (DCs). To reduce the total number of DCs and their collaborative travel distance, we propose an approach called Hybrid Collaborative Path Finder (HCPF). The targeted problem can be stated as *Given a set of DCs with a limited storage and predetermined trajectories under the coverage of a set of APs, find the minimum DCs count that can deliver these APs' data traffic while retaining storage capacity constraints and minimum travelled distances by DCs.*

To solve the above stated multi-objective problem, we use a hybrid metaheuristic path-finder approach, based on pure Genetic and Local Search plus (LS+) algorithms [12]. Below, we summarize the main contributions in this article toward solving the problem:

1. We propose a framework for PS in smart cities based on an IoT architectural model that integrates HetNets and data sources to support the smart grid project.

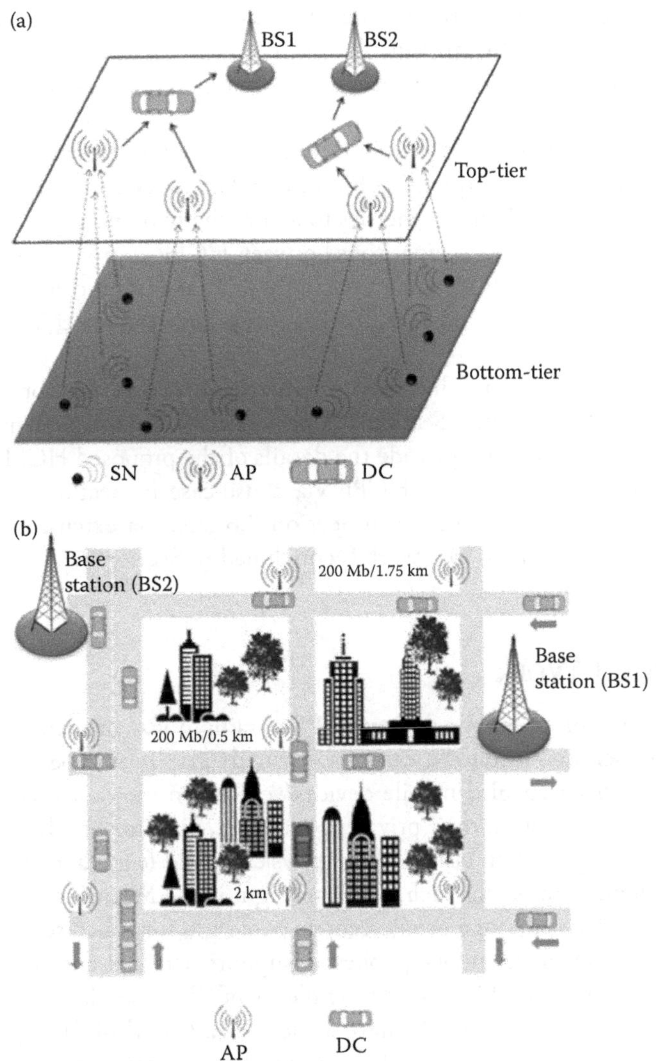

FIGURE 8.1 (a) Two-tier PS architecture and (b) selected DCs for data collecting in smart cities.

2. We propose a massive data-collecting network for DCs in smart cities. We do this by utilizing moving sensors that collect data in a smart environment efficiently in terms of the number of DCs involved and the total distant travelled by each DC.
3. For DCs that operate in competitive time complexities and experience traffic/memory capacity constraints, we propose a collaborative genetic-based approach, called HCPF.

4. We also propose a cost-based fitness function for DCs election in collaborative PS paradigm. In this function, we consider DCs resource limitation in terms of its DCs count, storage capacity, energy consumption, and the communication link quality.
5. We provide a dynamic two-tier scheme that adheres to the social welfare of the PS system by incorporating lifetime and capacity constraints while considering, delay, and quality metrics to assure the maximum gain.
6. We perform extensive comparisons between HCPF approach and other heuristics (e.g., EF and HS), and accordingly we make significant recommendations regarding such kind of heuristic approaches in smart cities.

The remainder of the chapter has been organized as follows. Section 8.2 reviews related work in the literature. Section 8.3 provides the utilized system models in this study. In Section 8.4, we provide the details of the proposed HCPF approach. Next, we elaborate more on the HCPF via a use-case in Section 8.5. The proposed HCPF approach is validated in Section 8.6 later via extensive simulation/experimental results. Finally, our work is concluded in Section 8.7.

8.2 Related Work

The recent explosion of mobile devices in PS paradigm viz., smartphones, tablets, and on board sensors, inspired a category of smart grid prototypes. These prototypes can use sensor-enabled mobile devices/vehicles to monitor their local environments via smart meters, their private spaces (e.g., monitor bodily vital signals), or create a binding between tasks and the physical world (e.g., take video or audio samples). A good example for such a PS paradigm is the MetroSense presented in Ref. [13] as a wider vision of a people-centric paradigm for urban sensing. It solely explores sensor-embedded mobile phones to support personal and public sensing. PS can be more effective while considering plenty of Cloud services as well. Joining the capabilities of a multitude of smart devices in a Cloud of Things (CoT) has been proposed in several works [14–16]. Proposals specifically address smart cities applications can be found in Ref. [17–19]. The authors of Refs. [17], for instance, highlighted how future cities need to collect data from an abundance of low-cost urban sensors including environmental sensors, GPS devices and building sensors. The key idea for getting high-quality services from such cheap sensors is the cross-correlation of sensed data from several sensors and their analysis with sophisticated algorithms.

The application of PS in smart cities has recently proven to be very effective with the IoT evolution [3,20]. In a paradigm such as the IoT, data collection is done by deploying sensors around the city, which periodically send data about the smart city variables (e.g., temperature, humidity, traffic conditions, etc.), to the processing center (BS) through a wireless link. To correctly collect data about a given sensor, it is advised to employ a number of mobile DCs and effectively utilize their path of interest. This DC are usually powered by low-power batteries and mounted on moving vehicles such as taxis and public transport means. Hence, the DCs in this framework will have low data transmission power, low memory capacity, and low processing power, which consequently leads to a major problem faced by most reliable smart city grid, where the main challenge is connecting with the main BS. Additionally, mobile DCs must be used due to the monitoring of non-stationary phenomenon such as animals, cars, people, etc. and the very limited network resources, such as communication range and energy budget, when it comes to large-scale applications like the ones found in smart cities. Mobile DCs were proposed in Ref. [21], given that centralized knowledge and decisions are made at the BS. Mobile SNs move along a predefined path in the sensing environment. There is a proven benefit of using DCs (mobile relays) over conventional WSN using static SNs. A network that uses the former nodes has more lifetime than a network that uses the latter ones. The authors in Refs. [22,23] state that DCs were first used to prolong the WSN lifetime. The lifetime of the network is divided into equal parts called rounds. Based on a centralized algorithm running at the BS, DCs are placed at the beginning of each round. The main goal was to minimize the average amount of energy used during one round. It was concluded that the optimal locations according to this objective function remain optimal, even after the objective is changed to minimizing the maximum energy consumed per SN. However, using this energy metric to find the optimal position of mobile DCs is not the most effective means, because the solution will not be in terms of time. Therefore, the DCs location found might be far from the optimal positions in a smart city setup.

Nevertheless, none of the aforementioned approaches introduces a comprehensive framework that addresses delivery, resource management, and cost challenges together. Most importantly and from the delivery perspective, they either apply simplistic communication protocols that are either basically cellular (e.g., GPRS) coupled with some WiFi or Bluetooth capabilities, or apply algorithms that are intended originally for mobile Ad-hoc networks such as Ad-hoc On demand Distance Vector Routing (AODV) [24] and Dynamic Source Routing (DSR) [25], which are short-path routing protocols based on minimum-hop count only. Genetic Algorithm (GA) does not break easily and is resistant toward noise unlike the mobile DC approaches mentioned earlier in Refs. [12,26]. Each chromosome contains information, and the population means a big data for the current generation. We can use GA to simultaneously search for candidate solutions; hence, it can be seen as a parallel searching approach. Numerous benefits come with the fact that GA is a powerful and robust optimization technique. The applicability of GA, as an artificial intelligence technique, comes from the fittest chromosomes along/over consecutive generations for solving a problem. With the progressive generation of solutions, chromosomes with potential solutions become more adaptable to the environment. As a result, successful results are obtained. According to Refs. [5,6,27,28], GA has a potential for solving problems such path planning and data gathering. The authors in Ref. [27] use the GA approach to solve the problem of point-to-point path for an offline autonomous mobile robot. The GA approach was used to effectively find the optimal path with reduced cost and computation time and smaller number of generations in a static field with known obstacles. The authors in Ref. [28] use GA to calculate an optimal path for autonomous mobile robots, where a robot is represented by its coaxial coordinates in its present environment. GA uses random searches, however, it actually directs searches into regions of better performance in the inside the search area [5,6]. Since the genes from good chromosomes go through the population, they are expected to handle better offsprings than the parent. In this research, we have mainly focused on genetic-based meta-heuristic approaches that consist of EF and HS algorithms. For comprehensive analysis, another non-genetic up-to-date approach, called LoRaWAN [7], which best fit big scenarios such as the PS in smart cities has been considered. LoRaWAN is a fully convergent technology based on open standards, low cost, and was designed from the start to build urban platforms for IoT. It is best suited to the big scenario of a smart city. Unlike other Mesh topology-based approaches, LoRaWAN has a star topology that simplifies the operation, and significantly reduces traffic on network destined to routing information. On the other hand, it does not count with the possibility of coverage extension through the relay on neighboring terminal device. In Ref. [5], EF considers the motion direction with some environmental effects to develop an EF function to be used in the genetic algorithm. In Ref. [6], the authors proposed a modified effective method, called HS, which takes into consideration speed of the motion to compute the composite fitness value. HS and EF are seeking the maximum fitness value for their designed fitness function, while others are seeking the minimum fitness that makes them less efficient in terms of the archived solution optimality. Accordingly, we focused on genetic-based approaches, which are efficient in terms of time complexity and optimality as well as practicality in solving the aforementioned path-planning problem. Unfortunately, some parameters such as the total travelled distance and the current application circumstances were not considered in the aforementioned collaborative approaches. Therefore, we propose a hybrid meta-heuristic approach that employs a modified version of the LS [29] in collaborative manner to improve the path search in smart cities based on application-specific characteristics, such as delay-sensitivity and bandwidth availability.

8.3 System Models

In this section, we discuss the main system models we assume in our HCPF approach. We start with the assumed notations, followed by the network, energy, and communication models, which are considered in this study.

8.3.1 Notations

The following is the list of the assumed GA components mapped to the aforementioned DC election problem:

- *Gene*: It is represented by a decimal value that can be between zero and the number of APs.
- *Chromosome*: It is a candidate solution to the given problem, and is composed of several genes. It is also called individual, and coded as a finite length vector of genes (represented in bit strings).
- *Search Space*: It is the area in which the search is performed to find the solution among the chromosomes.
- *Fitness value*: It is the associated with each chromosome and it represents quality of the candidate solution.
- *Population*: Consists of a number of chromosomes, and is maintained within a specific search space.
- *Generation*: It is the current population in any stage of the whole searching process.
- *Genetic Operator*: It is a variety of operations to be applied on a chromosome for producing better individuals (e.g., crossing over and mutation).

GA combines and passes the information of an individual to any other individual in order to produce new solutions with good information inherited from previously found solutions. It is expected that these new individuals, called offspring, are hopefully leading toward optimality; thus, the new population can be better than the previous one over successive generations toward the global optimum. This process continues until some condition (termination criteria), such as the maximum number of generations is reached, is satisfied.

8.3.2 Network Model

In this chapter, we consider a multi-tier telecommunication framework with three main components: (1) the BS, (2) DC, and (3) the WSN AP in a smart city (see Figure 8.1). Telecommunications infrastructure that supports the smart grid application in IoT-based PS is divided into outdoor RF devices, and indoor wired devices connected to the Internet via Network Interface Cards (NICs). RF devices are generally installed on GSM towers, top of buildings or even poles and are responsible to communicate with endpoints (end-users) in its covered area. The NICs are interfaces embedded on endpoints and are responsible to communicate with nearest RF device and make interface with the Internet. Outdoor RF infrastructure is usually divided into private and public networks. Public networks are those provided by third parties, generally the telecommunication companies that share their infrastructure with several other users to provide different services such as mobile network (GPRS, 3G, 4G), dedicated circuits, satellite services, and Internet. Private are those in which a network or a partition of network is dedicated to exclusive use of a single user and it can be own, assets acquired and operated by individuals or a specific company itself. There are several alternatives, which are currently available for utilities companies to construct a private telecommunication network for their implemented smart grid applications using the PS paradigm. Those alternatives can be employed at the backbone layer of the network such as the Optical Systems, and the Super High Frequency (SHF) Point-to-Point Radios, at the backhaul layer such as WiMAX, LTE, and Point-to-Multipoint Radios, and/or at the access layer such as RF Mesh and LoRaWAN. Although such wide range of technologies/alternatives can add further complexity to the overall data delivery process, it provides better financial situation for the proposed PS paradigm and makes it more flexible in satisfying the different users' categories/needs in such a heterogonous system. Furthermore, heterogeneity of the system will be a promising solution to satisfy multiple users' needs in public places such as downtowns or shopping malls, where the capacity of the cellular network can be overwhelmed. In such scenarios having WiFi as a backup network would be a solution to handle the gigantic incoming data requests from the smart cities' users.

In the assumed telecommunication network there are multiple BSs and this can be a typical situation in smart cities, where multiple service providers can co-exist [30,31]. Thus, different DCs can move in the same route and collaborate with each other via exchanging in-city routing plans. In this chapter, a two-tier hierarchical architecture is assumed as a natural choice in large-scale applications, in addition to provide more energy-efficient solutions. The lower layer consists of SNs that sense the targeted phenomena and send measured data to the APs in the top-tier, as shown in Figure 8.1a. Usually, these SNs have fixed and limited transmission ranges and do not relay traffic in order to conserve more energy. The top-tier consists of APs and DCs, which have better transmission range and communicate periodically with the base station to deliver the measured data in the bottom-tier. APs aggregate the sensed data and coordinate the medium access, in addition to supporting DCs in relaying data from other APs to the BS in the top-tier. A data packet consists of the data traffic that includes loads of a group of APs in the network. Each AP delivers its sensed data via multi-hop transmissions through other APs/DCs to a BS. DCs are equipped with wireless transceivers and are responsible for forwarding the APs' data load to the destination (AP or BS) once they are within their communication range. APs represent the target for a given DC its route. In our formulated path-planning problem, we assume the following inputs. Each DC has a specific/limited storage capacity. Different APs can communicate with each other for collaboration. Moreover, we assume the following constraints for the problem:

1. Every DC is usually associated with a predetermined route/trajectory in the city.
2. Each route includes a path. The path of any route of a DC includes a subset of the APs.

3. Each smart node/APs can be assigned to more than one DC (i.e., to be visited more than once).
4. The total traffic carried by a DC cannot be more than the DC storage/BW capacity.

We assume the outputs for this problem solution consists the following three characteristics: (1) the minimum required number of DCs; (2) the total travelled distance by them; and (3) the paths of all the optimally selected DCs. It's worth pointing out here that the total travelled distance represents the consumed time and energy in the found solution, where this distance is the total driven kilometers from the data source to distention.

8.3.3 Energy Model and Lifetime

The energy supply of a mobile DC can be unlimited or limited. When a DC has an unconstrained energy supply (rechargeable or simply have enough energy relative to the projected lifetime of the APs), the placement of DCs is to provide connectivity to each AP with the constraint of the limited communication range of APs. When the energy supply of DCs is limited, the allocation of DCs should not only guarantee the connectivity of APs but also ensure that the paths of mobile DCs to the BS are established without violating the energy limitation. In this research, we assume a fixed and limited power supply at the DCs. In order to measure the network lifetime, a measuring unit needs to be defined. In this work, we adopt the concept of a round as the lifetime metric. A round is defined as the time period over which every irredundant SNs and relay nodes in the network communicates with the BS at least once. It can also be defined as the time span over which each EC reports to the BS at least once. At the end of every round, the total energy consumed per node can be written as:

$$E_{\text{cons}}^{i} = \sum\nolimits_{\text{Per round}} J_{\text{tr}} + \sum\nolimits_{\text{Per round}} J_{\text{rec}}, \tag{8.1}$$

where $J_{\text{tr}} = L\left(\varepsilon_1 + \varepsilon_2 d^n\right)$ is the energy consumed for transmitting a data packet of length L to a receiver located d meters from the transmitter. Similarly, $J_{\text{rc}} = L\beta$ is the energy consumed for receiving a packet of the same length. In addition, ε_1, and β are hardware specific parameters of the used transceivers. In addition, if the initial energy E_{init} of each node is known, the remaining energy per node i at the end of each round is $E_{\text{rem}}^{i} = E_{\text{init}}^{i} - E_{\text{cons}}^{i}$. In order to consider the effect of packet retransmission at the MAC layer the extended energy model in Ref. [32] has been considered. Using this model, the effect of the wireless links reliability on the energy consumption of per DC is taken into consideration. Based on Ref. [32], the total energy consumed by a DC to receive the packet is

$$E_{DC} = x\frac{P_r L_d}{R_d} + y\left(P_t + \frac{P_{DC,SN}}{k}\right)\frac{L_a}{R_a}, \tag{8.2}$$

where L_d (bits) denotes the size of the data packet transmitted from SN to DC, R_d denotes the data rate with which the data packet is transmitted by SN, P_t is the power required to run the processing circuit of the transmitter, P_r is the power required to run the receiving circuit, $P_{DC,SN}$ is the transmission power from SN to DC, and k be the efficiency of the power amplifier. Also, L_a (bits) denotes the size of the acknowledgment, R_a denote the data rate by which the acknowledgment is transmitted by DC, and the value of x and y depends on reliability of the forward link between a SN and the DC for data packets and reliability of the reverse link (DC, SN) for acknowledgments.

8.3.4 Communication Model

Practically, the signal level at distance from a transmitter varies depending on the surrounding environment. These variations are captured through what we call *log-normal shadowing* model. According to this model, the signal level at distance from a transmitter follows a log-normal distribution centered on the average power value at that point [33]. This can be formulated as follows:

$$P_r = K_0 - 10\gamma \log(d) - \mu d, \tag{8.3}$$

where d is the Euclidian distance between the transmitter and receiver, γ is the path loss exponent calculated based on experimental data, μ is a random variable describing signal attenuation effects* in the monitored site, and K_0 is a constant calculated based on the transmitter, receiver and field mean heights. Let P_r equal the minimal acceptable signal level to maintain connectivity. Assume γ and K_0 in Eq. (8.3) are also known for the specific site to be monitored. Thus, a probabilistic communication model, which gives the probability that two devices separated by distance d can communicate with each other is given by:

$$P_c(d, \mu) = Ke^{-\mu d^{\gamma}}, \tag{8.4}$$

where $K_0 = 10\log(K)$. Thus, the probabilistic connectivity P_c is not only a function of the distance separating the SNs but also a function of the surrounding obstacles and terrain, which can cause shadowing and multipath effects (represented by the random variable μ).

* Wireless signals are attenuated because of shadowing and multipath effects. This refers to the fluctuation of the average received power.

8.4 Hybrid Collaborative Path Finder (HCPF)

Although there has been improvement on the computing and energy resources of connected devices in smart cities, they are not enough for the unrestricted deployment of ambitious IoT systems. The mobile edge computing nowadays highlights that such resources ought to be wisely employed for shifting and distributing computing and sensing tasks toward the network edge (user). Accordingly, we based our proposed HCPF solution on opportunistic PS due to its inherent ability in sensing the surrounding environment and providing a response to detected changes (events) in timely and cost-effective manner.

PS in IoT is a capable framework to provide a convergent telecommunications infrastructure, interoperable between different manufacturers that enables investments optimization and the provision of services at a low cost. Based on open standards, it can meet all the demands of IoT in smart cities, including the smart grid application. It can be viewed as a dynamic networked system composed of a large number of smart object that can communicate with each other and/or with users through heterogeneous wireless connections. Mobile data collectors (DCs) are used as communication devices for the smart objects in the smart city to access the Internet and reach the end user. The average speed per DC while it moves in the smart city is assumed be between 15 and 60 km/h. By using such DCs, a significant amount of traffic can be offloaded from cellular networks and other similar global ones, which can be overloaded and inefficient during peak times of the year when dense wireless devices are served simultaneously. Therefore, the operator's cost can significantly be reduced. In addition, due to the short distance between receivers and transmitters in the proposed PS framework, the power consumption and battery of the mobile user devices can be saved and better utilized. As much as we increase the number of deployed DCs, they consume much more energy to work and this can increase the overall cost as well. Consequently, the network will not be financially affordable to be in service. Thus, the HCPF approach aims at optimizing the count of DCs through a multi-objective fitness function that takes into consideration the aforementioned smart city characteristics. This fitness function considers the available resource limitations in terms of delay, capacity, and price on the data providers' side, as well as user's quality and trust requirements from the requesters' side. In characterizing the smart city network, we use a set of locations that belong to a number of APs. We use some Vehicle Routing Problem (VRP) instances from the literature [34] for referencing, and thus, we can know the exact deployed APs' locations. To minimize the total energy of the network, we try to minimize both the total number of DCs, which is required to serve all the APs in the network, and the total travelled distance by the minimized set of DCs. Now, after elaborating a bit more on the used genetic-based notations, we propose our HCPF framework to optimize the cost, calculated by DC count and travelled distance, in a smart-city setup.

8.4.1 Chromosome Representation

To solve the aforementioned problem using genetic-based algorithms, we represent each feasible solution/path by one chromosome that is a chain of integers where each integer value corresponds to an AP or a BS in the network. Each DC identifier uses a separator (*zero* value) between a route pair and a string of AP identifiers to represent the sequence of deliveries on a route.

Example 1

Assume we have nine APs and three DCs are in the city shown in Figure 8.1b. Then, in the sequence below, we give a potential solution for the problem in order to demonstrate the encoding of a chromosome representation. Note that "0" represents the BS, and the number of zeros indicates the number of required DCs in the solution.

Gene sequence: 0 — 5 — 6 — 3 — 0 — 7 — 8 — 2 — 1 — 6 — 0 — 8.
Path 1: BS 5 6 3
Path 2: BS 7 8 2 1 6
Path 3: BS 8

These will be the paths found for the three DCs that are employed by the BS in this example.

8.4.2 Generating Initial Population Strategy

The strategy for generating initial population for a GA is an imperative part. In this step, we choose the closest nodes for a specific node in creating the location sequence. The idea is to determine the close by nodes of each node before the creation of an initial set in the genetic pool and prevent it to be completely random. This information is accumulated over several communication rounds and leads to a Knowledge Base (KB) at the DC, which can be looked up by reasoning mechanisms to make quick decisions about the data delivery sub-path. This KB facilitates learning from feedback in the network about the actual values of desired QoS attributes as seen at receiver nodes. This helps in reducing the communication and energy overhead in the network. In fact, the initial population strategy does not wait for the establishment of an end-to-end path from source to BS. Instead, it relies on the learning from the feedback provided at each hop via Quality-aware Cognitive Routing (QCR) algorithm. A pseudo code description of the request dissemination and information gathering is shown in Algorithm 1. Steps 1–2 represent the beginning of stage 1, where the request is multicast to all DCs. Steps 3–5 indicate the action taken when an *exact-match* is found at a DC, at any stage of the query dissemination. Steps 6–14 indicate the actions involved in disseminating the request through nearby distributed SNs till all the information required to update a *partial-match* to *exact-match* is found. At the end of stage 1, a DC containing data that is an *exact-match* with the *Query* is identified and the QCR algorithm is initiated. Steps 16–18 indicate the steps taken if no match was found or the request timed out. A pseudo code description of the QCR algorithm is shown in Algorithm 2. Step 1 starts with acknowledging the DC with an *exact-match* to the *Query* from Algorithm 1. In steps 2 and 3, a next-hop routing path is chosen from the KB, based on the traffic type and QoS attribute priorities associated with it. Steps 4–11 show the actions taken at the next-hop nodes depending on whether they were SNs or DCs, and their location relative to the BS. Steps 12–15 indicate that the once data is received successfully at the receiving node, an "Ack" is sent back to the transmitter with information about the QoS of the received data.

Algorithm 1: Query Dissemination and Data Gathering

---

1. **If** BS has a new *Query* from the user
2. **Then** multicast *Query* to all DCs in the network
3. **If** an *exact-match* is found in the cache of any of the DCs
4. **Then** initiate Algorithm 2 (QCR) from that DC
5. **Goto** Step 18
6. **If** *partial-match* is found at a DC
7. **Then** multicast *Query* from that DC to SNs in its cluster
8. **If** *partial-match* is found at any of the SNs
9. **Then** multicast *Query* from DCs to SNs in their cluster
10. Gather *Query*-relevant data from SNs at each DC
11. **If** *exact-match* found at DC
12. **Then** transmit data from DC to the DC that issued the *Query*
13. Aggregate the data received from SNs at the DC
14. **If** *exact-match* is verified for this data at DC
15. **Then Goto** Step 4
16. **If** no match found or request timed out
17. **Then** abort current request
18. **End**

Algorithm 2: QoS Aware Cognitive Routing (QCR)

1. **If** a DC with data that is an *exact-match* to *Query* is identified
2. **Then** identify the priorities with respect to QoS attributes and Energy Cost based on the attribute: "Traffictype"
3. Look-up the *KB* to identify one or more next-hop paths that satisfy the QoS requirements for the traffic type
4. Transmit data from source SN to the next-hop receiver node(s) listed in the KB
5. **If** a DC located at one-hop from the BS received the data
6. **Then** continue to relay data from that DC to the BS
7. **If** an intermediate DC received the data
8. **Then** look-up DC's routing table (RT) to identify the next-hop DC **and** goto step 2
9. **If** a neighboring DC received the data
10. **Then** goto step 2
11. **If** any node other than the BS received the data
12. **Then** calculate the QoS of received data and send this information back to the source SN with an "Ack"
13. The SN stores this information about the receivers and their associated QoS in its KB
14. **If** BS received the data
15. **Then** BS computes the QoS for the data received and sends it back to the transmitter(s) along with an "Ack"
16. **If** "Ack" transmitted by BS
17. **Then** data has been received for the current request
18. **End**

Lines 16–18 indicate that when the sink transmits an "Ack," it means that the requested data was successfully received and that ends the current transmission cycle. Nodes that fail to send an "Ack" in a previous transmission cycle will be chosen for upto three times without receiving an "Ack." Beyond this, the transmitter presumes the HCPF genetic algorithm described in the following subsection.

8.4.3 Path Planning in HCPF

HCPF is an event-driven and collaborative-based optimization algorithm to select the best chromosome from the search space. In this section, we formulate a data path based on the available set of touring DCs in the city. We assume that each AP location is a target for DCs on its route where they stop (or satisfy speed and channel characteristics based on Eq. (8.3)) for the minimum amount of time required for exchanging data via *wifi* connections. APs are assigned to DCs based on their passing time and requested service specifications and requirement (e.g., security level, maximum delay, etc.). This service is quantified usually using a QoS factor that can be archived in collaboration with other BSs and APs. This QoS represents the successful arrival of bit-rate at the final destination (i.e., BS), where some services (e.g., VoIP and video streaming) require a higher quality level in terms of transmission rates as opposed to data/information about road-traffic updates, for instance. This is predicted in this work via log-based function running on the top-level of the system (e.g., a set of accessible BSs/APs in the cloud). We formulate a multi-objective fitness function for our genetic-based approach. The fitness function $F(x)$ for the xth chromosome in our GA population aims to find solutions that takes in to consideration cost, QoS, and DC motion characteristics such as speed and travelled distances. This function formulated as follows:

$$F(x) = \lambda * RDistance\ (x) + \mu * RCount(x) + \beta * QoS(x) + P_c, \quad (8.5)$$

where $RDistance(x)$ computes the total travelled distance when chromosome x is applied, and the $RCount(x)$ function returns the minimum required number of DCs. This count is proportional to total cost. Furthermore, $QoS(x)$ is used to calculate the achievable bit-rate of a generated chromosome x. And P_c is the average probabilistic connectivity between the neighboring nodes on the same path, which is calculated based on Eq. (8.4). These four variables are weighted via tuning parameters (λ, μ, β, and P_c) that makes the proposed framework adaptable to the heterogeneous nodes/applications in a typical smart-city's PS. Total distance is weighted by λ value equal to 0.001 and the number of DCs is weighted by μ value equal to 100 based on preliminary studies in Refs. [11,12]. Since total distance has more importance than the number of DCs, the coefficient of it is more sensitive, and thus, λ value is set to be 0.001. These tuning parameters formulate our fitness function that directs the search space to find chromosomes that cover all APs interested in exchanging their data loads, and then it finds the shortest path with the least possible DCs. Meanwhile, for the creation of a new generation we apply four main steps: (1) ***Selection***, which is used for choosing an individual pair to apply the genetic operators (*crossover* and/or *mutation*), and in this study we use *tournament* selection method, (2) ***Crossing-Over***, which is used to add the genes of the two different parents into the child, (3) ***Mutation***, which is applied for modifying some genes in the individuals, randomly selects an AP in the chromosome (test) to be flipped from one digit to another within the range of the available AP count without affecting the test logical structure, and (4) ***Reproduction*** which is used for the construction of the next generation using the current parents. In this approach, we used a variant version of the ***Reproduction*** step, where almost every new population replaces an old one per iteration. Unlike Ref. [29], we apply an improved 2-opt LS algorithm as a ***Mutation*** operator. We combine our genetic algorithm in Ref. [26] with the improved LS+ (applied at the end of each created generation) for better converging results. We point out here, that neighborhoods of a potential solution are searched for better convergence as well.

Algorithm 3: Pseudo-Code of the Improved LS+ Approach

Function: Improved_LS+()

While (local minimum is not achieved)

Select best AP pair: (*i*, *i*+*1*) and (*j*, *j*+*1*)

If distance(*i*, *i*+*1*) + distance(*j*, *j*+*1*) > distance(*i*,*j*) + distance (*i*+*1*, *j*+*1*)

Exchange the edge with 2-opt

End If

End While

Algorithm 4: Pseudo-Code of the HCPF Approach

Function: HCPF()

Inputs

Max_generations: the maximum number of iterations to run the algorithm.

Initial and *max_population* size.

Crossover and *mutation* probabilities.

The *AP sequences* in a smart city.

Begin

Generate initial population chromosomes from *AP sequences*.

Compute the fitness of each chromosome using Eq. (8.5).

Do{

While (offspring population != parent population size)

1st parent=Tournament_selection(2 andom chromosomes)

2nd parent=Tournament_selection(2 andom chromosomes)

Crossover(1st parent, 2nd parent) based on crossover probability.

Apply all three mutation types on the two newly-generated offspring based on *mutation probability*, using **Improved_LS+().**

Delete/Repair any logically invalid offspring.

Replace the parents by offspring if the parent has worse fitness. Otherwise, let parent propagate to the new population.

}While (Max_generations)

if (population converged && population size < *max_population)*

Double the population size and add extra new randomly generated chromosomes to the population.

End

Return best solution in current generation.

In terms of time complexity analysis, we use big-O notation. This means that the time taken to solve a problem of size n can be described by $O(f(n))$. Since traditional LS+ algorithm has complexity of $O(n)$ when non-enumerative (data-dependent) search is allowed [29], where n is the total count of APs in our proposed two-tier architecture. Our improved LS+, in Algorithm 3, has a time complexity of $O(n\log n)$. And thus, the total time complexity of the proposed HCPF approach, in Algorithm 4, end up being $O(n^2\log n)$.

For verification purposes, the running time and number of populated solutions have been recorded while HCPF is running. The recorded runtime is then plotted and polynomial fitting has been used to estimate the growth function of the running time. The theoretical value of the population growth of the proposed HCPF algorithm is evaluated as:

$$P_{\max} = C_P P_0 (1 + r_c)(1 + r_m). \qquad (8.6)$$

Equation (8.6) is derived from the trend of the population growth with every iteration of the HCPF algorithm. Initially, the population size is a positive value P_0. In worst case, the whole population is selected during the *selection* process. This means that P_0 number of solutions is considered for the *crossover* operation. A portion of the population is selected for the *crossover* process, which reproduces additional chromosomes/solutions to be added to the population. Let us assume r_c is the *crossover* rate where $0 \le r_c \le 1$; the number of additional chromosomes would be $P_0 r_c$. This makes the total number of chromosomes in the population so far equal to $P_0 + P_0 r_c$. by factorization, this can be simplified to

$$P_0 = P_0(1 + r_c)\ (1 + r_m). \tag{8.7}$$

The expression in Eq. (8.7) is then multiplied with a coefficient of population C_P, to enable some amount of control over the maximum population. The time complexity of the HCPF algorithm is then determined from the fitted growth function to be $O(n^2 \log n)$ as well. It is worth pointing out also that HCPF implementation is not straightforward, as it might produce computational and energy overheads on resource-constrained devices. Accordingly, event-driven processing is applied as a natural choice to efficiently orchestrate the synchronous operation of the different stages in HCPF without incurring substantial overheads. We opted to follow an event-driven approach in which system components are only activated upon message notification for carrying out a time-bounded task. Such approach plays well with power management facilities of the employed PS paradigm in smart grid. However, as the HCPF framework heavily relies on public mobile relays, it could become vulnerable to ambient data manipulations, which brings up security issues. Moreover, considering the criticality of smart grid applications is an aspect that needs to be increasingly observed in the construction of infrastructures of telecommunications technology in order to avoid security issues. Accordingly, it is recommended to maintain the selection of the telecommunication technology to meet the requirements of high availability, and security under the control and responsibility of the company that owns/rent the infrastructure equipment. In addition, encryption methods can be applied while utilizing unregistered DCs in relying security-sensitive data. Also, privacy and anonymity issues are expected due to the personal exchanged information in the smart grid project. Therefore, it is of great relevance to keep HCPF artifacts reliable and trustworthy to be able to secure the core parts of it.

8.5 Use-Case

In this section we elaborate more on our proposed HCPF approach via a typical delay-tolerant communication scenario in rural areas where people and/or public transportation can reach but the infrastructures of WiFi/cellular networks are not deployed [33]. We illustrate an example where three regions in a city has the APs: *A*, *B*, and *C*, which have established one or more connecting paths, through some DCs, to the BS1 shown in Figure 8.2. Note that each DC path has its end-to-end storage capacity and travelled distance characteristics as speculated by the HCPF approach given in the format (capacity/distance) in Figure 8.2. These characteristics are based on routing table exchanges between DCs and the APs. This is achieved via our newly developed fitness function in Eq. (8.5), which adopts these

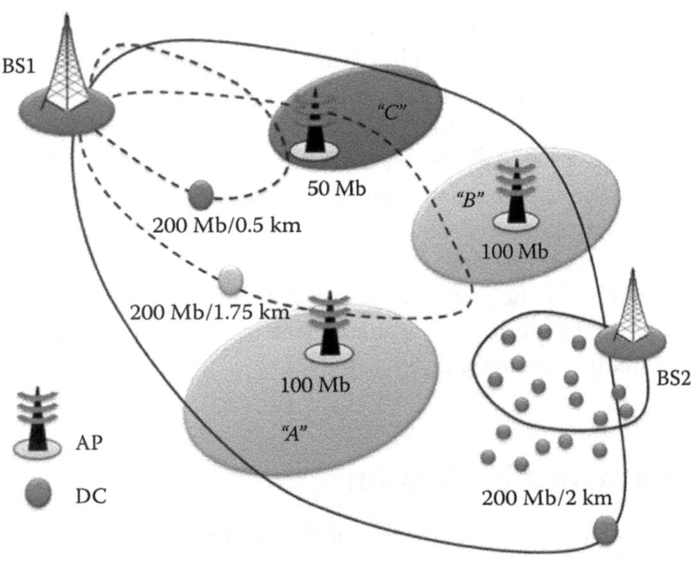

FIGURE 8.2 A use-case for a group of regions' APs serving multiple BSs.

characteristics while the hybrid genetic algorithm in HCPF is applied. The HCPF approach aims at selecting the paths (shown in dashed-lines) that guarantees minimum total travelled distance without violating the DC capacity constraints. The decision of utilizing these DCs depends on its resources, which are stored, as well, at the AP's routing table. For example, assuming that AP in region *A* in Figure 8.2 has to transmit packets with a data-load that requires a minimum DC capacity of 100 MB, the HCPF detects two different paths connecting *A* to the BS. Both of them conform the capacity constraints, and the one among these two that provides the minimum total travelled distance is ultimately chosen. It will be also able to serve the AP at *B* as the remaining storage of that DC is still satisfying the capacity constraint. Then, the second route (dash-line) passing by city *C* will be utilized to deliver its corresponding data. Hence, the selected paths will be:

Path 1: BS1 A B
Path 2: BS1 C

Note that the non-dashed path satisfies the DC capacity constraint, however, it does not provide the minimum travelled distance in total if it has been chosen for *A* and *B*. The selection of the minimum DC count in the proximity of each AP in the network is achieved by periodically exchanging routing tables and/or registration records with BSs to deliver delay-tolerant data packets. This selection process is repeated at the beginning of each triggered round as we aforementioned. The mobility history of the DCs is examined against the communication range of the corresponding destination (AP). Based on the results, DC candidates are defined according to best (i.e., minimum) travelled distance in total per round. Amongst the most significant processing implications is the mobility factor. Since the average experienced speed per DC in the assumed smart city setup is between 15 and 60 km/h, there will be enough time to exchange messages using the IEEE 802.15 standard. However, we classify the received signal quality into real-time (RT) and non-RT traffic for more user satisfaction. Where RT users can be charged extra fees for this type of data traffic. Meanwhile, traffic load, the velocity of mobile DCs, and the multi-interface DC are all crucial parameters to achieve the best average energy consumption by the grid-system. For instance, increasing the DCs' request rate in the smart city can lead to an increment in the average waiting time, and energy consumption per data unit.

8.6 Performance Evaluation

In this section, we assess the correctness of the proposed HCPF approach first via preliminary study relying on realistic route instances experiments. In this experimental work, we compare HCPF against the pure GA and LS algorithms. Then, we evaluate the performance of the proposed HCPF against the LoRaWAN, EF, and HS algorithms in terms of varying design aspects that affect the total cost in a medium scale city per the depicted parameters values in Tables 8.1 and 8.2 Detailed description of our experimental and simulation setup is given in the following sub-sections.

TABLE 8.1 Parameters of the Simulated Networks

Parameter	*Value*
τ	70%
nc	110
ψ	0.001 ms
D_{max}	500 ms
ω	200,000 km/s
γ	4.8
δ^2	10
P_r	−104 dB
K_0	42.152
R	100 m

TABLE 8.2 Components of the Simulated Networks

Simulator Component	*Value*
AP radius	100 m
DC velocity	Low, medium, high
AP transmit power	20 mW
Transmission bandwidth	5 MHz
AP count	100
Expected failure rate	0.001/h

8.6.1 Simulation Setup

Using MATLAB R2016a and Simulink 8.7, we simulate randomly generated HetNets to represent the PS environment in a smart city. A discrete event simulator is built on top of the aforementioned MATLAB platforms for more realistic results.

In this simulator, an event-based scheduling approach is taken into account, which depends on the events and their effects on the system state. One of the most commonly used stopping criterion, called relative precision, is employed in this simulation to be stopped at the first checkpoint when the condition $\beta < \beta_{max}$, where β_{max} is the maximum acceptable value of the relative precision of confidence intervals at $100(1-\alpha)\%$ significance level, $0 < \beta_{max} < 1$. Accordingly, our achieved simulation results are within the confidence interval of 5% with a confidence level of 95%, where both default values for β and α are set to 0.05. Our Simulink simulator supports wireless channel temporal variations, node mobility, and node failures. The simulation last for 2 hours and run with the lognormal shadowing path loss model. Based on experimental measurements taken in a site of dense heterogeneous nodes [21], we adopt the described signal propagation model in Section 8.3, where we set the communication model variables as shown in Table 8.1, and μ to be a random variable that follows a log-normal distribution function with mean zero and variance of $[\delta^2]$. The parameters of the utilized network components in this study are mainly driven from Refs. [35,36], and summarized in Table 8.2.

8.6.2 Experimental Setup & Baseline Approaches

We verify our proposed HCPF approach via a number of real experimental VRP instances taken from Ref. [34]. This step is essential to validate the correctness of the proposed genetic-based algorithm. To do this, we compare it with pure GA and LS algorithms. To use GA without LS, we set the mutation probability to 0. Also in order to use LS without GA, we set the crossing-over probability to 0. We execute our demo 100 times for each experiment and take the averages of the results. Then, we extend our assessment section via simulation-based scenarios in which we vary the aforementioned parameters (in Tables 8.1 and 8.2) for more comparative analysis about the targeted path-planning problem. We assume a network of vehicles in a smart city, where DCs are a subset of the moving public transportation vehicles. We assume up to 100 total APs with one BS. For some experiments, we keep data traffic or DC capacity fixed, and for other system components' values as depicted in Table 8.2 The assumed networks in this study are random in terms of their nodes' positions and densities. In order to select the most appropriate DC trajectory in these randomly generated networks, we apply our HCPF scheme. The output of the HGPF scheme is compared to output of another set of baseline approaches in the literature. These baseline approaches address the same problem tackled in this research; however, they use different path-planning strategies; the EF, the HS, and the LoRaWAN. In addition, we adopt three different versions of the proposed HCPF approach. These versions vary in the utilized stopping criteria of the proposed HCPF algorithm, which is the maximum number of generations. This value is set to be 200 in HCPF_1, 150 in HCPF_2, and 100 in HCPF_3.

8.6.3 Performance Metrics and Parameters

In this research, we assess our proposed HCPF in terms of three main metrics:

1. **Throughput:** It is the total arrived data to the main BS via DCs per the time unit (second). It is measured in Mbps and it reflects QoS in the system.
2. **DC count:** It is the number of DCs to be used for visiting all the devices in the network.
3. **Travelling distance:** It is the Euclidean distance between consecutive APs to be visited by a DC while travelling on the route and it is measured in (km). This parameter has been chosen to reflect the consumed time and energy in every solution.

In the following, we also list the varying simulation parameters:

1. **AP count:** It is the number of APs in the network, and represents the network size. This metric has a direct effect on DC count and total distance and it reflects the complexity and scalability of the exploited path-planning approach.
2. **DC count:** It is the number of DCs to be used for visiting all the devices in the network.
3. **AP Load:** It is the amount of data to be delivered from an AP to the BS via a DC. This parameter is measured in Mbytes it reflects the generated data traffic by these smart nodes in a smart-city setup.
4. **DC capacity:** It is the maximum storage capacity for the utilized DCs measured in Mbytes. This parameter represents the limited hardware on the

selected DC, where the sum of the collected data loads by a DC cannot be more than the capacity of that DC.

5. **Area size:** This parameter represents the targeted region area measured in km^2. It has been chosen to assure the proposed solution efficiency in terms of scalability and different scale deployments.
6. **Solution cost:** It is the fitness value computed by the fitness function in Eq. (8.5) and it represents the overall cost of the achieved solution for the proposed DC allocation problem.

8.6.4 Experimental Results

In this section, we examine the correctness of HCPF in comparison to the aforementioned GA and LS algorithms. Genetic-based parameters of these three algorithms are given in Table 8.3.

Experimental results are performed with the related location set, APs demands and the DC capacity, for each symmetrical capacitated VRP instance given in literature [34]. These instances are summarized in Table 8.4. N is the number of APs, M is the number of DCs, and Q is the DC capacity. As shown in Table 8.4, the minimum number of DC with the minimum total distance is obtained with complete convergence in each run for all the instances. The average error and standard deviation values are zero, which indicates stability in the proposed solutions. The results obtained are consistent as well with Refs. [37,38].

As depicted in Figures 8.3–8.5, while increasing the data load, the total travelled distances are increasing. In Figure 8.3, we are applying only global search (GA). Convergence is not sufficient in GA. In Figure 8.4, we are applying only LS.

TABLE 8.3 The Assumed Parameters' Values for Genetic Algorithms

	Approach Values		
Parameter	*LS*	*GA*	*HCPF*
Population size	100		
Maximum number of generations	100		
Selection method	Tournament selection		
Probability of mutation	5%	0%	5%
Probability of crossing-over	0%	80%	80%
Crossing-over operator	Permutation		
Mutation operator	2-opt LS		
Elitism	2		

TABLE 8.4 VRP Instances Used for Testing the HCPF Approach

Problem Instance	*Best-Known Value [34]*	*N*	*M*	*Q*
att-n48-k4	40002	47	4	15
A-n34-k5	778	33	5	100
An80k10	1763	79	10	100
B-n39-k5	549	38	5	100
E-n22-k4	375	21	4	6000
E-n23-k3	569	22	3	4500
E-n30-k3	534	29	3	4500
E-n51-k5	521	50	5	160
En101k8	815	100	7	200
F-n45-k4	724	44	4	2010

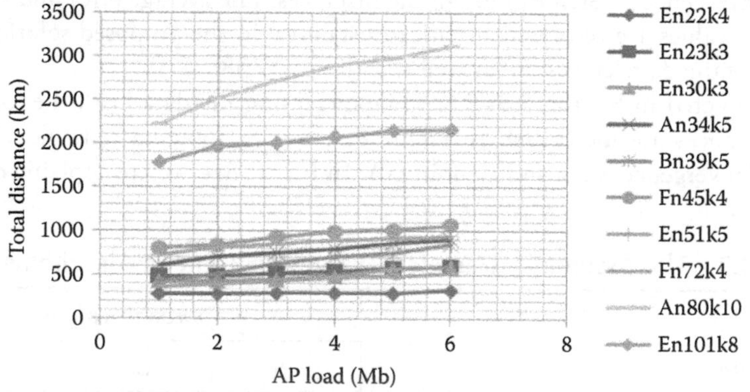

FIGURE 8.3 Data traffic vs. the total travelled distance (GA).

In this figure, convergence can be achieved immediately with LS and the values are better in comparison to Figure 8.3. But using pure LS is not a good way to solve path-planning problems, because LS searches the neighborhoods of the current solutions in the same pattern without any randomness in contrast to GA which leads mostly to local minima/maxima.

In Figure 8.5, we are applying the proposed meta-heuristic approach (HCPF). Saturation is obvious for our hybrid approach after a number of iterations when genetic algorithm is improved by the LS algorithm. For data load test, we should state that the best approach is the hybrid one and then LS and later GA. To observe

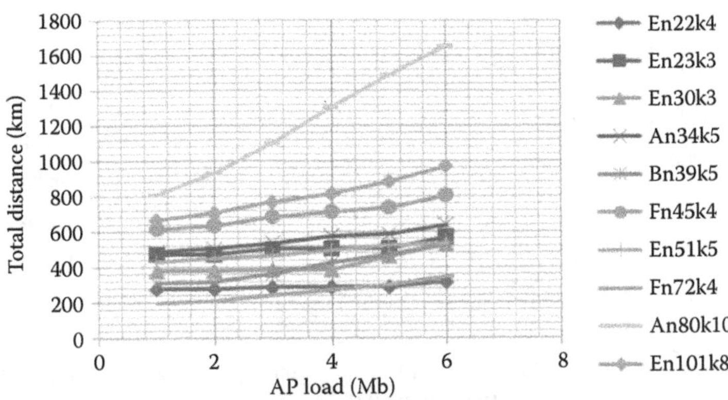

FIGURE 8.4 Data traffic vs. the total travelled distance (LS).

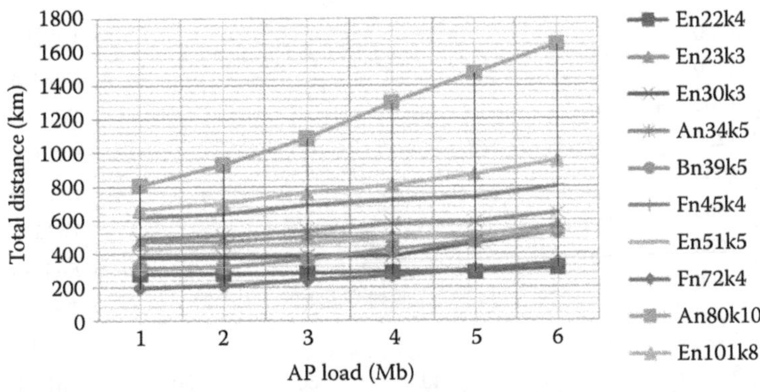

FIGURE 8.5 Data traffic vs. the total travelled distance (HCPF).

the change in DC capacity, the data load for all APs are set to 1 and DC capacity is chosen to be 10, 20, 30, 40, 50 and 60 respectively for all instances.

Figures 8.6–8.8 show that while increasing the DC capacity, the total travelled distances decreases. In Figure 8.6, we are applying only global search (GA). As can be seen in this figure, convergence is not sufficient for the global search without the LS algorithm. Figure 8.7 is identical with Figure 8.6 but we are applying only LS. As shown in Figure 8.7, convergence can be achieved immediately with the LS approach and results are better in comparison to Figure 8.6. However, using pure LS algorithm without any improvement is not a good way to solve the proposed path-planning problem as we have shown in Figure 8.4. In Figure 8.8, everything is identical with Figures 8.6 and 8.7; but we are applying the proposed hybrid approach (HCPF).

In Figure 8.8, the saturation is obvious for our hybrid approach after a number of iterations when the genetic algorithm is improved by the LS algorithm. For the DC capacity, we should also mention that the best approach is the hybrid one and

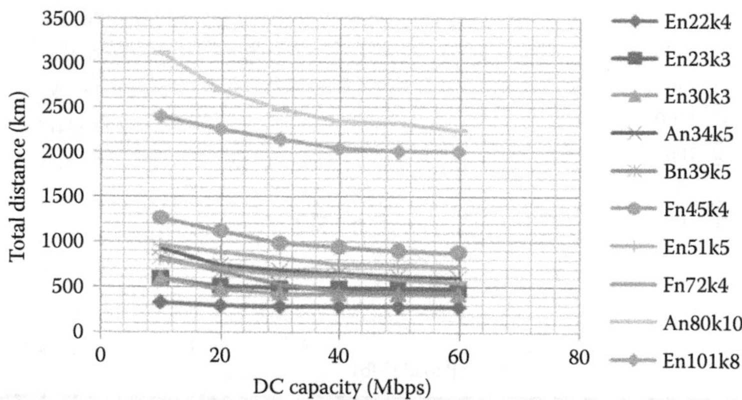

FIGURE 8.6 DC capacity vs. the total travelled distance (GA).

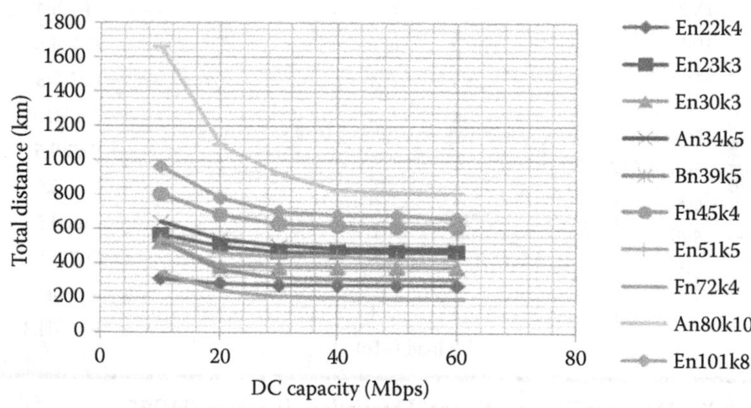

FIGURE 8.7 DC capacity vs. the total distance (LS).

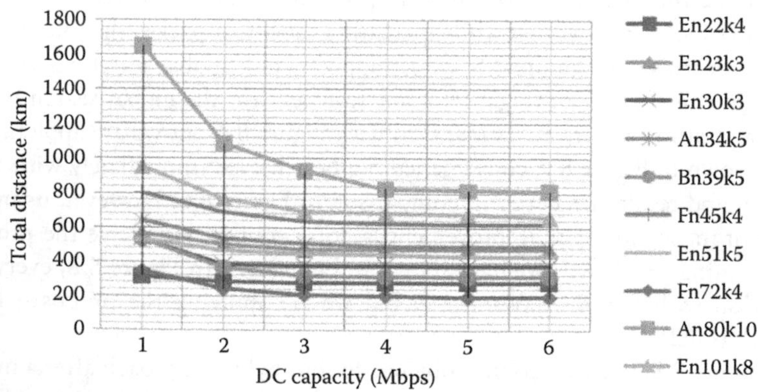

FIGURE 8.8 DC capacity vs. the total distance (HCPF).

then pure LS and later GA. To observe the change in total travelled distances under the proposed approach, we vary the data load values which have been used in the literature for all the AP devices. In Figure 8.9, we use the last instance (En101k8) in Table 8.4 which also includes DC capacity values. This instance has 100 APs, and device locations are random from the set of {10, 20, 30, 40, 50, 60, 70, 80, 90 and 100}. We also fixed the data load (traffic), the DC capacity and location information. As depicted in Figure 8.9, we compare the three different approaches with the same instance for random locations. It is noted that pure GA is not as good as pure LS, and also pure LS is not as good as HCPF.

In Figures 8.10 and 8.11, we examine the effect of the total cost instead of the total travelled distances with the same configurations used above for the three different approaches. To observe the effect of varying data load values, the total cost values are given in Figure 8.10. In this figure, we can observe the cost increase for

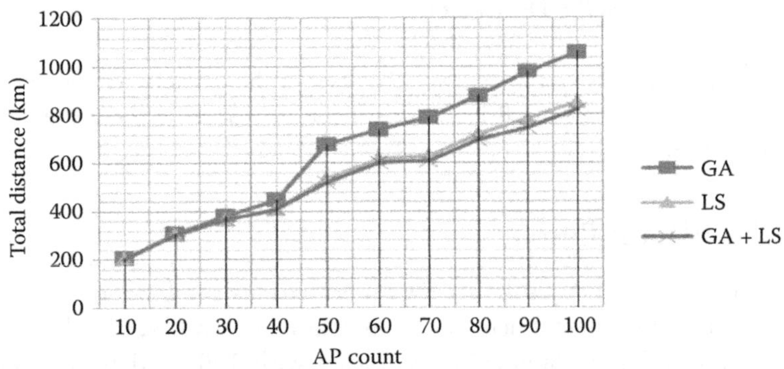

FIGURE 8.9 The three different approaches—total distance.

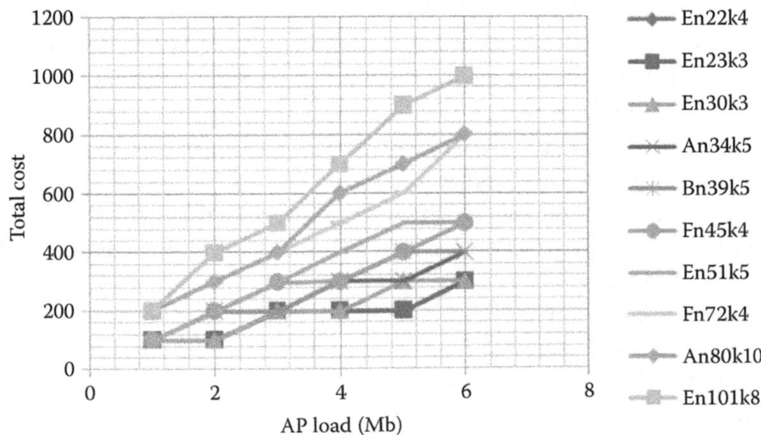

FIGURE 8.10 Cost vs. the AP data load.

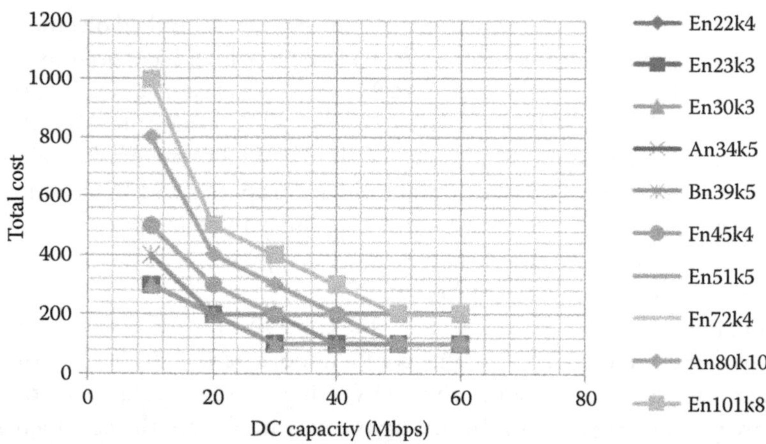

FIGURE 8.11 Cost vs. the DC capacity.

all instances as the data load increases. To observe the capacity effect, the overall cost values are given in Figure 8.11. In this figure, we observe the decrease in cost for each instance.

8.6.5 Simulation Results

In this section, simulation results are shown for various combinations of data load and DC capacity for different heuristic approaches; LoRaWAN, EF, HS, HCPF_1, HCPF_2, and HCPF_3. Simulation is started with 100 random locations and repeated 10 times with the same location set. Averaged results are plotted in the following figures.

In Figure 8.12, it is observed that total travelled distance is decreasing for all methods (i.e., the EF, HS, LoRaWAN, and HCPF), while increasing the DC counts. However, the HCPF approach outperforms LoRaWAN, EF, and HS by at least 50% of their best performance. HCPF_1 approach achieves the least total travelled distance with respect to all other baseline approaches. It is worth pointing out here that LoRaWAN outperforms typical meta-heuristic approaches when they are applied in a non-event-driven fashion such as EF and HS. Unlike these meta-heuristic approaches, the HCPF is an event-driven system that takes into consideration the different network characteristics as we aforementioned. Thus, it outperforms the LoRaWAN technique.

Moreover, HCPF_1, HCPF_2, and HCPF_3 demonstrate the same performance when the total count of DCs is greater than or equal to 150. Similarly, HCPF_1 and HCPF_2 approaches demonstrate the same performance when the DC count is greater than or equal to 100. Meanwhile, when the DC count is greater than or equal to 170, all methods cannot improve anymore in terms of the total travelled distance.

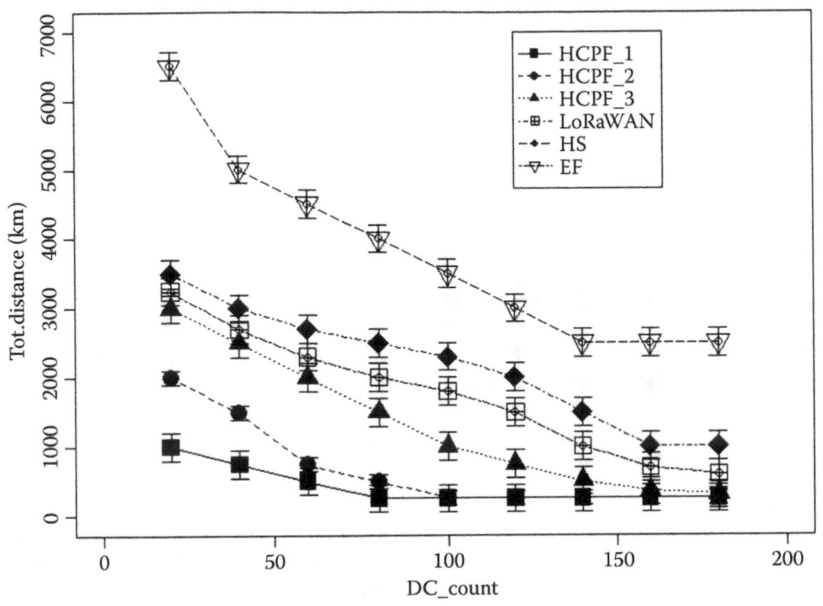

FIGURE 8.12 Travelled distance vs. the DC count.

In Figure 8.13, as AP Count is increasing we notice that all approaches, except the EF based approach, are converging to the same total travelled distance (~1800 km). It is worth pointing out here that the total travelled distance will reach an upper limit that cannot be exceeded as long as the total DC count and traffic load is fixed. Unlike LoRaWAN, HS, and EF approaches, HCPF_1, HCPF_2, and HCPF_3 are monotonically increasing in terms of the total travelled distance as the DC count is increasing. This is because of the systematic steps has been followed in the proposed HCPF approach toward optimizing the search performance. We remark also that the HCPF_1 has still the lowest total travelled distance and it is the fastest approach in terms of convergence. Meanwhile, HCPF_2 and HCPF_3 approaches demonstrate the same performance when the AP count is greater than or equal to 30. While the AP count is greater than or equal to 30 all methods cannot improve anymore in terms of the total travelled distance as we remarked before.

In Figure 8.14 we study the AP traffic load effect on the total travelled distances by the occupied DCs. As depicted by Figure 8.14, the total travelled distance is increasing for all approaches, except the HCPF_1, while increasing the AP load from 10 to 80 MB. This indicates a great advantage for the proposed HCPF approach while using a bigger max_generation value, where no more travelled distance is required for a few more extra Mbytes at the APs. It is obvious as well how the HS and EF approaches are monotonically increasing with a larger slope than HCPF_2 and HCPF_3, which means that HS and EF are more sensitive to the AP loads to be delivered. This can be a significant drawback in the IoT and big data

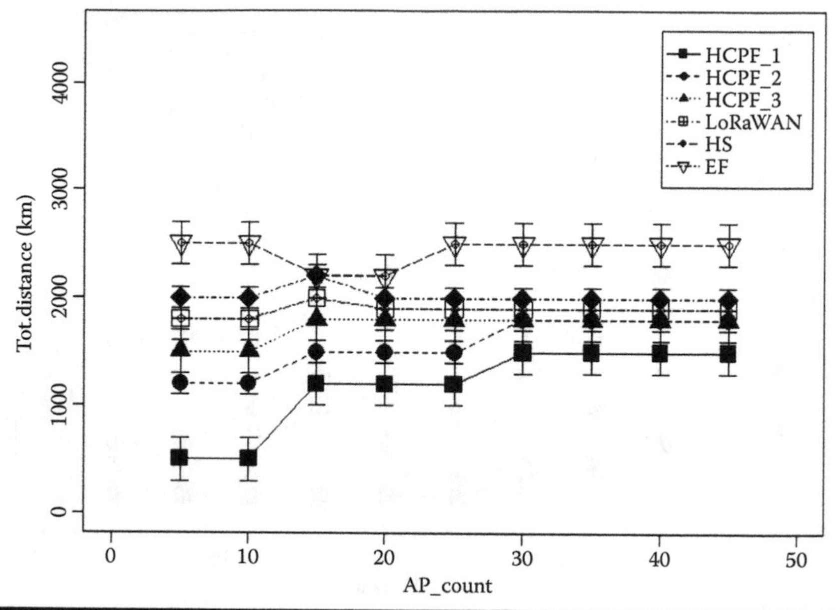

FIGURE 8.13 Travelled distance vs. the AP count.

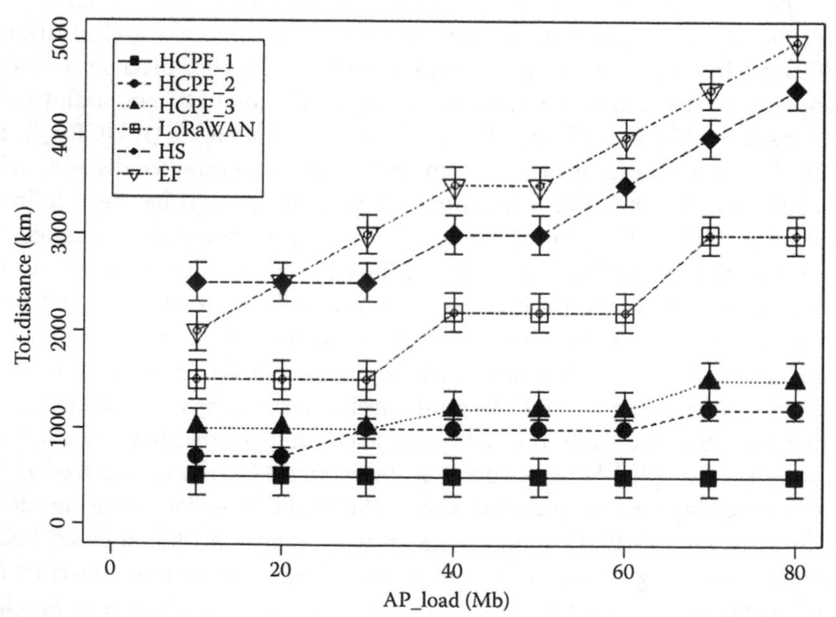

FIGURE 8.14 Travelled distance vs. the AP load.

era. Obviously, the HCPF_1 experiences the lowest total travelled distance while the AP load is increasing. Also, we notice that when the AP load is greater than or equal to 70 MB, the HCPF_2 and HCPF_3 cannot improve anymore in terms of the total travelled distance.

On the other hand, the total travelled distance is decreasing for all experimented approaches while the DC capacity is increasing as shown in Figure 8.15. In this figure, HCPF_1, HCPF_2, and HCPF_3 demonstrate similar performance when the DC capacity is greater than or equal to 60. HCPF_1, HCPF_2, and HCPF_3 are very close to each other in terms of the total travelled distance as the DC capacity increases. Again, HCPF_1 outperforms the other approaches in terms of the total distance. On the other hand, EF and HS are the worst in terms of the total travelled distance as the DC capacity increases.

In Figure 8.16, we study the AP count effect on the total required DCs. Obviously, the DC count is increasing monotonically for all approaches while the AP count is increasing. Surprisingly, HS, LoRaWAN, and HCPF_3 approaches demonstrate the same performance when the AP count is less than or equal to 25. This can be returned for the small number of generations that has been used with the HCPF_3 approach. Meanwhile, HCPF_1 necessitates the lowest DC count while increasing the AP count. EF necessitates the most DC count while the AP count is increasing.

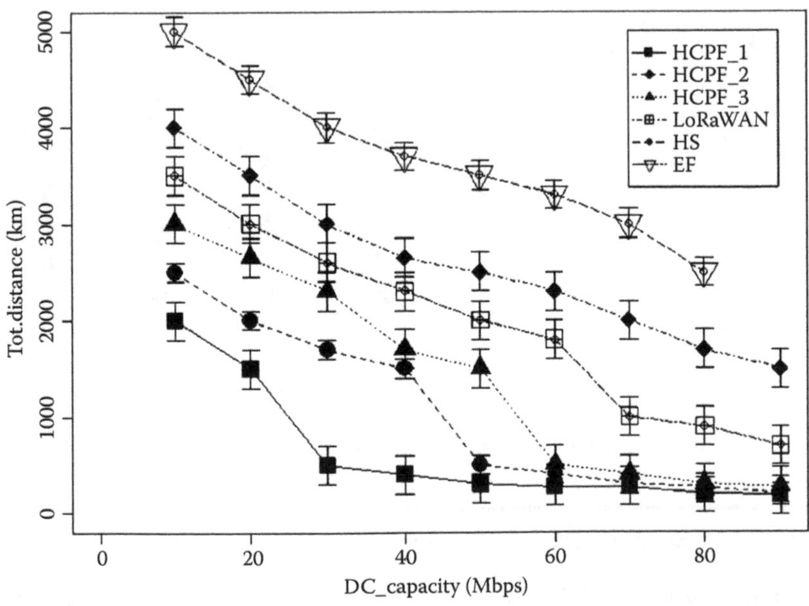

FIGURE 8.15 Travelled distance vs. the DC capacity.

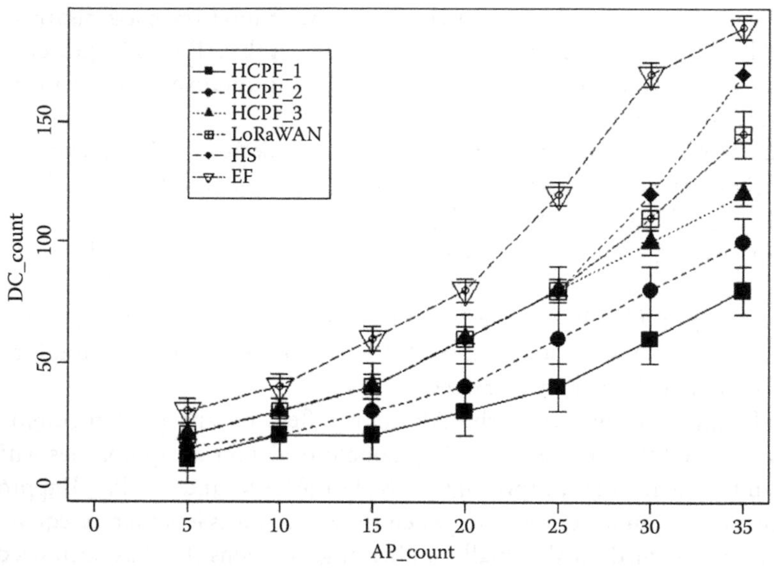

FIGURE 8.16 Required DC count vs. the AP count.

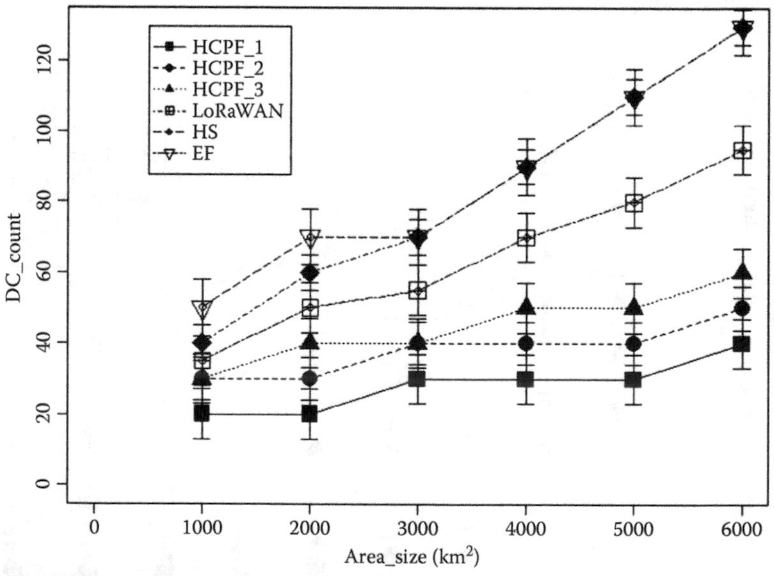

FIGURE 8.17 Required DC count vs. the targeted region size.

In Figure 8.17, the DC count is increasing monotonically for all methods while the deployment area size is increasing. EF and HS approaches demonstrate the same performance when the DC count is greater than or equal to 3000 km^2 and these approaches necessitate the largest DC count while the area size is increasing. Meanwhile, HCPF_1 necessitate the lowest DC count as the area size is increasing.

In Figure 8.18, we study the cost effect on the average system throughput measured in Mbps. As depicted by Figure 8.18, overall average system throughput is increasing for all methods while the cost parameter is increasing. However, they saturate after reaching a specific cost value (~1000) in terms of their overall achieved throughput. For example, HCPF_3 is saturating at cost equal to 1000 and its corresponding throughput is not enhancing anymore once it reaches 127 Mbps regardless of the cost amount. HCPF_2 and HCPF_3 demonstrate the same performance once the system cost reaches a specific value (~1000). On the other hand, EF and HS approaches demonstrate almost the same performance over all cost values. However, HCPF_1 outperforms all approaches under all examined cost values. EF and HS has the lowest throughput.

In this section, we have compared total travelled distances and overall cost for the different input samples as shown in the above figures for our experimental work. As a result, we remark that the proposed hybrid approach outperforms HS and EF approaches in terms of quality and time complexity. Additionally, it is worth mentioning that under the different parameters' values and configurations, the proposed hybrid method outperforms EF or HS. However, it should be noted also that parameters and configurations have a significant impact on the achieved solutions quality and the approach convergence as elaborated in Sections 8.6 3 and 8.6.4. Another noteworthy implication of the proposed HCPF framework is the ability to simply estimate the overall framework cost for a small to medium scale city. Where the average counts of sensing nodes per path can be controlled and

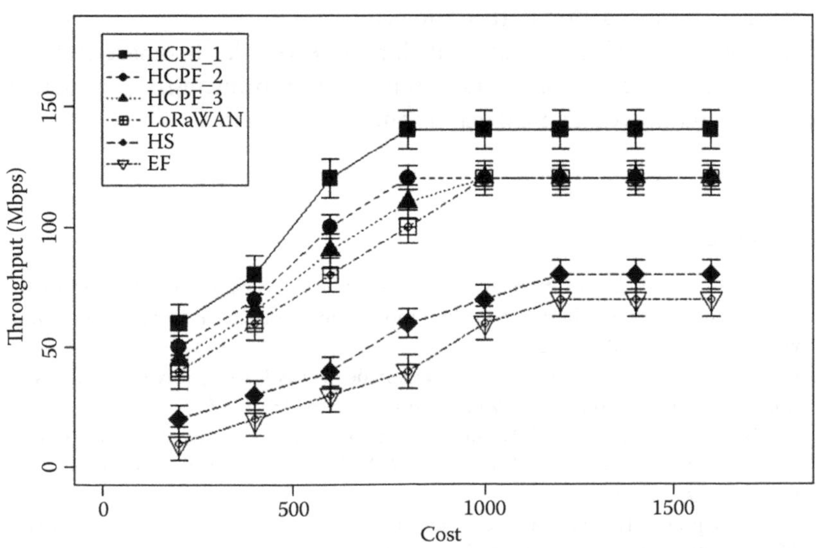

FIGURE 8.18 Throughput vs. the cost.

assessed through Eq. (8.5). Moreover, based on Eqs. (8.5) and (8.6), QoS can be adjusted and manipulated based on the available operator/service provider budget. This can be returned of course to the emphasized edge computing principle where a considerable amount of data traffic can be offloaded from the core of the already deployed telecommunication infrastructure to those tiny easy to deploy/access SNs.

8.7 Conclusion

We introduce in this work the HCPF—a hybrid IoT PS framework for smart cities. This framework is based on a multi-tier architecture that gratifies heterogeneous data sources (e.g., sensors) with mobile data collectors in urban areas, which can be isolated form the BS. According to this framework, APs at the top-tier of the proposed architecture receive sensor readings and initiate delivery requests. The data delivery approach implements a genetic-based algorithm, which realizes distance- and cost-sensitivity in competitive time complexity. At the top tier, an IoT-specific cost model employs a fitness function that maximizes the network operator gain according to the limited DCs count, storage capacity, and total travelled distances. We provide results from simulations and experimental instances in practice showing the efficiency of our framework when compared to two prominent heuristic approaches; EF and HS. Our simulation results show that the HCPF framework exhibits superior performance for different network sizes, storage capacities, DCs counts, and end-to-end travelled distances. Taking into account what was studied in this work, it is recommended that utility companies that have not invested yet in a smart metering network based on genetic-based solutions should consider HCPF approach as a possible option in their projects.

Future work would investigate utilizing same vehicles in smart city settings with non-deterministic mobility trajectories. It is also interesting to look at the same application where APs are dynamic and not only static.

References

1. G. Singh, and F. Al-Turjman, A data delivery framework for cognitive information-centric sensor networks in smart outdoor monitoring, *Elsevier Computer Communications*, vol. 74, no. 1, pp. 38–51, 2016.
2. F. Al-Turjman, Information-centric sensor networks for cognitive IoT: an overview, *Annals of Telecommunications*, vol. 72, no. 1, pp. 3–18, 2017.
3. J. Sahoo, S. Cherkaoui, and A. Hafid, A novel vehicular sensing framework for smart cities, *2014 IEEE 39th Conference on LCN*, 8–11 Sep. 2014, Edmonton, Canada, pp. 490–493, 2014.
4. F. Al-Turjman, Impact of user's habits on smartphones' sensors: an overview, *HONET-ICT International IEEE Symposium*, Kyrenia, Cyprus, pp. 70–74, Oct. 2016.

5. C. Shih, Y. Yang, M. Horng, T. Pan, and J. Pan, An effective approach to genetic path planning for autonomous underwater glider in a variable ocean, *Proceedings of International Forum on Systems and Mechatronics*, Tainan, pp. 1–6, 2014.
6. M. Soulignac, Feasible and optimal path planning in strong current fields, *IEEE Transactions on Robotics*, 2011, doi:10.1109/tro.2010.2085790.
7. H. G. S. Filho, J. Pissolato Filho, and V. Moreli, The adequacy of LoRaWAN on smart grids: a comparison with RF mesh technology, *Proceedings of the IEEE Smart Cities Conference (ISC2)*, Trento, 2016.
8. D. Turgut and L. Bölöni, Heuristic approaches for transmission scheduling in sensor networks with multiple mobile sinks, *Computer Journal*, vol. 54, no. 3, pp. 332–344, Mar. 2011.
9. F. Al-Turjman and H. Hassanein, Towards augmented connectivity with delay constraints in WSN federation, *International Journal of Ad Hoc and Ubiquitous Computing*, vol. 11, no. 2, pp. 97–108, 2012.
10. D. Said, S. Cherkaoui, and L. Khoukhi, Scheduling protocol with load management for EV charging, *GLOBECOM 2014*, 8–12 Dec. 2014, Austin, pp. 362–367, 2014.
11. M. Biglarbegian and F. Al-Turjman, Path planning for data collectors in precision agriculture WSNs, *Proceedings of the International Wireless Communications and Mobile Computing Conference (IWCMC)*, 4–8 Aug. 2014, Nicosia, Cyprus, pp. 483–487, 2014.
12. J. Tiu and S. X. Yang, Genetic algorithm based path planning for mobile robots, *IEEE Conference on Robotics and Automation (ICRA)*, Taipei, Sep. 2003.
13. A. T. Campbell, S. B. Eisenman, N. Lane, E. Miluzzo, and R. Peterson, People-centric urban sensing, *Proceedings of the 2nd Annual International Workshop on Wireless Internet, (WICON '06)*, Boston, pp. 18–31, 2006.
14. H. Patni, C. Henson, and A. Sheth, Linked sensor data, *International Symposium on Collaborative Technologies and Systems (CTS)*, Illinois, pp. 362–370, 2010.
15. R. Golchay, F.L. Mouel, S. Fŕenot, and J. Ponge, Towards bridging IoT and cloud services: proposing smartphones as mobile and autonomic service gateways, *CoRR abs/1107.4786*, 2011.
16. F. Hao, T.V. Lakshman, S. Mukherjee, and H. Song, Enhancing dynamic cloud-based services using network virtualization, *SIGCOMM Computer Communication Review*, vol. 40, no. 1, pp. 67–74, 2010.
17. M. Naphade, G. Banavar, C. Harrison, J. Paraszczak, and R. Morris, Smarter cities and their innovation challenges, *IEEE Computer*, vol. 44, no. 6, pp. 32–39, Jun. 2011.
18. J. Lee, S. Baik, and C. Lee, Building an integrated service management platform for ubiquitous cities, *IEEE Computer*, vol. 44, no. 6, pp. 56–63, Jun. 2011.
19. K. Su, J. Li, and H. Fu, Smart city and the applications, *International Conference on Electronics, Communications and Control (ICECC)*, Ningbo (China), pp. 1028–1031, 2011.
20. F. Al-Turjman, Cognition in information-centric sensor networks for IoT applications: an overview, *Annals of Telecommunications*, vol. 72, no. 3, pp. 209–219, 2017.
21. G. Solmaz, M. I. Akbas and D. Turgut, A mobility model of theme park visitors, *IEEE Transactions on Mobile Computing (TMC)*, vol. 14, no. 12, pp. 2406–2418, 2015.
22. F. Al-Turjman, H. Hassanein, and S. Oteafy, Towards augmenting federated wireless sensor networks, *Proceedings of the IEEE International Conference on Ambient Systems, Networks and Technologies (ANT)*, Niagara Falls, pp. 224–231, 2011.

23. L. Bloni, D. Turgut and S. Basagni and C. Petrioli, Scheduling data transmissions of underwater sensor nodes for maximizing value of information, *Proceedings of IEEE GLOBECOM'13*, Atlanta, GA, pp. 460–465, Dec. 2013.
24. F. Al-Turjman, Price-based data delivery framework for dynamic and pervasive IoT, *Elsevier Pervasive and Mobile Computing Journal*, vol. 42, pp. 299–316, 2017.
25. D. Johnson and D. Maltz, Dynamic source routing in ad-hoc wireless networks, *Mobile Computing*, vol. 353, ch. 5, pp. 153–181, Aug. 1996.
26. S. Al-Harbi, F. Noor, and F. Al-Turjman, March DSS: a new diagnostic March test for all memory simple static faults, *IEEE Transactions on CAD of Integrated Circuits and Systems*, vol. 26, no. 9, pp. 1713–1720, Sep. 2007.
27. W. Parvez and S. Dhar, Path planning of robot in static environment using genetic algorithm (GA) technique, *International Journal of Advances in Engineering & Technology*, Jul. 2013.
28. O. Castillo and L. Trujillo, Multiple objective optimization genetic algorithms for path planning in autonomous mobile robots, *International Journal of Computers, Systems and Signals*, vol. 6, no. 1, pp. 269–279, 2005.
29. C. Papadimitrio and K. Steiglitz, On the complexity of local search for the travelling salesman problem, *SIAM Journal of Computing*, vol. 6, no. 1, pp. 337–346 Mar. 1977.
30. Y. Luo, D. Turgut, and L. Boloni, Modeling the strategic behavior of drivers for multi-lane highway driving, *Journal of Intelligent Transportation Systems*, vol. 19, no. 1, pp. 45–62, 2015.
31. D. Turgut and L. Bölöni, Heuristic approaches for transmission scheduling in sensor networks with multiple mobile sinks, *Computer Journal*, vol. 54, no. 3, pp. 332–344, Mar. 2011.
32. J. Vazifehdan, R. V. Prasad, M. Jacobsson, and I. Niemegeers, An analytical energy consumption model for packet transfer over wireless links, *IEEE Communications Letters*, vol. 16, no. 1, pp. 30–33, 2011.
33. E. Ever, F. Al-Turjman, H. Zahmatkesh, and M. Riza, Modelling green HetNets in presence of failures for dynamic large-scale applications: a case-study for fault tolerant femtocells in smartcities, *Elsevier Computer Networks Journal*, vol. 128, pp. 78–93, 2017.
34. Networking and Emerging Optimization, Capacitated VRP instances | vehicle routing problem, http://neo.lcc.uma.es/vrp/vrp-instances/capacitated-vrp-instances/. [Accessed on Feb. 14 2016].
35. J. Zhang, G. De la Roche, et al., *Femtocells: Technologies and Deployment*, Wiley Online Library, 2010.
36. V. Chandrasekhar, J. G. Andrews, T. Muharemovic, Z. Shen, and A. Gatherer, Power control in two-tier femtocell networks, *IEEE Transactions on Wireless Communications*, vol. 8, no. 1, pp. 4316–4328, 2009.
37. E. Uchoa, D. Pecin, A. Pessoa, M. Poggi, A. Subramanian, and T. Vidal, New benchmark instances for the capacitated vehicle routing problem, Technical Report – ArXiv 2014-10-4597, unpublished, 2014.
38. M. Stanojević, B. Stanojević, and M. Vujošević, Enhanced savings calculation and its applications for solving capacitated vehicle routing problem, *Applied Mathematics and Computation*, vol. 219, no. 20, pp. 10302–10312, 2013.

Chapter 9

Cognitive Caching in Fog-Based IoT*

Fadi Al-Turjman
Antalya Bilim University

Contents

* F. Al-Turjman, "Cognitive Caching for the Future Sensors in Fog Networking", *Elsevier Pervasive and Mobile Computing*, vol. 42, pp. 317–334, 2017.

9.1 Introduction

The growing demand for efficient distribution of content/data over the cloud has motivated the development of future Internet architectures based on Named Data Objects (NDOs), such as web pages, videos, documents, or other pieces of information. The approach of these architectures is commonly called Information-Centric Networking (ICN). In contrast, current networks are host-centric where communication is based on named hosts, for example, web servers, PCs, laptops, mobile handsets, and other devices. ICN serves as a content-based model, which focuses on client's demands disregarding of the data's address or the origin of distribution.

ICN is the next generation model for the Internet that can cope with the user's requests/inquiries regardless of their data-hosts' locations and/or nature. The current Internet model is suffering from the exchange of huge amounts of data while still relying on the very basic network resources and IP-based protocols. Meanwhile, ICNs promise to overcome major communication issues related to the massive amounts of distributed data in the Internet. ICNs adopt a content-centric architecture, which focuses more on the networked data itself rather than the metadata. This kind of network architectures is known usually by the Content-Oriented Networks (CONs) term [1]. Luckily, these CONs architectures match a lot with the emerging communication trend that aims at exchanging Big-data over tiny and energy-limited Wireless Sensor Networks (WSNs) in order to realize numerous attractive projects such as the smart planet and the Internet of Things [2,3]. Thus, a new platform is needed to meet these requirements. A new platform, called Fog Computing [4], or, simply, Fog, because the fog is a cloud close to the ground is proposed to address the aforementioned requirements. Fog is a Mobile Edge Computing (MEC) that puts services and resources of the cloud closer to users to be facilitated in the edge networks.

Unlike Cloud Computing, Fog Computing enables a new breed of light applications and services that can be run at particular edge networks, such as WSNs. In order to enable WSNs to support this trend in communication and function in a large-scale application platform, such as the *Fog Computing*, we proposed the cognitive framework in our previous work [5]. In Ref. [5], an information-centric scheme is proposed for the future WSNs using *cognitive* in-network devices that makes dynamic routing decisions based on specific *Knowledge-* and *Reasoning-* observations in WSNs. Analytic Hierarchy Process (AHP) is applied on Quality of Information (QoI) attributes in next generation WSNs such as reliability, delay, and network throughput observed over the communication links/paths [6,7]. This cognitive Information-Centric Sensor Network (ICSN) framework is able to significantly outperform the *non-cognitive* ICSN paradigms. However, this cognitive ICSN framework did not consider yet the in-network caching feature. Caching in multitude of nodes in ICNs has pivotal role in enhancing the network performance in terms of reliability and response time. In this chapter, we propose a novel caching framework in fog-based ICSNs, called CCFF. It identifies the most suitable data to be replaced in order to maintain prolonged data availability time periods while enhancing the network performance.

The cache replacement strategies in the literature [8,9], have been strategically designed so far for IP-based networks, which have significant variations against the targeted vision of the future Fog networks. Where Fog is emerging as one of the most promising content-oriented networks that necessitates massive changes in the core architecture of the system. Foremost thoughts must be specifically put into the cache replacement strategy, as it counts to dramatic influence on the network performance. To this end, a novel cognitive caching framework, which can serve several kinds of applications in the cloud with content-oriented requirements is needed.

Up to this point, we propose a CCFF framework for in-network caching that computes a qualitative value for each content in the fog. This value is used to determine which content shall be removed from the cache of an edge device, when it experiences a limited memory size. Our CCFF replacement approach is compared to existing major replacement categories, namely the Node Functionality-based Caching (CF), the Content-based Caching (CC), and the Location-based Caching (LC). The comparison counts on comprehensive performance metrics in Fog computing under a variety of parameters including cache size, data popularity, in-network cache ratio, and network connectivity degree. To strengthen the proposed CCFF framework, a formal trust analysis has been applied as well using Burrows–Abadi–Needham (BAN) logic [10]. This theoretical analysis emphasizes the effectiveness of CCFF in Fog paradigms, where the edge devices can be always under security threats and prone to untrusted data.

The remainder of this chapter is organized as follows. Section 9.2 provides a literature review on the caching techniques and categories in ICSNs. Section 9.3, introduces our ICSN-specific system models. In Section 9.4, we elaborate in details about the proposed CCFF framework. Section 9.5 presents detailed theoretical analysis in edge computing from the perspective of data fidelity. Moreover, a use-case and simulation results are proposed in Section 9.6 for more elaboration on suitability of CCFF framework in Fog computing paradigm. Section 9.7 concludes our work.

9.2 Related Work

At the core of the Fog paradigm, which comes with plenty of heterogeneous edge devices, data has to be located close to the device requesting it. This is the mission of the caching scheme usually running over the cloud nodes. Since the environment, the type and amount of data received, the energy of the devices and many other factors varies from one device to another over the cloud, it is hard to come up with a single caching strategy that fits everyone's needs. Therefore, several caching strategies have been proposed under various assumptions for different scenarios. However, efficient caching strategies most often demands two main properties: first, an updated copy of the requested data must be always close to the area of interest. Second, the data copy shall stay "alive" as long as it might still be used.

Caching is usually coupled with data delivery and naming architectures. Consider DONA [5] as an example, due to the coupling of naming tuples, it is possible to enable any in-network caching in the framework that hold a valid copy. There are several other architectures and attempts with different strategies on how to decide which entity shall provide the data copy and how it should be retrieved. A two-fold-metric method for NDO caching based on the time to live metric was recently proposed in Ref. [11]. This two-fold method assumes NDOs, which are located at the edge of the network or has a higher popularity are cached for longer time periods. Entities that hold replicas can collaborate in turning the age counter to manifest such factors. The inherent need for book-keeping is a major challenge in today's cashing systems; as a result, the massage overhead cannot scale to the internet and at the same time claim efficiency. Therefore, we are bound to analyze caching schemes under conditions like communication overhead per NDO, storage requirement, NDOs with different priorities, and granularity in assessing requests' frequencies, types, locations, etc. Consequently, we review the caching methods in cloud-related ICSNs and identify the technique that best suits the appropriate caching decision in Fog. Network caching in the literature can be categorized into the following: (i) *Functionality-based caching*, (ii) *Content-based caching*, and (iii) *Location-based caching*.

9.2.1 Node Functionality-Based Caching (FC)

To maximize the full potential of the CON, we must consider which content should be stored in the control level rather than guessing it at the data level. Consequently, the authors in Ref. [12] claimed that there are side-effects for handing over the caching decision to the data level, and they propose an approach to handle the data at the control level. However, their approach is not applicable in large-scale scenarios. The authors in Ref. [13] propose an information-centric caching algorithm known as LocalGreedy, where a cluster of caches is considered, with different leafs either connected directly or indirectly via a parent node. An inter-level cache cooperation is utilized in order to fetch the data from a specific node called the parent node and not from any other node. This approach employs a strong conceptual similarity check to provide more cost-effective solutions especially when the idea of "access cost" is adopted. The access cost is represented either by the accrued cost of latency when fetching data from remote caches, or by the consumed bandwidth when retrieving content from peer nodes. However, this method necessitates the knowledge of the in-network nodes' capabilities and this contradicts with the Fog vision. The authors in Ref. [14] have studied the trade-off between caching the content in distributed IP-based networks and the new emerging content-oriented architectures in Fog systems. The sturdy is applied to a mixture of real traffic from sources like the web, file sharing, and multimedia streaming, and it concludes that caching real-time contents in routers can increase the cache hits. Nevertheless, this caching system is not sufficient for Fog systems, where other content types would likely be more efficiently handled at more capable devices such as smartphones/ laptops at the edge of the network.

9.2.2 Location-Based Caching (LC)

In order to achieve better network performance, the authors in Ref. [15] recommends caching less, and argues against caching everywhere in ICSNs. This caching policy states that data should only be stored in a node with the highest probability of getting a cache-hit. Cache Aware Target identification (CATT) is a proposed caching policy in Ref. [16] where a node is selected for caching as long as the node has the highest connectivity-degree based on its geographical location. However, this makes the node a geographical bottleneck in the network. Another location-based caching has been proposed in Ref. [17], where a probabilistic caching method combined with geo-factors is utilized. The proposed method is called pCASTING and a probabilistic Least Recently Used (LRU) approach, as well as data freshness, device energy, and storage capability were considered. Authors show how these metrics are employed in a utility function to decide the caching option. They compare their results with no caching, with 50% probability to cache, and with the everything and everywhere caching strategy. As a result, they conclude that there is at least 10% difference in favor of pCASTING for the cache hit ratio, and a higher value of successfully received packets as well as lower delay.

Moreover, a topology-based replacement approach is proposed in Ref. [18] where replicas are placed at the intermediate routers of the Internet. The authors have found out that the router fan-out based replacements need to be carefully designed to maximize the cache efficiency in content-oriented systems, where self-organized caches are required. Self-organized caches are caches, which make consistent decision about caching. The authors in Ref. [19] have considered this strategy and concluded that it is has better average delay compared with traditional methods (Cache at intermediate nodes), using smaller per-node caches.

Meanwhile, the authors in Ref. [20] have proposed another caching method, namely the selective neighbor approach, which selects the most appropriate neighboring proxies for minimum mobility overhead in terms of average delay and caching cost. This approach is based on proactively caching data requests and the corresponding metadata to a subject of proxies that are one hop away from the proxy. The authors in Ref. [21] suggests a probabilistic approach for ICNs. They claim that the probability of a file being cached should be increased as it travels from source to destination by considering the following parameters: (i) The distance between source and current node, (ii) Distance between destination and current node, (iii) Time-to-Live for the routed data content, and (iv) the Time-Since-Birth. The authors also suggest redundancy in caching on a single path between source and distention. However, this degrades the ICN performance dramatically while experiencing limited caching spaces. Moreover, in Ref. [21], authors assume that all the network nodes has the capability of caching, which is not the case in practice with Fog systems. The proposed approach is weak as well due to considering static data requests' frequency from a subnet where that data can exist. Nevertheless, we believe caching should be based on dynamic frequencies and location-independent.

9.2.3 Content-Based Caching (CC)

CC is another category for caching in content-oriented fog networks, in which the data replacement decision is taken based on the content of the exchanged data. For example, an automatic cache management system that dynamically assigns data to distributed caches over the network has been proposed in Ref. [22]. In this reference, distributed file managers make the data replacement decision based on the observed request patterns. This approach assumes that every cache manager has access to all the caches and data request patterns. Consequently, this approach experience reduction in access time at the cost of additional massage exchange and computational overhead, which can dramatically degrade the fog performance.

Another approach, proposed by Hail et al. (2015), suggests that LRU would most likely be the best candidate strategy for caching to be used in a heterogeneous information-centric network such as the cloud [23]. The authors in Ref. [23] compare the pure LRU strategy with three other strategies: the pure randomness, the probabilistic LRU and the probabilistic caching method pCASTING. According to their results, the probabilistic LRU performs better than the other methods in terms of energy, and environmental metrics.

The authors in Ref. [24] aims at minimizing the data publisher load and maximizing the in-network cache hit by storing frequently requested data on selected routers over the network. The authors have presented two popularity-based caching algorithms from the basis of optimal replica replacement. However, this work may not be practical, as authors have only considered one gateway in the network, while plenty of them can be utilized in Fog paradigms. Where Availalable Void Exist (WAVE) is another caching approach in which the cache size is adjusted according to data popularity features [25]. In this approach, a master upstream node recommends the count of data chunks to be cached at the downstream slave nodes. This count increases exponentially with the number of requests in order to avoid any unnecessary overhead. WAVE distributes data over the network edge while considering popularity to distance relationships. However, the different volumes of data chunks have not been deliberated in this work.

Authors in Ref. [26] propose an age-based caching approach, which aims at reducing data publisher load and the in-network delays. This approach provides the techniques where ages of the contents/data are dynamically updated. It distributes popular content to the edge of the network while eliminating unnecessary replicas at the intermediate routing nodes. However, this approach fails to handle frequently changing contents, and hence, in-network devices which are far away from the data publisher may experience extended delays.

Accordingly, the design of the cache replacement policy must be a dynamic one based on the user request pattern and the implemented application. In this chapter, even though a content-oriented caching policy can resolve several challenges in the Fog paradigm, typical information-centric strategies cannot be applied without extra manipulations. That is simply because of the energy-constrained in-network devices, the uncertainty of the wireless medium, and the disparate need for user requirements-awareness in Fog-specific ICNs. Power supply, storage capacity, and heterogeneity in terms of sensors and other utilized enabling technologies at the edge devices are further challenges to be considered. Additionally, there are some factors that the replacement policy should be taken into consideration. Unlike other related woks, this chapter proposes a dynamic caching framework, called CCFF, based on knowledge- and reasoning-elements in the Fog. We provide a vast utility function that sets the value of each data item based on their application specifics like required trust, delay, and data age. This makes our proposed CCFF framework able to cope with Fog networks as a new trend in communication.

9.3 System Models

In this section, we list our content-oriented system models and abridge our Fog-based ICSN assumptions accordingly.

9.3.1 ICSN Network Model

The main components in the assumed ICSN network in this research are the sensor/edge nodes (SNs), which are stationary/mobile nodes that send/respond to data requests via relay nodes (RNs). These RNs forward the data requests through the closest Local Cognitive Node (LCN) to the cloud. LCNs/gateways are sophisticated processing devices, which are connected to the cloud via the Internet or any other backhaul network (e.g., WiFi, LiFi, 4G, etc.). In our proposed ICSN architecture, gateways play an intermediary role by replying to users with the requested data in addition to providing values for the processed/requested data based on trust factor and other application-specific requirements. In fact, these LCNs/gateways are cognitive nodes, which mean that they have knowledge, reasoning, and can learn about the status of the network and user requirements regularly. This helps them to better handle user requests and requests' responses. They interact with SNs, RNs, and the network Sink. The Sink is where all the collected data is delivered. It is also fitted with cognitive elements to help in managing the data and improve the overall performance of the system, and thus, we call it a Global Cognitive Node (GCN). Accordingly, our framework is based on the following two prominent scenarios. First, a network with a stationary edge device located at the edge of the cloud. Here, all data packets are destined to it via multi-hop transmissions. Second, a network with a mobile edge device that moves along the boundary of the fog (i.e., the perimeter) of the cloud. These mobile edge devices are represented in a Fog paradigm by people handheld devices, cars/trains equipped with sensors, etc.

9.3.2 ICSN Traffic Model

There are three types of data traffics that an ICSN can handle in the proposed Fog framework; these types are as follows: on-demand, periodic, and emergency traffics. Each of them is associated with a different QoI requirement based on the served application. The QoI requirement for the cached data can be decided by the following attributes: (i) the network reliability (R), (ii) the end-to-end delay/latency (L), (iii) the energy consumed (E), and (iv) the average node throughput (T). As shown in Table 9.1, we are note allocating absolute numbers on the attribute; rather we associate priorities with each attribute and let the priorities decide on the importance of the attribute absolute value. Values in Table 9.1 show the associated priority with the attribute. Number 1 indicates the "highest priority," while number 3 indicates "the lowest." The "x" symbol in Table 9.1 indicates "do not care." This means that there is no strict requirement on the value of the QoI, and its value does not affect the caching decision.

9.3.3 Delay Model

Different sensors and edge devices experience varying latencies to be exposed to the environment and to effectively send back their data. This in turn, also affects the node and network lifetime [27]. In order to prolong the lifetime of the ICSN node, we have to store data for extended periods of time when the experienced

TABLE 9.1 QoI Priority for Different Data Traffic Types

Request Type	*Latency (L)*	*Energy (E)*	*Reliability (R)*	*Throughput (T)*
Type I: On-demand	x	3	1	2
Type II: Periodic	1	2	4	3
Type III: Emergency	1	1	x	2

delay in gathering data from the surrounding environment is increasing; this is called the sensing delay. Additionally, propagation delay would be added if every time that data is requested, it has to be moved from SNs/edge nodes to LCNs, especially if the edge nodes are located far away from the Sink/data publishers in the cloud. Hence, the delay components we have to consider are the sensing delay δ, and the propagation delay τ. Where δ is a function of the maximum delay among all the accessed edge devices to provide fresh/new data. The propagation delay τ a packet encounters is generally proportional to the number of hops between the packet's source and destination. The number of hops between these two points can be approximated to be a linear function of the Euclidean distance between them. Therefore, the propagation delay a data packet encounters can be expressed by:

$$\tau = \alpha L, \tag{9.1}$$

where α is a constant and L is the Euclidean distance between the packet's source and destination. With a stationary data requester located at the center of the edge of the cloud, the maximum propagation delay a packet may encounter occurs when the source is at the center of the cloud; this results in a delay of αR, where R is the radius of the user cloud. On the other hand, the maximum delay in a network with a mobile data requester is significantly worse. With a mobile data requester, the worst case occurs when a packet arrives to a relaying node which has just been left by the requester/user; such a packet needs to wait for the data requester to complete a full round along the perimeter of the cloud which depends on the speed of the data requester and on other factors. It is obvious that such a delay is much longer than that of a network with a stationary data requester/user. The total delay (Δ), which involves delivering fresh data to the user is a combination of the sensing and propagation delay, given by Eq. (9.2) below:

$$\Delta = \tau + \delta. \tag{9.2}$$

9.3.4 Age Model

The recommended age factor in CCFF makes use of the following two conditions to decide which content shall be removed from the edge device cache memory. The first is based on the validity duration of the periodic request (Type I traffic type in Table 9.1), and second, when the edge device cache is full. We make use of this duration interval, because freshly retrieved data has to be provided at the start of each periodic request cycle [28]. Thus, when the cache is full at the end of each periodic request cycle, old data can be replaced with more recent ones. Thus, the age attribute-value pair is represented by its Time-To-Live (TTL), which is based on the periodicity of the request for each application type. Since we are not considering the use of historic data, our model implies that cached contents may be refreshed after every periodic time interval, as long as the data is being transmitted to the edge device at the end of each cycle (Figure 9.1).

$$TTL_c \propto T_{periodic} \tag{9.3}$$

Equation (9.3) represents that the TTL_c of the cached data (*c*) represented as attribute-value pair, is directly dependent on the periodicity of a request in Type I traffic flow. In case the application requires that the periodic data is stored for a prolonged duration of time, for example 24 h, before making a single transmission to the sink, then the cache retention period becomes a function of the duty cycle's periodicity.

9.3.5 Popularity of On-Demand Requests

Traffic flow generated in response to on-demand requests have been classified as Type 2 traffic. More number of users may be interested in a particular type of sensed data, or a specific sensed data may be requested more number of times by one or more users. Such sensor data is said to be popular, and can be retained for prolonged periods in the edge device/LCN's cache. Thus, the popularity of the cached data attribute-value pair can be represented by Eq. (9.4) below:

$$Popularity_c \propto Req_c / Req_{total}, \tag{9.4}$$

where Req_c is the total number of requests for an attribute-value pair received at an LCN, and Req_{total} is the total number of requests received by that LCN, within

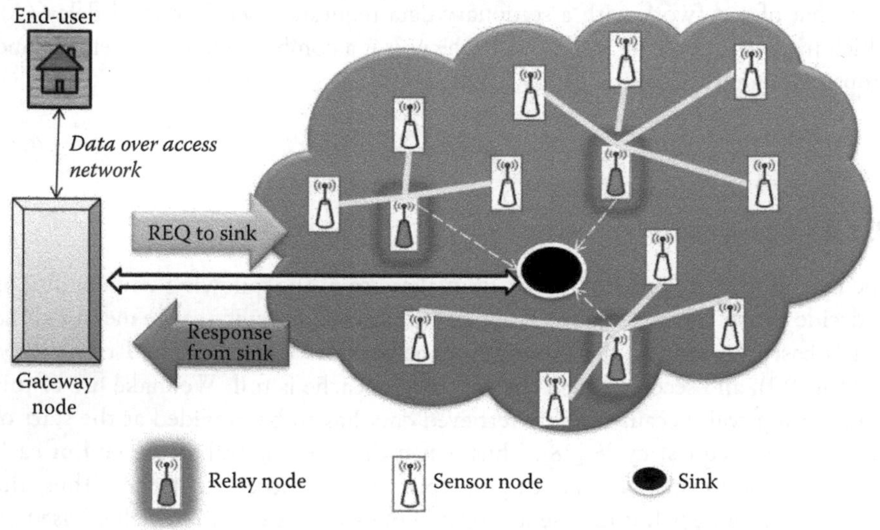

FIGURE 9.1 Network model with sensor/edge nodes and cognitive nodes.

a particular periodicity cycle. In addition, when the in-network devices' batteries start to be depleted, LCNs should store the data for prolonged time periods to maintain their availability. When the edge device/LCN storing such data starts to die out, neighboring LCNs can be good alternatives to store the data and/or provide extra storage for guaranteed availability.

9.3.6 Trust Model

Since we are relying heavily on distributed resource, which can be owned by different parties in the Fog paradigm, it is necessary to verify the fidelity of the retrieved data in order to avoid any unsecure/improper access. Thus, a Trust factor (T) is introduced in this model. This factor is calculated at the gateway/LCN per user to represent a GW_j fulfillment measure. A higher T_{GW_j} indicates that previous data exchanges between $user_j$ and GW_j have been fulfilled according to attributes promised by GW_j. Hence, we mark some data to be trusted while other data may not. The calculation of T_{GW_j} could follow a function similar to the fuzzy reputation formula presented in Ref. [26]. In this chapter, we assume an arbitrary value between 0 and 1 to express the trust parameter according to the function

$$T = (T_{GW_j})^{\gamma}, \tag{9.5}$$

where $\gamma > 0$ is a weight variable to control the slope of the function by giving more emphasis to the trust parameter.

9.3.7 Communication Model

The highest achievable transmission rate R is directly proportional to the communication channel capacity [29]. For a single hop, this can be simply expressed using Shannon's formula by

$$R = \log_2\left(1 + \frac{E_s}{N_0}\right), \tag{9.6}$$

where E_s is the average signal power in watts and N_0 is the noise power ratio. Assuming N hops instead of a single one can have the following consequences. The transmitted energy per symbol remains same, but the received energy at each hop is N^{α} times higher due to reduced attenuation over shorter distances. Additionally, each hop must accommodate the transmission of the same number of information bits in $1/N$ of the channel usage in the single hop mode, and thus increasing the required per-hop spectral efficiency. Accordingly, the transmission rate for the multihop case can be derived by the following:

$$R = \frac{1}{N}\log_2\left(1 + \frac{E_s}{N_0}N^{\alpha}\right). \tag{9.7}$$

9.4 CCFF Framework

In this section, we detail and discuss the prioritization process of data replacement over ICSN-based systems using the Cognitive Caching approach for the Future Fog (CCFF). This approach aims at making the most frequently requested and valuable data the edge of the cloud. It employs elements of cognition such as learning and reasoning techniques in achieving in extending the data hit ratios while maintaining the least delay to find and locate data. To this end, we classify the existing data/content types in the fog into the following; (i) delay sensitive data, (ii) demand-based data, and (iii) age-based data. The usability of delay sensitive data is usually associated with the delay time a consumer needs to wait before being able to locate the requested data. For example, collected data in emergency situations such as disaster and fire detections must be delivered and cached quickly and in timely manner to be fruitful for the end-user. Meanwhile, in the demand-based category, the priority of data is based on its popularity in the fog. In other words, how frequent it can be requested per time unit from the edge nodes. This is usually the case with file sharing and web browsing systems where same content is mostly to be requested again if it has been associated with a high-demand record. Finally, in the age-based category, the exchanged data over the fog is more sensitive to its age. For instance, if we are interested in displaying the current temperature in a specific geographical location over the earth, the retrieved data (i.e., temperature value) shall be stored and updated momentarily for the validity of this kind of applications. Accordingly, the choice of which data to replace shall base on a combination of factors that takes into consideration the aforementioned data categories and this what we call here, the data/content value. Therefore, we propose a new component for the fog system, called LCN, which act as a gateway toward deeper nodes in the cloud. This LCN/gateway is responsible to implement the cognition elements, which are used in mentoring the fog content conditions. These conditions are represented by a value (V) which can be calculated based on a utility function that takes into consideration all the aforementioned criteria and data attributes. This value is not only dependent on the history of the data itself, but also on the fog node status and conditions. It is secured via a secret key SK_i that is generated/assigned by the end user/device i. At the beginning of every interactive round with the fog and based on the abovementioned models, each user i resets its V_i value based on the following utility function:

$$V_i = (\alpha \cdot \Delta + \beta \cdot TTL_c + \gamma \cdot Popularity_c + \lambda \cdot T) \oplus SK_i, \tag{9.8}$$

where α, β, λ, and γ are priority-tuning weights. The key advantage to CCFF framework stems mainly from its ability to prioritize between the trusted/secured cache contents of the edge node based on the assumed application. This utility function aims at increasing the users' gain from the coming gateway/LCN replies according to the user's requirements. To improve the basic priority-caching method, the weights of the parameters can be adjusted to find the most efficient approach. In the following, Algorithm 1 provides the steps to be executed by each node if its cache is full in order to drop the content with the least value *V*.

Algorithm 1: Best Value Replacement Approach

```
Function CCFF (data, node type)
Input
    data: A content item within the fog.
Begin
if node-type == LCN
for each LCN node, do
  for each interactive round with the fog, do
     Set value of each V in the cache based on Eq. (9.8)
     if cache-is-full
         Check the history of requested data
         Check the current node/link status in the network
         Apply learning element
         Apply reasoning element
         Drop the data content of the least V
       End if
   End for
 End for
 End if
 if node-type == GCN
     store data
 End if
 if node-type == SN
     Call CCFF(data, LCN)
 End if
 End
```

This Algorithm serves as a function of a data content. It computes the value V for each data in the cache based on Eq. (9.8) and it applies the requested data comparison at each node for each cycle during the network lifetime. Then, it checks whether the cache memory is full or not. If it is full, it decides the content replacement based on the history of the requested content and drops the old least valuable content from the cache. Moreover, elements of cognition are implemented at the LCNs as we aforementioned. These elements are, namely, the *learning* element and *reasoning* element.

9.4.1 Learning Element

Supervised, unsupervised and reinforcement learning are the major categories of learning techniques. Among these, reinforcement-learning techniques are mostly applied to improve data delivery in resource-limited networks such as the one at the edge of the cloud because of their ability to learn incremental changes, without the aid of any prior information about the environment in which they are applied. In CCFF, learning is used to find the best paths that can be used by data/requests to reach destination. It is choosing the nodes that can make the path between the source and destination as short as possible and store it in LCNs in order to learn the best paths in the next data-exchange rounds. In other words, it selects best RNs that can lead to the GCN in the ICSN-based fog. Since the learning process might be too slow to respond/converge before further changes occur in the fog, we choose a planning strategy based on a relatively fast heuristic search algorithm, called A*, to identify the RNs that can be used for data caching and delivery to the end user [3]. This assists in making quicker decisions, with the help of a model of the original network topology combined with the currently observed changes.

Example 1:

Assume S_1 and S_2 have the same data to be sent to destination nodes D_1 and D_2. RNs are all the available relays toward the destinations. Out of these relays, it is determined that two RNs are in the way from S_1 and S_2 to D_1 and D_2 as shown in Figure 9.2a, respectively. Therefore, S_1 initiates routing data to these two RNs and data will be typically cached on these RNs for future retrieval. Meanwhile, S_2 also forwards and caches the same data to another two RNs (depicted by the arrows in Figure 9.2). Nevertheless, by introducing the learning elements at the edge LCNs, they will be able to identify the common RNs at which the data shall be cached for more efficient retrieval as shown in Figure 9.2b. Sharing this observation with neighboring nodes, the cognitive fog would be able to adapt to severe circumstances proactively.

9.4.2 Reasoning Element

Concepts of reasoning have been applied to networks using neural network models and software agents to implement distributed intelligence [30]. However, most reasoning algorithms are applied to application specific deployments, and usually

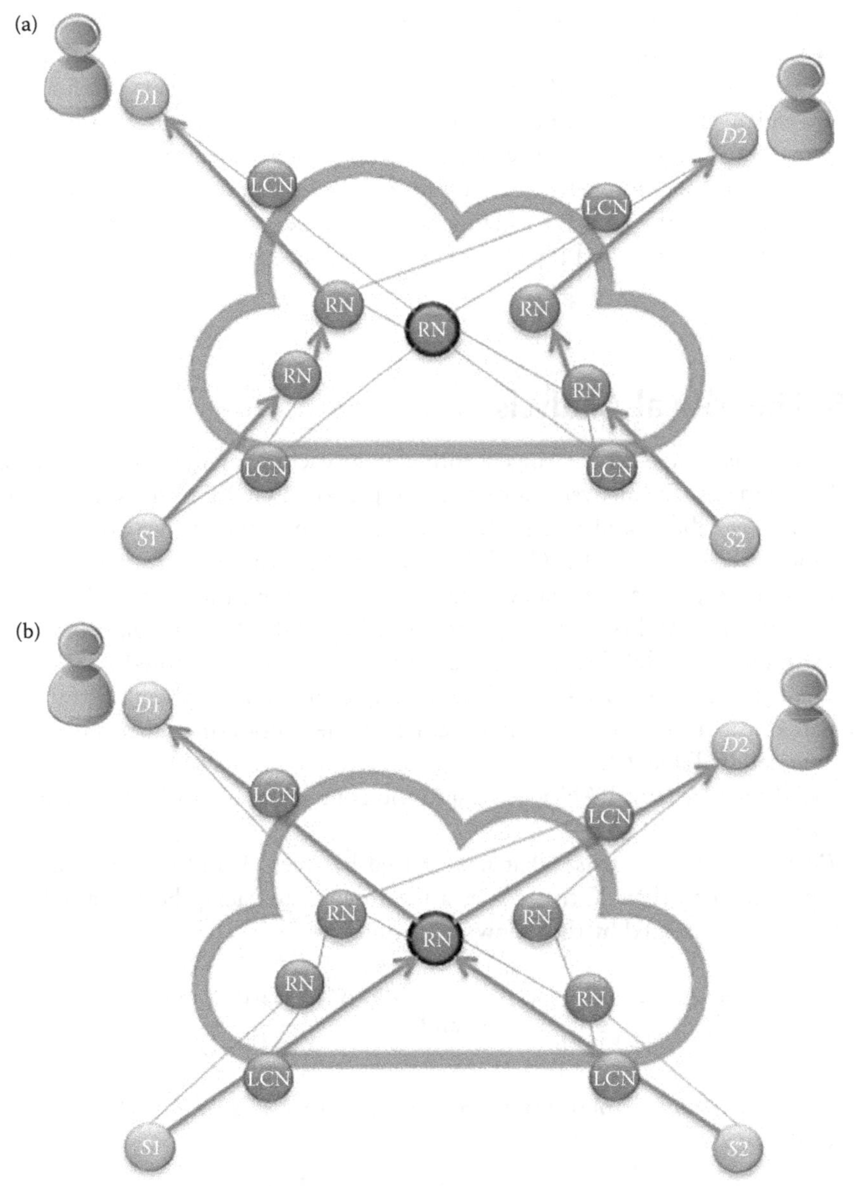

FIGURE 9.2 (a) Typical on route caching and (b) cognitive caching at common relay node RN.

take too long to converge. This does not suit the resource limited nature of the edge devices in the fog, which may change in its resource availability and topology even before the reasoning from previously observed changes is applied to the network to make a positive impact. Consequently, CCFF approach assumes a modified version of the AHP reasoning approach [5]. AHP supports multiple-criteria decision making while choosing the data path. For example, if we have a delay-sensitive data, AHP assists in searching for the node that supports the least latency. If it finds two nodes, which provide the same latency, it will check for a second parameter like the remaining energy for instance and prioritize between the two available nodes. In this way, the reasoning element can provide the best next node to consider and so on. In general, the CCFF reasoning helps to decide on the immediate actions to be taken by the fog, and learning is used to achieve the long-term goals of the fog network.

9.5 Theoritical Analysis

This section is composed of stringent formal edge caching analysis of the proposed CCFF approach. The analysis shows that the proposed CCFF framework not only offers data availability and accessibility, but also prevents the various potential data loss, and unsecure access. The CCFF framework offers a secure agreement between the customer C_i, and a relay/gateway LCN_i and it is proven using the hierarchical Kura architecture and BAN logic proposed in Refs. [31,32], and [10], respectively. Where gateways in Ref. [32] are equivalent to LCN in our proposed ICSN architecture. Let us assume that X and Y be the requesting devices of data, P and Q be the data to be cached and s_k be the secret key while considering other notations as summarized in Table 9.2.

Based on Ref. [10], BAN logic assumptions/procedures can be formulated as follows:

Procedure 1: If X trusts that s_k is shared between X and Y and observes P encrypted with s_k, then X trusts Y as a legal device to cache the requested data. This can be represented by the following formulas:

$$\frac{X|\equiv X \xleftrightarrow{s_k} Y, X \triangleleft \{P\}s_k}{X|\equiv Y|\sim P} \text{ and } \frac{X|\equiv X \xleftrightarrow{Q} Y, X \triangleleft \{P\}Q}{X|\equiv Y|\sim Q}, \tag{9.9}$$

Procedure 2: If only Y observes P, then X trusts that Y is sure of P as shown in Eq. (9.10).

$$\frac{X|\equiv\neq P, X|\equiv Y|\sim P}{X|\equiv Y|\equiv P} \text{ and } \frac{X|\equiv\neq Q, X|\equiv Y|\sim Q}{X|\equiv Y|\equiv Q}. \tag{9.10}$$

Procedure 3: If X trusts P and Q, then X can cache P and Q. And thus,

$$\frac{X|\equiv PX|\equiv Q}{X|\equiv (P,Q)}. \tag{9.11}$$

TABLE 9.2 Notation Used in Our Analysis

Notation	*Description*
$X \mid\equiv P$	X trusts P
$\neq P$	P fidelity is assured
$X \mid\Rightarrow P$	X takes the authority over P
$X \lhd P$	X recognizes P
$X \mid\sim P$	P formerly trusted as X
(P, Q)	P or Q is an individual part of (P, Q)
$\{P\}_{sk}$	P is encrypted using s_k
$\langle P \rangle_{sk}^{Q}$	P is cached at Q
$X \xleftrightarrow{sk} Y$	X and Y uses s_k to establish a link. Besides, s_k is totally secure; and thus can not be discovered by any in-network device except X and Y.
$User_i$	ith user
PID_j	Identity of jth user
SK_j	Secret key of jth user
LCN_i	ith local cognitive node
CID_i	Unique identity of ith LCN
H_1:{0, 1}*	Map to point hashing functional operation
H_2:{0, 1}*	Secure collision free one way cryptography hashing function
x	Secret random integer controlled by GC_S
$E_{S_k}(.)$	Symmetric key encryption function
ΔTS	Expected delay transmission time
TS_S	Timestamp
$\Vert$	Concatenation operator
$\oplus$	Bitwise X-OR operator
S, r, y, z	Random integers $\in Z_q^*$
S_{k1}, S_{k2}	Secure session key
q	Prime order integers

(*Continued*)

TABLE 9.2 (*Continued*) Notation Used in Our Analysis

Notation	*Description*
H_1, H_2	Hashing operators
s_k	Secret key
GC_S	Global cognitive sink
RN	Relay node

Procedure 4: If X trusts the key of P, then Y trusts (P, Q) as follows:

$$\frac{X|\equiv\neq P}{X|\equiv\neq (P,Q)}. \tag{9.12}$$

Procedure 5: If X believes that Y can affect the cached data P and X trust Y in terms of P validity, then X trusts P. This can be formulated as follows:

$$\frac{X|\equiv Y \Rightarrow P\ X|\equiv Y|\equiv P}{X|\equiv (P,Q)}. \tag{9.13}$$

In order to satisfy the security factor in Fog computing, the proposed CCFF framework must be able to meet the following objectives:

$$\textbf{\textit{Objective}}_1 : User_i|\equiv GC_S|\equiv LCN_i \xleftrightarrow{Sk} RN$$

$$\textbf{\textit{Objective}}_2 : User_i|\equiv LCN_i \xleftrightarrow{Sk} S_C$$

$$\textbf{\textit{Objective}}_3 : RN|\equiv User_i|\equiv LCN_i \xleftrightarrow{Sk} RN$$

$$\textbf{\textit{Objective}}_4 : RN|\equiv LCN_i \xleftrightarrow{Sk} RN$$

Where, the structural flow of BAN logic is as follows:

1. Messages to be exchanged/cached:

$$M_1 : User_i \rightarrow GC_S : \langle H_2(x \oplus SK_j)\rangle, \langle PID_j, H_2(x \oplus SK_j)\rangle_{x\in Z_q^*}$$

$$M_2 : GC_S \rightarrow RN : \langle (Certify_j = S.H_1(PID_j \parallel H_2(x \oplus SK_j)); TS_j$$
$$= H_2(PID_j \parallel y); H_j = H_2(TS_j); V_j = TS_j \oplus H_2(x \oplus SK_j))\rangle_{x\in Z_q^*}$$

$M_3 : User_i \rightarrow RN : \langle Certify_j, V_j, H_j, x \rangle_{x \in Z_q^*}$, where $V_j = TS_j \oplus H_2(x \oplus SK_j)$ and $H_j = H_2(TS_j)$.

2. Transmitted/cached messages in cognitive form:

$$T_{M1} : User_i \rightarrow GC_S : \langle PID_j, H_2(x \oplus SK_j) \rangle_{User_i \xrightarrow{PID_j} GC_S}$$

$$T_{M2} : GC_S \rightarrow RN : \langle Certify_j, V_j, H_j \rangle_{User_i \xrightarrow{PID_j} GC_S}$$

$$T_{M3} : User_i \rightarrow RN : \langle Certify_j, V_j, H_j, x \rangle_{User_i \xrightarrow{PID_j} GC_S}$$

3. Cached data in Hypotheses form:

$$H_{M1} : User_i \mid\equiv \neq (CID_i),\ LCN_i \mid\equiv \neq (TS_1, TS_2)$$

$$H_{M2} : GC_S \mid\equiv \neq (PID_i),\ GC_S \mid\equiv \neq (TS_3, TS_4)$$

$$H_{M3} : User_i \mid\equiv GC_S \mid\equiv LCN_i \xleftrightarrow{sk} RN$$

$$H_{M4} : User_i \mid\equiv LCN_i \xleftrightarrow{sk} RN$$

$$H_{M5} : RN \mid\equiv User_i \mid\equiv LCN_i \xleftrightarrow{sk} RN$$

$$H_{M6} : RN \mid\equiv LCN_i \xleftrightarrow{sk} RN$$

Consequently, the proposed cognitive CCFF framework's security and fidelity can be examined and proved based on BAN logic assumptions' and objectives' as follows:

Proof. Based on the transmitted/cached message T_{M1}, the CCFF has $P_1 : GC_S \triangleleft \langle PID_j, H_2(x \oplus SK_j) \rangle_{User_i \xrightarrow{PID_j} GC_S}$. And from H_{M2}, P_1 and Eq. (9.9), the CCFF acquires $P_2 : GC_S \mid\equiv User_i \mid\sim \langle PID_j, H_2(x \oplus SK_j) \rangle$. And since the transmitted/cached message T_{M2}, the CCFF has $P_3 : RN \triangleleft \langle Certify_j, V_j, H_j \rangle_{User_i \xrightarrow{PID_j} GC_S}$. Now based on H_{M5}, P_3 and Eq. (9.9), the CCFF acquires $P_4 : LCN_i \mid\equiv RN \mid\sim \langle Certify_j, V_j, H_j \rangle$. And from H_{M1}, P_4, Eqs. (9.10) and (9.12), the CCFF obtains $P_5 : User_i \mid\equiv GC_S \mid\equiv LCN_i \xleftrightarrow{sk} RN$. This validates our target in $\langle \boldsymbol{Ojective_1} \rangle$. And based on H_{M5}, P_5 and Eq. (9.9), the CCFF gets $P_6 : User_i \mid\equiv LCN_i \xleftrightarrow{sk} RN$. This validates our target in $\langle \boldsymbol{Ojective_2} \rangle$. Now from the transmitted/cached message T_{M3}, the CCFF has $P_7 : GC_S \triangleleft \langle Certify_j, V_j, H_j, x \rangle_{User_i \xrightarrow{PID_j} GC_S}$. And from H_{M1}, P_7 and Eq. (9.9), the CCFF acquires $P_8 : GC_S \mid\equiv User_i \mid\sim \langle Certify_j, V_j, H_j, x \rangle$. From H_{M5}, P_8, Eqs. (9.10) and (9.12), the CCFF obtain $P_9 : RN \mid\equiv User_i \mid\equiv LCN_i \xleftrightarrow{sk} RN$. This validates our target in $\langle \textbf{\textit{Objective}}_3 \rangle$. Finally, from H_{M6}, P_9 and Eq. (9.11), the CCFF eventually achieves $P_{10} : RN \Big| \equiv LCN_i \xleftrightarrow{s_k} RN$. This validates our target in $\langle \textbf{\textit{Objective}}_4 \rangle$. ■

Provided the aforementioned proved objectives $\langle \textbf{\textit{Objective}}_1 - \textbf{\textit{Objective}}_4 \rangle$, the CCFF framework states that it uses a common s_k to cache data; and hence the proposed CCFF framework achieves the proper data fidelity and trust.

9.6 Use-Case and Performance Evaluation

As a practical test-bed for our fog computing CCFF solution, we have developed and deployed a simple use-case application in the domain of cognitive ICSN. In our ICSN application, a fog node (Raspberry Pi 2 model B, with 900 MHz quad-core ARM Cortex-A7 CPU, and 1 GB RAM) connected to a smart device (e.g., smartphone, tablet, etc.) and following pre-determined paths assists the end-users in retrieving their cached data to show the performance of our fog-based CCFF framework. The fog node in this framework acts as an LCN or a gateway to the Internet that aggregates/caches the data gathered from multiple heterogeneous sources/publishers. This fog node/LCN keeps track of gathered data by storing critical QoI-related attributes, e.g., priorities, locations, security keys, and/or reliability-based measures. The fog node application is implemented in Golang, which is supported by Docker to provide different performance measures in terms of both computing and communication as performed in other related studies [32,33]. In this study, the performance of containers-based solutions (e.g., Docker [34] and Kura [31]) in comparison to our proposed CCFF execution is performed, by assessing the overall overhead and delays. In addition, we have thoroughly evaluated the usage of multiple concurrent caching levels over the fog nodes to verify the capability of multilevel caching in edge devices to the Cloud. Furthermore, we evaluate our CCFF framework against the general categories we summarized in Section 9.2, namely the FC, the CC, and the Location-based Caching (LC) techniques using a discrete event simulator in NS3. We simulate our fog-based network with the detailed physical layer built-in specifications in NS3 with system features of 32-bit Linux-based Operating Systems. The employed caching schemes, FC, CC, LC, and CCFF, at the edge of the cloud are applied on randomly generated 100 wireless heterogeneous network topologies in order to get statistically stable results. The average results hold confidence intervals of no more than 5% of the average values at a 95% confidence level. The examined cache is implemented in both: a single storage level and a hierarchical storage of two levels. Simulation results are generated after applying the CCFF, LC, CC, and FC methods. These results have been derived from a discrete event simulator with a cache size per fog node (LCN) ranging from 10 to 100 MB, and 100 different requests from randomly selected end nodes (devices) in the simulated fog network.

9.6.1 Performance Metrics and Parameters

To assess the performance of the proposed CCFF framework, we target fog-specific metrics in order to achieve subjective conclusions. We consider the following performance metrics:

- Time-To-Hit (TTH): The amount of time spent looking for the requested data in the Cloud. In this study, the fog-based ICSN is proposed to minimize this amount per request on average.
- Hit-ratio: It is the average times a data is found in a targeted node cache.
- Delay: This metric represents the end-to-end delay in retrieving a cached data over the Cloud as described above. It can be measured in seconds, minutes, hours, or days in specific circumstances.
- ARP: The average number of requests per publisher, measured in requests per hour (req/h) and it represents the average load per publisher in the proposed fog-based ICSN.

Meanwhile, the varying parameters in our simulation can be summarized as follows:

- PoC: The percentage of in-network nodes, which are equipped with caches. This parameter is used to study our system sensitivity to fog-specific metrics such as the TTH and delay.
- Speed: It is the available Internet speed at the user device (i.e., laptop, smartphone, etc.) which is used to access data over the Fog. It is measured in *bytes per seconds*.
- Connectivity degree: It represents the average number of edges getting in and out a node in the ICSN fog. It is used to show the complexity of network while examining the aforementioned caching methods.
- Popularity (%): It is the average percentage a specific data is requested per time unit. This parameter is randomly generated per experiment for more realistic results.

9.6.2 Simulation and Results

In this section, we compare our proposed CCFF approach against traditional and new emerging trends in caching data over the Cloud and/or the Fog. Our benchmarks and data sets are provided by enterprise solutions in practice such as Mozy [35], Evernote [36], and Extended-Kura [33]. Mozy is an online service that retrieves and backup data over the edge of the cloud. It backups deleted files from personal computers based on the user-request. Moreover, it provides safe and easy access to recovered data. Evernote a google cloud platform that aims at caching and retrieving data as well. Recently has made a few improvements in performance, security, efficiency, and scalability. On the other hand, Kura [31] is an open-source framework for IoT applications at the gateway-level. It aggregates and controls device information, as well as it supports the simplification of the overall development and deployment process in fog computing. Furthermore, it comes with self-contained pluggable bundles suitable for IoT applications. Kura framework has been extended in Ref. [33] to avoid unnecessary overheads at the gateway level for more compatibility with fog computing.

In comparison to statistics provided in Refs. [35,36], our CCFF approach is not only safer as elaborated above in the use-case section but also faster in terms of accessing and restoring the cached data over the Fog as shown in Table 9.3.

In Table 9.3, we highlight the average experienced delay associated with the most fog-relevant operations performed while running our application at the fog nodes, including I/O operations (represented by reading a file of size 100 MB–10 GB) and CPU operations (represented by a key word search algorithm). As expected, the execution of non-typical container-based approaches (e.g., Extended-Kura and CCFF) outperforms the container-based ones. This is due to the extra time spent in creating the container at the gateways. Among the examined container-based approaches, Mozy has shown the worst performance in terms of delay due to the

TABLE 9.3 Available Internet Speed vs. the Delay to Retrieve Data at the User Device

Approach	*Data/ Speed*	*56 Kbps*	*256 Kbps*	*1.5 Mbps*	*10 Mbps*	*100 Mbps*
Mozy [35]	100 MB	4 h	1 h	10 min	2 min	9 s
	1 GB	2 days	10 h	2 h	15 min	2 min
	10 GB	18 days	4 days	16 h	3 h	15 min
Evernote [36]	100 MB	5 h	1 h	10 min	2 min	9 s
	1 GB	3 days	9 h	2 h	15 min	2 min
	10 GB	19 days	4 days	16 h	3 h	15 min
Extended-Kura [33]	100 MB	3 h	45 min	10 min	2 min	9 s
	1 GB	10 h	5 h	2 h	15 min	2 min
	10 GB	2 days	1 days	16 h	3 h	15 min
CCFF	100 MB	1 h	30 min	2 min	1 min	2 s
	1 GB	8 h	4 h	1 h	4 min	10 s
	10 GB	1 days	10 h	6 h	1 h	1 min

strong dependency on the used file size. Meanwhile, Extended-Kura and CCFF have almost similar performance except at the beginning while experiencing very low internet speeds.

In the following, we elaborate on the different caching categories simulated and examined against our CCFF approach. Figures 9.3–9.12 show the achieved results.

In Figure 9.3, we compare the hit ratios against varying cache volumes while applying a single cache level. Four caching strategies have been compared. We

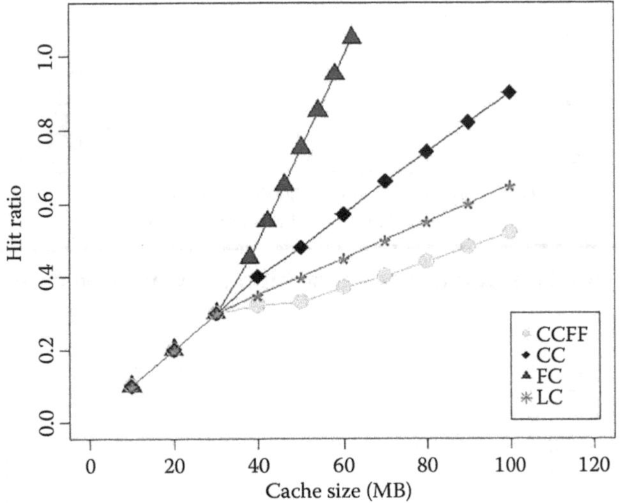

FIGURE 9.3 Cache size vs. the hit ratio with single caching level.

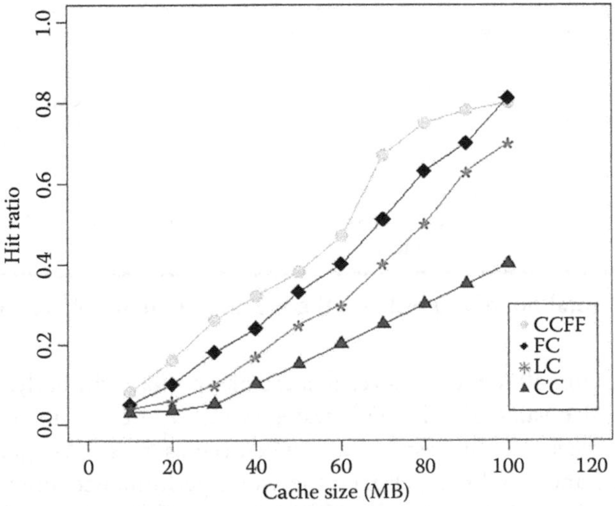

FIGURE 9.4 Cache size vs. the hit ratio with double caching levels.

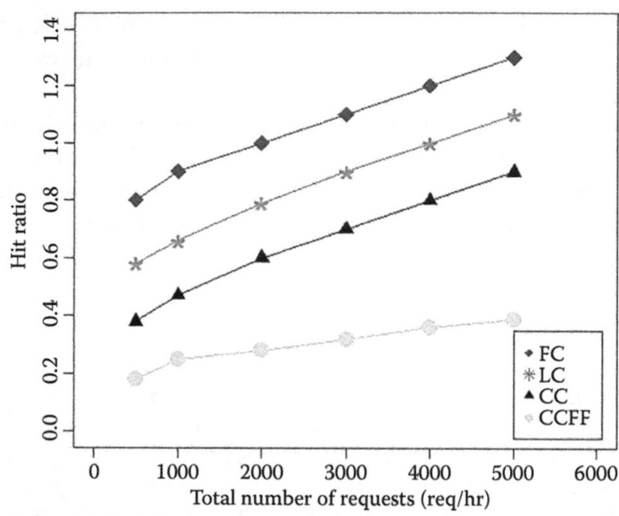

FIGURE 9.5 Total no. of requests vs. the hit ratio with single caching level.

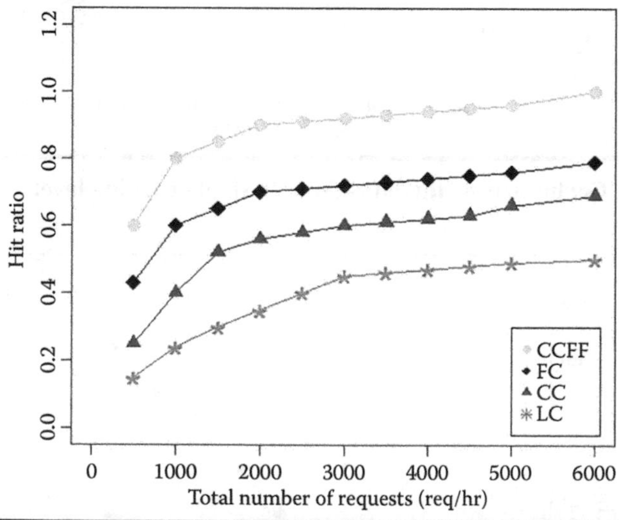

FIGURE 9.6 Total no. of requests vs. the hit ratio with double caching levels.

remark that up to ~40 MB cache size, all strategies perform similarly. After exceeding the cache size value 40 MB, FC strategy's hit ratio curve increases dramatically and it achieves the highest hit ratio when the cache size is equal to 60 MB. On the other hand, CCFF experience the worst performance among the others. Accordingly, we can deduce that the CCFF is not efficient in one level caching, however, FC and LC caching techniques perform equally well in this scenario.

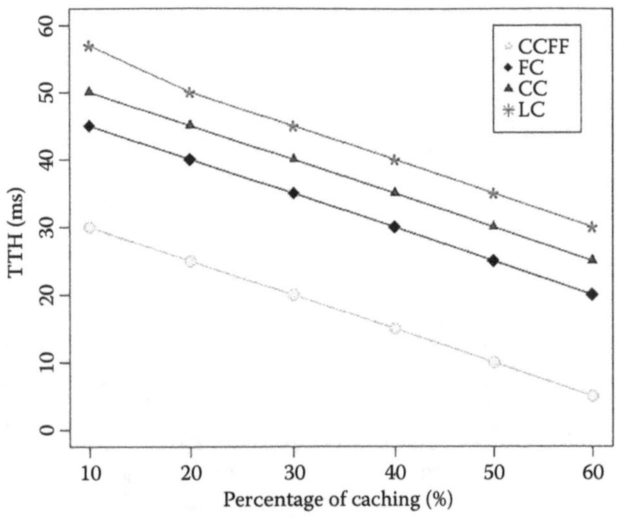

FIGURE 9.7 Time to hit ratio vs. the PoC (with connectivity degree = 30).

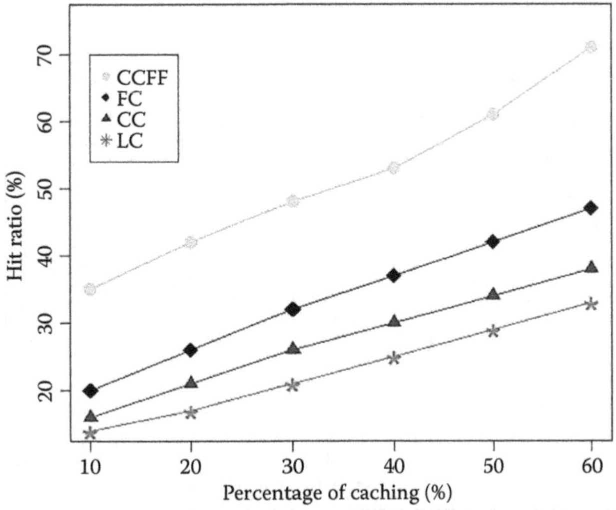

FIGURE 9.8 Hit ratio vs. the PoC (with connectivity degree = 30).

It confirms that increasing the number of caching levels in ICSN fog can reduce the total spent time to meet the requested data.

In Figure 9.6, where we have two levels of caching, we find out that CCFF outperforms the other caching techniques. Again, it compares the hit ratios against the varying cache volumes. The four caching strategies are compared while increasing the cache size and recording the effects on the hit ratio. Achieved results are very

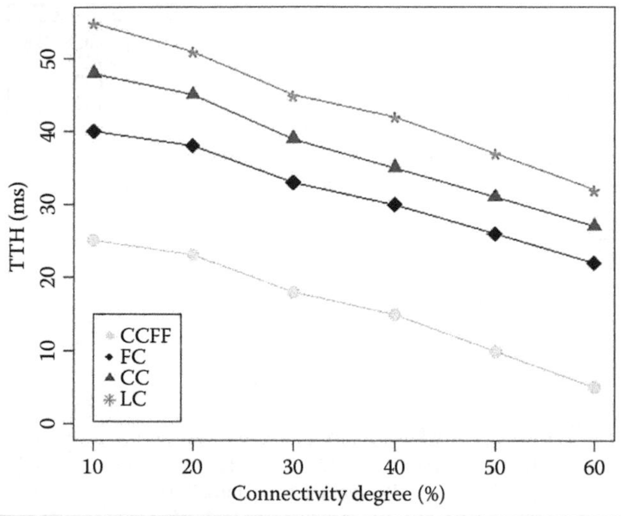

FIGURE 9.9 Time to hit ratio vs. the connectivity degree percentage.

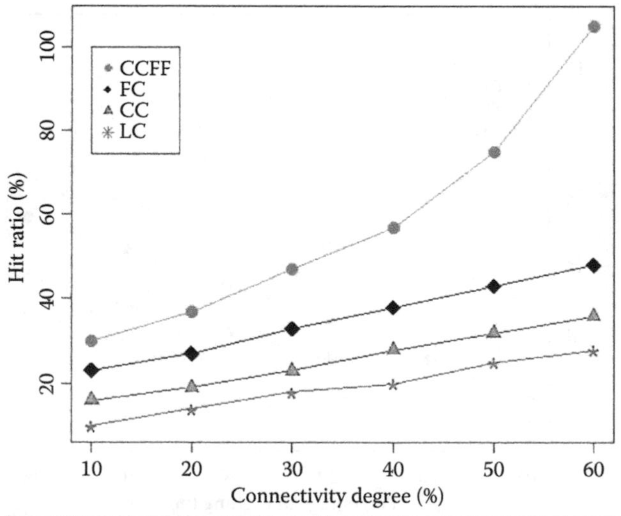

FIGURE 9.10 Hit ratio vs. the connectivity degree percentage.

close to each other except for the CC category, where it experiences the worst results among the others. In addition, CCFF shows the best performance among the others up to ~100 MB cache size after which it goes into a steady state with the FC approach. We further remark that CC cache category is the worst one while applying the double caching level in terms of the hit ratio. Where it reaches almost ~0.4 hit ratio as the cache size increase to 100 MB. In addition, we notice that there is

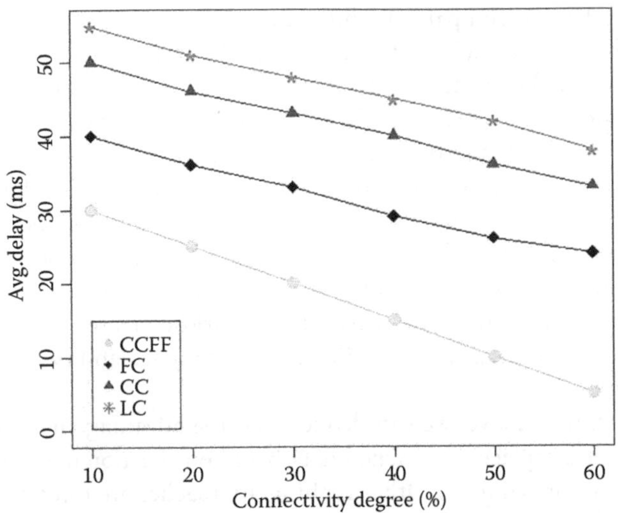

FIGURE 9.11 Avg. in network delay vs. the connectivity degree.

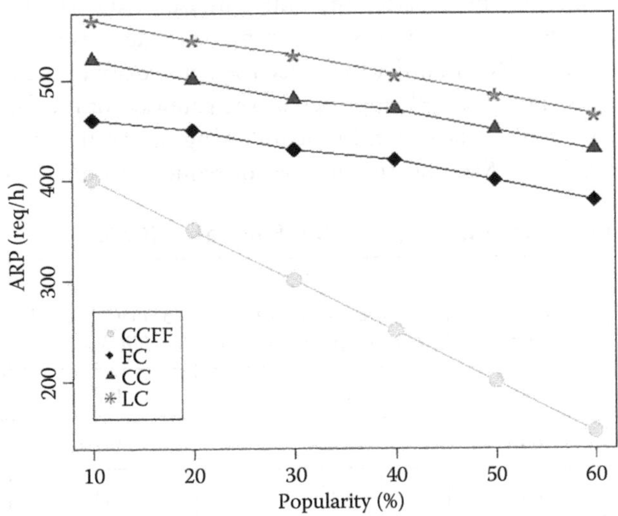

FIGURE 9.12 Publisher load vs. the data popularity.

no significant gain when the cache size is increased beyond 30 MB. This is because we have only one type of requests considered in this experiment.

The next set of simulations, depicted in Figures 9.5 and 9.6, analyze the caching performance while increasing the number of requests from 600 to 6000 request per hour. The total number of requests' types is increased to four different types and the cache size per node is set to be 100 MB.

In Figure 9.4, we compare the hit ratio while varying the total number of requests with a single caching level. Again, FC cache category gives the best results in this case. However, FC, LC, and the CC categories increase linearly in terms of the hit ratio whereas CCFF increases linearly but with a very small slope compared to others. In addition, LC and CC algorithms achieve very close and similar results' trends. Consequently, we can conclude that FC algorithm is the most appropriate choice in terms of hit ratio when the total number of requests is high. In Figure 9.5, we further compare the hit ratio against the varying counts of requests while applying the double caching level. As a result, CCFF framework gives the best results in this case where the hit ratio increases up to ~99% as the number of requests increases. This indicates the strong scalability feature in CCFF versus the others.

From the figures above, we can deduce that the advantage held by the CCFF approach is that it replaces the cached data based on the aforementioned fog metrics and the user prerequisite. Other caching approaches are only concerned with the match of the requested data, without considering the age, popularity or even the delay associated with this data. Nevertheless, the use of CCFF approach takes into consideration the age of data, value and how popular the data is. Unlike other caching techniques, CCFF replaces the oldest unused data based on the utility function in Eq. (9.8). Consequently, we recommend the use of two levels of caches at the fog gateways/LCNs. In order to reduce the complexity of the computations, we suggest employing the CCFF approach at the gateways of the fog while applying FC or LC on regular relays in it. In the following, we set the size of level 1 and level 2 caches to be 10 MB and the number of requests to be 20,000. Table 9.4

TABLE 9.4 Two-Level Caching over the Fog-Based ICSN

First Level Caching Approach	*First Level Hit-Ratio*	*Second Level Caching Approach*	*Second Level Hit-Ratio*	*Total Hit-Ratio*
CCFF	0.86	LC	0.87	0.865
CCFF	0.57	FC	0.79	0.71
CCFF	0.91	CCFF	0.1	0.921
LC	0.77	CCFF	0.4	0.58
LC	0.91	FC	0.05	0.91
LC	0.82	LC	0.49	0.65
FC	0.74	LC	0.01	0.81
FC	0.84	CCFF	0.02	0.81
FC	0.84	FC	0.0	0.84

depicts that the best possible combination for the two level caching over the fog network is while applying the CCFF approach at the 1st level cache and CCFF or LC at the 2nd level cache. Despite having a good hit ratio sometimes, FC is not reliable to serve the user needs based on the hit-ration metric and the delay required to retrieve the required data.

In the following figures (Figures 9.7 and 9.8), we elaborate more on applying the two level caching while examining the four targeted caching methods. According to Figures 9.7 and 9.8, data availability increases proportionally with the increase in the percentage of fog caching nodes in the network. In Figure 9.7, we observe the decrement in the time to meet data (i.e., TTH) while that percentage increases from 10% to 60%. In particular, CCFF approach outperforms the other benchmarks due to learning elements in the gateways/LCNs, which promptly discover and learn the most appropriate and empty storage in the Fog. On the other hand, LC algorithm gives the worst results amongst the others. In fact, CCFF still achieves competitive results with respect to LC, CC, and FC while experiencing low caching percentages.

In Figure 9.8, we compare the hit ratio against the varying percentage of caching while applying connectivity degree equal to 30. While increasing the percentage of caching nodes, we analyze the changes in the hit ratio. We observe the increment in the hit ratio for all approaches; however, it is the most obvious while applying CCFF approach. This can be returned to the effectiveness of Eq. (9.8) in addressing and replacing the most unnecessary data at the available fog storage/cache. This is achieved via identifying the most appropriate data attributes at the network LCNs. Again, LC category gives the worst results. Moreover, the CCFF reacts much faster in response to the increment in the percentage of caching nodes when compared to the others. Therefore, CCFF can be considered the most appropriate choice in terms of the hit ratio when the percentage of caching is high or low.

In the following figures, we examine the connectivity degree effect on the four evaluated approaches. Figure 9.9 shows the increment in time to hit as the fog connectivity increases. Nevertheless, we remark here that CCFF can tolerate denser fog topologies than other competitive alternatives such as the FC, CC, and LC approaches. In fact, denser topologies support the CCFF approach in converging toward the best storage in the fog in more competitive time periods. It gives better learning feedback that can help Algorithm 1 to converge faster toward more optimized solutions. Therefore, when the connectivity degree increase, the time to hit decreases dramatically for the CCFF approach. Hence, CCFF algorithm is the most appropriate choice in terms of time to hit when the percentage of connectivity degree is high.

Figure 9.10 depicts the caching performance in terms of hit ratios versus the varying connectivity degrees. In this figure, we observe the exponential increment in the hit ratio metric while applying the CCFF approach. CCFF approach gives the best results in this figure and LC approach performs the worst amongst the other caching categories. When we check the figure, we observe that the CCFF performance increases exponentially whereas the others increase linearly. In addition, it shows that the connectivity degree is a very important metric while searching better hit ratios.

In Figure 9.10, we perform a comparison between the average delay metric against the varying percentage of connectivity degree. We increase the percentage of connectivity degree and analyze the influence on the hit ratio percentage. Generally, when we increase the percentage of the connectivity degree, we observe a decrease in delay for all strategies. However, CCFF approach achieves the best performance in comparison to others and the LC approach achieves the worst. We remark as well that the CCFF curve decreases linearly but with a higher slope than others do. In addition, Figure 9.11 shows that the connectivity degree is inversely proportional to the average experienced delay in general. This can be better utilized among the densely connected edge devices of the Fog.

In Figure 9.12, we compare the four caching approaches in terms of the average publisher load while varying the percentage of data popularity. Obviously, when we increase the percentage of data popularity, we observe significant reductions in publishers' loads, especially while applying CCFF approach. In general, CCFF outperforms the other alternatives in this in this figure. In addition, it shows that popularity is inversely proportional with the publisher load.

9.7 Conclusions

This work investigates the most appropriate in-network caching approach, which can cope with the advances we are experiencing nowadays in the cloud, and more specifically the fog-computing trend. There have been several attempts so far in the literature to come up with an efficient caching approach, which can enhance the requested/processed data accessibility and availability. These attempts have been classified in this chapter into: (i) *Location-based caching*, (ii) *Content-based caching*, and (iii) *Functionality-based caching*. However, these attempts suffer from critical aspects in fog computing, including data latency, availability, and scalability. Consequently, a value-based caching framework, called CCFF, is recommended to be used in the fog and especially in ICSNs where the sensed data is mostly processed/requested at the edge of the network. In this framework a new component, called Local Cognitive Node (LCN), is introduced to retain information about data popularity and other parameters that affect data availability and network scalability. Since the cached data information can be provided from the gateways/LCNs, the rapidly changing edge nodes, which are varying significantly over short time periods sometimes will not degrade the performance of the fog. Furthermore, CCFF addresses the need for the delay-tolerant caching in edge networks. In addition, it maximizes the gain of the data publishers by reducing their load. Our presented simulation results in this work show the efficiency of CCFF when compared to other similar approaches in the literature such as FC and LC. It outperforms these approaches in terms of the required time to find data, fidelity, and the experienced publisher loads.

References

1. B. Ahlgren, C. Dannewitz, C. Imbrenda, D. Kutscher, and B. Ohlman, "A survey of information-centric networking", *IEEE Communications Magazine*, vol. 50, no. 7, pp. 26–36, 2012.
2. IBM. "smarter planet, smarter cities". Available: http://www.ibm.com/smarterplanet/us/en/smarter_cities.
3. F. Al-Turjman, "Cognition in information-centric sensor networks for IoT applications: An overview", *Annals of Telecommunications*, pp. 1–16, 2016. doi:10.1007/s12243-016-0533-8.
4. F. Bonomi, "Connected vehicles, the internet of things, and fog computing", *Eighth ACM International Workshop on Vehicular Inter-Networking (VANET)*, Las Vegas, NV, 2011.
5. G. T. Singh and F. M. Al-Turjman, "A data delivery framework for cognitive information-centric sensor networks in smart outdoor monitoring", *Elsevier Computer Communications*, vol. 74, no. 1, pp. 38–51, 2016.
6. F. Al-Turjman, A. Alfagih, W. Alsalih, and H. Hassanein, "A delay-tolerant framework for integrated RSNs in IoT", *Elsevier Computer Communications Journal*, vol. 36, no. 9, pp. 998–1010, May 2013.
7. F. Al-Turjman, H. Hassanein, and M. Ibnkahla, "Efficient deployment of wireless sensor networks targeting environment monitoring applications", *Elsevier Computer Communications Journal*, vol. 36, no. 2, pp. 135–148, January 2013.
8. A. Chankhunthod, P. Danzig, C. Neerdaels, M. Schwartz, and K. Worrell, "A hierarchical internet object cache", *Proceedings of USENIX*, 1996.
9. M. Gritter and D. R. Cheriton. *TRIAD: A New Next-Generation Internet Architecture*. Stanford, CA: Stanford University, July 2000.
10. M. Burrows, M. Abadi, and R. Needham, "A logic of authentication", *ACM Transactions on Computer Systems*, vol. 8, no. 1, pp. 18–36, 1990.
11. A. Al-Fagih, F. Al-Turjman, W. Alsalih, and H. Hassanein, "A priced public sensing framework for heterogeneous IoT architectures", *IEEE Transactions on Emerging Topics in Computing*, vol. 1, no. 1, pp. 135–147, October 2013.
12. W. Yaogong, K. Lee, B. Venkataraman et al., "Advertising cached contents in the control plane: Necessity and feasibility", *Proceedings INFOCOM Workshop on Computer Communications*, Orlando, FL, 2014.
13. S. Borst, V. Gupta, and A. Walid, "Distributed caching algorithms for content distribution networks", *In Proceedings of the IEEE INFOCOM*, San Diego, CA, 2010.
14. C. Fricker, P. Robert, J. Roberts, and N. Sbihi, "Impact of traffic mix on caching performance in a content-centric network", *INFOCOM Workshops*, Orlando, FL, pp. 310–315, 2012.
15. W. K. Chai, D. He, I. Psaras, and G. Pavlou, "Cache 'less for more' in information-centric networks (extended version)", *Elsevier Computer Communications*, vol. 36, no. 7, pp. 758–770, 2013.

16. S. Eum et al., "CATT: Cache aware target identification for ICN", *IEEE Communications Magazine*, vol. 50, no. 12, pp. 60–67, 2012.
17. A. Hail, M. Amadeo, A. Molinaro, and S. Fischer, "Caching in named data networking for the wireless internet of things", *Proceedings of the International Conference on Recent Advances in Internet of Things (RioT)*, Singapore, April, 2015.
18. P. Radoslavov, R. Govindan, and D. Estrin, "Topology-informed internet replica placement", *Proceedings of WCW'01: Web Caching and Content Distribution Workshop*, Boston, MA, June, 2001.
19. S. Bhattacharjee, K. L. Calvert, and E. W. Zegura, "Self-organizing wide-area network caches", *In IEEE Infocom*, San Francisco, CA, pp. 752–757, April, 1998.
20. X. Vasilakos, V. Siris, G. Polyzos, and M. Pomonis, "Proactive selective neighbor caching for enhancing mobility support in information-centric networks", *Proceedings of the ICN Workshop on Information-Centric Networking*, New York, pp. 61–66, 2012.
21. I. Psaras, W. K. Chai, and G. Pavlou, "Probabilistic in network caching for information centric networks", *Proceedings of the 2nd Edition of the ICN Workshop on Information-Centric Networking*, pp. 55–60, 2012.
22. V. Sourlas, P. Flegkas, L. Gkatzikis, and L. Tassiulas "Autonomic cache management in information centric networks", *Proceedings of the IEEE Network Operations and Management Symposium (NOMS)*, Helsinki, Finland, August 2012.
23. M. A. Hail, M. Amadeo, A. Molinaro, and S. Fischer, "On the performance of caching and forwarding in information-centric networking for the IoT", *International Conference on Wired/Wireless Internet Communication*, Malaga, Spain, pp. 313–326, 2015.
24. J. Li, H. Wu, B. Liu, X. Wang, Y. Zhang, and L. Dong, "Popularity-driven coordinated caching in named data networking", *ACM/IEEE Symposium on Architectures for Networking and Communications Systems(ANCS)*, Austin, TX, pp. 200–211, 2012.
25. K. Cho, M. Lee, K. Park et al., "WAVE: Popularity-based and collaborative in-network caching for content-oriented networks", *Proceedings of INFOCOM WKSHPS*, Orlando, FL, pp. 316–321, 2012.
26. J. Carbo, J. M. Molina, and J. Davila, "Trust management through fuzzy reputation", *International Journal of Cooperative Information Systems*, vol. 12, no. 1, pp. 135–155, 2003.
27. F. Al-Turjman, H. Hassanein, and M. Ibnkahla, "Towards prolonged lifetime for deployed WSNs in outdoor environment monitoring", *Elsevier Ad Hoc Networks Journal*, vol. 24, no. A, pp. 172–185, January 2015.
28. F. Al-Turjman, "Cognitive routing protocol for disaster-inspired internet of things", *Elsevier Future Generation Computer Systems*, 2017. doi:10.1016/j.future.2017.03.014.
29. M. Z. Hasan et al., "Optimized multi-constrained quality-of-service multipath routing approach for multimedia sensor networks", *IEEE Sensors Journal*, vol. 17, no. 7, pp. 2298–2309, 2017.
30. K. Shenai and S. Mukhopadhyay, "Cognitive sensor networks", *Proceedings of the IEEE 26th International Conference Microelectronics (MIEL)*, Nis, Serbia & Montenegro, pp. 315–320, May 2008.
31. Kura. Available: https://eclipse.org/kura.
32. P. Bellavista, A. Corradi, and A. Zanni, "Integrating mobile internet of things and cloud computing towards scalability: Lessons learned from existing fog computing architectures and solutions", *3rd International IBM Cloud Academy Conference (ICA CON)*, Budapest, Hungary, May 2015.

33. P. Bellavista and A. Zanni, "Feasibility of fog computing deployment based on docker containerization over Raspberry Pi", *Proceedings of the ACM International Conference on Distributed Computing and Networking*, Hyderabad, India, 2017.
34. Docker. Available: https://www.docker.io.
35. Mozy-Backup Times. Available: http://support.mozy.com/articles/en_US/Documentation/mozy-c-lotsofdata-howlong-faq.
36. Evernote Pro. Available: https://blog.evernote.com/blog/ 2016/09/13/evernotes-future-cloud/.

Chapter 10

Secure Access for Mobile Applications in Industrial IoT

Fadi Al-Turjman
Antalya Bilim University

Contents

10.1 Introduction

Following the remarkable success of 2G and 3G mobile networks and the fast growth of 4G, the next generation mobile networks, including the 5G and Industrial Internet of Things (5G/IIoT) was proposed aiming to provide endless networking capabilities to mobile users. Industrial Internet of Things (IIoT), also known as industrial internet, brings together smart machines, innovative analytics, and people at work. It is an interconnection of many devices through a diverse communication system to bring forth a top-notch system capable of monitoring, collecting, exchanging, analyzing, and delivering valuable information. These systems can then help manage smarter and faster business resolutions for industrial companies. As the author in Ref. [1] puts it, IIoT is more advanced than commercial IoT, simply due to the dominance of the connected sensors in the industrial platform. Sensor interface is a key factor in industrial data collection, the author in Ref. [2] states that the present connect number, sampling rate, and signal types of sensors, are highly restricted by the sensing device.

To advance beyond the traditional mobile networks, intelligence, and secure network access need to be diffused, empowering even the smallest connected sensors. IIoT can revolutionize the ubiquitous computing with multitude of applications built around various types of "smart" sensors enabled with intelligence and machine learning techniques. Smart sensors evolve our knowledge and conception of the world as a hyper-connected environment that has raised new requirements for making the IIoT ready for large-scale deployments. This new paradigm is all about "sensing," an evolution, or may be revolution, that will take the mobile users into new territories. Secure communication and sensing techniques enable a participatory approach for achieving integrated solutions and creating novel applications related to industry and especially to healthcare. With the proliferation of intelligent identification methods, sensors are expected to lead further innovation in IIoT. Sensors play an undeniably key role in driving the IIoT revolution by incorporating 5G networks that provide functions such as data conversion, spectrum sensing, digital processing, and communication to external devices. 5G is expected to be more than a new generation of mobile communications. Instead, it is already considered as the unifying fabric that will connect billions of devices in some of the fastest, most reliable and most efficient ways possible. Of course, the impact of such an enabling technology is expected to be revolutionary. The new infrastructure for communication is expected to transform the world of connected sensors and reshape industries. Such a revolution would of course require research and development for the co-existence and device inter-operability for sensors with 5G networks. Integrated IIoT is an intelligent communication network capable of acquiring knowledge about its users and its surroundings, and uses such knowledge to help users achieve their goals in a context-based manner. This definitely improves users' quality of life, and helps in optimizing and controlling the dramatically increasing consumption rates of resources in IIoT environments. Inhabitants (users) of IIoT environments could be people, devices, services, or systems, occupied with smart enabling technologies such as Radio Frequency Identifications (RFIDs), sensors, 5G smartphones, etc. in varying applications in our daily life. A key application for the 5G-based IIoT is the e-Healthcare, which aims at maintaining the patient's medical information in electronic environments such as the Cloud via up to date telecommunication paradigms. In e-Healthcare application system,

Wireless Medical Sensor Networks (WMSNs) have become a prominent technology for Wireless Sensor Networks (WSNs) [2]. For the early diagnosis, the hospitals have developed several tiny medical sensors that are used to sense the patients' body to collect the health data, such as heart beat, blood pressure, temperature and so on. Besides, these sensing data are broadcasted via the profession handheld devices, namely PDA and smart phone to do further analysis. Various medical research communities have begun their patient health monitoring using WMSNs [1,3–5]. Therefore, user authentication scheme is necessitated to protect the illegal access in the medical information system.

WSNs' have recently been a pioneer for the source of data unification. Further, it can be able to provide an executable solution to the critical mission applications. The deployment of WSNs' can be a *proviso* for the various mission critical environments such as surveillance in military, fire detection in forest, medical health care system and in monitoring a wild life. Recently, several research works have proven that the heterogeneous sensor networks have better network performance, more reliable, scalable, transparent, load-balance, lifetime, and cost-effective [6–9]. Hence, the heterogeneous types of the sensor networks have been a salient system for the practical related real-time applications. Though it has more prominent features [6–9], it still requires a promising user authentication scheme to offer mutual authenticity, secure session key sharing, and privileged-insider resiliency. Thus, the real-time mission-critical application necessitates a secure cum mutually reliable authentication scheme to protect the network system from the anomalies. In the past, numerous user authentication protocols have been proposed for the heterogeneous sensor network system [10–13]. Since the existing mechanisms focus on unilateral user authentication, it is thus not able to deduce whether the connection is legitimate or not; further the trustworthy between the communication entities becomes zero. Le et al. in Ref. [14] presented a mutual-authentic based secure authentication using Elliptic-Curve Cryptography (ECC). However, the Le et al. scheme is susceptible to information leakage attack. In 2009, Das in Ref. [15] discovered a two-factor user authentication scheme for WSNs. Since the Das scheme uses a one-way hashing to encrypt and decrypt the sensed data, several authentication schemes [16–19] prove that the Das scheme is insecure to offline password guessing, key-impersonation, privileged-insider and gateway (node) bypassing attack. Lately, He et al. [20], Yoon et al. in [21], and Chuang et al. in Ref. [22] have proposed the user authentication schemes, where the pitfalls are as follows: (1) The computational cost is expensive; and (2) The security weaknesses, like mutual authentication and session-key establishment and vulnerabilities, like privileged-insider attack are intangible.

To address the above issues in the real time based WSNs' application, this chapter proposes a secure cum mutual authentication scheme that draws a prominent feature of the symmetric crypto-system to provide strong mutual authenticity, secure-key agreement and resilient to privileged-insider attack to use in Wireless Multimedia Medical Sensor Network (WMSN) system. We propose Context-sensitive Seamless Identity Provisioning (CSIP), a secure context-sensitive seamless multi-modal identity-provisioning framework for smart environments. CSIP builds an encrypted and compressed 360-Degree Inhabitant Profile [23,24] by using his activities' history and usage patterns of the environment's resources, based on that, it can build Disposable Customized Virtual Inhabitant Profile (DCVIP) then, it creates an identity proxy to perform the verification required during the interaction. Moreover, the paradigm is lightweight, and hence perfect for 5G-based technologies. We can divide the CSIP into two parts: client-based part and cloud-based part. The first part considers the client and it is responsible for gathering inhabitants' access data into blocks, then compresses these blocks and sends it to the cloud-side. The second part is related to the cloud; it receives inhabitants' data from the first part, applies deep analysis to gather the information required for the interaction, and then classifies and saves this information. We believe CSIP reduces the risks for identity theft/loss, as it does not depend on static identifiers, and it takes into consideration the dynamics of the inhabitants' behaviors as well as their access and usage patterns to handle incoming requests. Extensive simulation results and security analysis show that CSIP is practical as it provides acceptable performance when compared with a basic static identity proxy approach. In the following, we overview the attempts related to this system in the literature.

In e-Healthcare, WSNs represent the infrastructure, which extract the information of smart sensing object [25]. WSNs are one of the essential components of the infrastructures employed for smart e-Healthcare in IoT-based applications. Recently, security issues in WSNs have gained much attention of the researchers not only to satisfy the security properties of Authentication and Key Agreement (AKA) protocol but also to mitigate the computation and communication cost of the system. For the achievement of minimum overhead, several lightweight authentication schemes have been proposed [26–32]. Watro et al. proposed the lightweight two-factor user authentication based on RSA cryptosystem for WSNs. However, the Watro et al. scheme [27] is vulnerable to replay, denial of service and key impersonation attacks. Wang et al. [28] presented a lightweight user authentication scheme for WSNs, which only demands the computation of a hashing function. Later on, Srinivas et al. [30] shows that the Wang et al. scheme [28] is vulnerable to stolen verifier and many logged-in users with the same login identity attack. Tseng et al. [31] improved the version of Wong et al., which does not offer mutual authentication between the base-station and sensor-node. To overcome the security weakness of mutual authentication, Lee [32] presented a novel password based dynamic user authentication scheme, which also fails to satisfy mutual authentication between the base-station and the sensor-node.

Meanwhile, Cloud-based solutions provide significant long- and short-term benefits over an internal computing grid. Over the long-term, Cloud storage, analysis and archival are scalable for the streaming, real-time data volume that is expected to be generated by users over an extended period of time. In the short-term, Cloud solutions provide more cost-effective deployment. Maintenance costs are also optimized, since the Cloud infrastructure is kept up to date by the Cloud services providers. The proposed CSIP framework addresses the identity verification problem by only accumulating data related to the user's access to resources. Collecting such information does not necessitate the patients to do extra disorderly activities in the process of confirming their individuality. Moreover, only activity types are used in the confirmation process, instead of the activities themselves, which makes the proposed framework less conspicuous than biometric and context-aware methods [9,10]. As the data needed by the framework is composed anyhow and no further information is obligatory, the verification choices are made more rapidly.

The rest of the chapter is devised as follows. Section 10.2 describes the system models. Section 10.3 presents the CSIP secure mutual authentication scheme for the real-time e-Healthcare application. Simulation results and analysis of the security properties are discussed in Section 10.4. Finally, Section 10.5 concludes this chapter.

10.2 System Models

This section will discuss the system model of WMSN and related assumptions.

10.2.1 Network Model

The considered network model for the proposed WMSN is illustrated in Figure 10.1 in which the patient may use medical sensor/smart phone to sense and update health metrics, such as current blood pressure, body movement, heart bit rate, pulse rate and so on for the doctor/medical technical expert.

The doctor/medical expert may access the current health information of the patients via smart phone system, such as Personal Digital Assistant (PDA) and laptop. The WMSN system model has three actors, namely patient, medical expert, authentic-gateway access.

Since the patient and medical experts share the confidential information over an Ad hoc network, this chapter thus interpolates an authentic-gateway system on the Ad hoc network to provide proper user authentication, secure-session key sharing and minimum computation overload as the information is sent fast with little or no slow human intervention. Additionally, it is user friendly and resistant to potential attack of a privileged-insider, from unauthorized personal access to confidential information, by providing a secure gateway between the physician and the patient.

In addition, a smart PnP sensor like IEEE 1451.7 (incorporated with the medical sensor) is used to define the hardware and software transducer interface (sensor or actuator). The specific objective of IEEE 1451.7 [17] is the sensor security that is interfaced between sensor and user, like patient/medical expert to achieve the features of mutual authentication and session-key establishment. Thus, it is preferred for security operations and secure data (multimedia) transmissions. To simplify the sensor descriptive characteristics, a smart sensor like medical sensor is incorporated

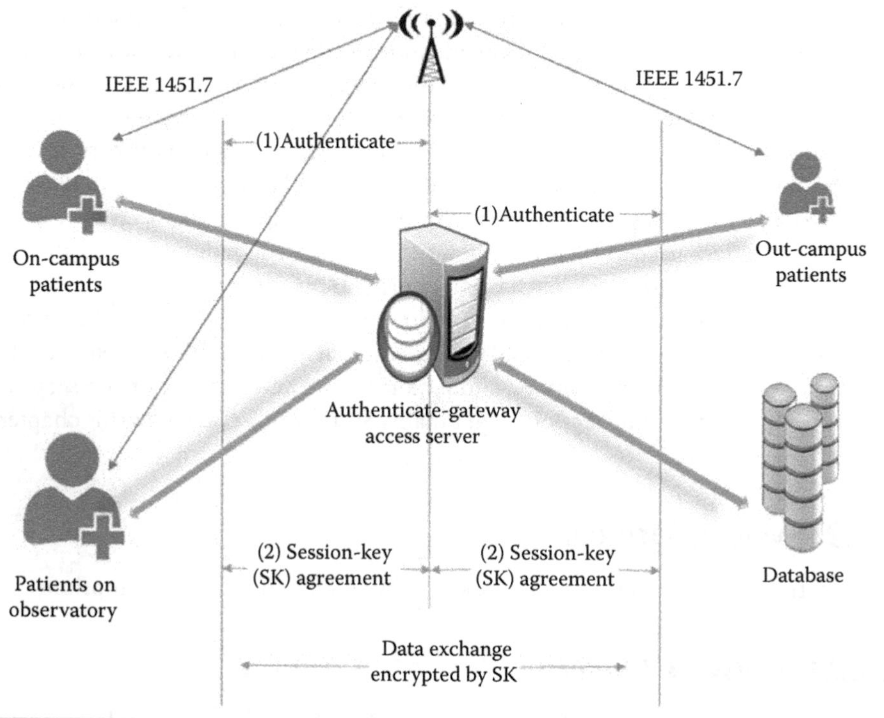

FIGURE 10.1 Network model of the proposed WMSN.

as a Transducer Electronic Data Sheet (TEDS). The main purposes of this technology are as follows:

1. Assisting the host-device to identify the sensor related parameters;
2. Providing the wireless connection to the host-device provided self-descriptive sensor parameters.

10.3 Proposed CSIP Approach

In this section, this chapter presents a secure-cum-mutual user authentication scheme for the WMSN system using medical-sensor/smart phone and the prime objective of this scheme is to ensure that the medical transmission of information between the patient and doctor are mutually authenticated or not. To execute the authentication scheme, the following assumption is considered: (1) the authentic-gateway is believed to be a trustworthy node; (2) two long-term master key should be used between the entities (based on existing identity-based key agreement protocols such as SCK-1, and SCK-2 [25]); and (3) a long-term secret-session key should be shared to ensure mutual authenticity $ssk_{a-gw} = H\left(S_N \oplus ID_{gw}\right)$.

10.3.1 Three-Phase CSIP

The proposed scheme is comprised of three phases: system registration, system login, and system authentication.

Phase I - System Registration

In this phase, the medical expert enters the credentials into the authentic-gateway system and its related execution flows are as follows:

Step 1: The medical expert chooses M_{id} and s_k and then he/she submits into the authentic-gateway node via secure channel.
Step 2: Upon receiving the medical expert M_{id} and s_k, the authentic-gateway determines the followings: $C = E_J\left[M_{id} \| ID_{gw}\right]$ and $N_i = H(M_{id} \oplus s_k \oplus S_{key})$.
Step 3: Thenceforth, the authentic-gateway provides a secure-ware to the medical expert with the configuration of the following parameter $\left\{H(\cdot), C, N_i, S_{key}\right\}$. Herein, S_{key} is a long-term gateway secret key that is securely bound between the entities.

Phase II - System Login and User-Authentication Phases

This phase may be invoked, when the medical expert visits the patients' ward and wishes to review the current information status of the patient. To access such confidential data, the medical expert should enter a proper long-term secret key into the smart phone system. Upon receiving the login-request, the authentic-gateway verifies the long-term secret key into the system database to execute the following operations: $N_i^* = H\left(M_{id} \oplus s_k \oplus S_{key}\right)$; Compare: $N_i^* = N_i$; Compute: $H\left(M_{id}\right)$ and $CID_i = E_k[H\left(M_{id} \| M \| S_N\right]$; and eventually, generate $\left\{CID_i, C, T'\right\}$, then sends to the authentic-gateway node. Herein, M is a random nonce determined by the medical expert to establish secure session-key. Upon receiving the medical expert's message, the authentic-gateway executes the following tasks to authenticate his/her access-request:

Step 1: Initially, the authentic-gateway node validates the current access time T_c: validate whether $\left(T_c'' - T_c'\right) \geq \Delta T_c$, if the expression holds, then the authentic-gateway node refuses the access-request and terminate the process. Otherwise, the authentic-gateway executes the further steps. Herein, T_c'' is the current request-time of the authentic-gateway and ΔT_c is the delay time interval.
Step 2: After the successful validation, the authentic-gateway executes the following tasks: Compute: $D = D_J\left[M_{id}^* \| ID_{gw}^*\right]$ from the M_{id}^* and ID_{gw}^*; Compute: $H\left(M_{id}^*\right)$; Compute: $D_k\left[CID_i\right] = E_k[H\left(M_{id} \| M \| S_N\right]$ from the M_{id}', M and S_N. Compare $H\left(M_{id}^*\right) = H(M_{id}')$ and $ID_{gw}^* = ID_{gw}$; if the condition is satisfied, then the request is authentic; Otherwise, terminate the rest of the process; Compute $V_i = E_{SK_{gw}}\left[M_{id} \| S_N \| M \| T_c'''\right]$; generate the

request-message $\{V_i, T_c'''\}$ and then sends the request-message to the nearby medical sensor/access point wherein the medical expert is available to access the patient info.

Step 3: Upon receiving the authentic-gateway message, the medical sensor/smart phone executes the following tasks: S_N verifies the time T_c''': validate whether $T_c'''' - T_c''' \geq \Delta T_c$, if the validation is successful, the medical sensor node rejects the authentic request-message of the authentic-gateway and terminate the further process. Herein, T_c'''' is the current execution time of the medical sensor node and ΔT_c is the delay time interval.

Step 4: After the successful validation, the medical sensor node executes the following tasks: Compute $D_{Sk-gw} = D_{gw}\left[M_{id}^* \| S_N^* \| ID_{gw}^* \| M^*\right]$ from the $M_{id}^*, S_N^*, ID_{id}^*, M^*$; Compare $S_N^* = S_N$, if the condition holds, then terminate the request-message of authentic-gateway node; otherwise continue the further process; S_N computes the secure-session key $ss_k = H\left(M_{id}^* \| S_N \| M^*\right)$; Compute: $L = E_{ssk}\left[S_N \| M^* \| T_c^*\right]$; Generate the response-message $\{L, T_c^*\}$ and then sends the response-message to the medical expert.

Step 5: Upon receiving the response-message from the medical sensor node, the medical expert executes the following tasks: Medical expert validates the time interval T_c^*. Validate, whether $T_c^{**} - T_c^* \geq \Delta T_c$, if the condition is satisfied, then refuses the request and terminate the process. Otherwise, continue the rest of the process. Herein, T_c^{**} is the current process time of the medical expert's system and ΔT_c is the delay time interval.

Step 6: After the successful validation, the medical experts' system executes the following tasks: Compute: $SK = H(M_{id} \| S_N \| M)$; Decrypt: Using L, to obtain the valid SK, S_N^* and M^*; Compare: $S_N^* = S_N$ and $M^* = M$, if the condition is satisfied, then the secure session key has been established between the medical sensor and experts successfully; Otherwise not established successfully.

Phase III - Session-Key Update Phases

The session-key update is called forth while $User_i$ wishes to update his/her password. The working procedure for secret-update is as follows:

Step 1: $User_i$ puts his/her smart-card in the terminal to enter his/her credentials, namely M_{id} and s_k.

Step 2: Upon the credentials validation and verification, the smart-card carries out the operation of $N_i' = H\left(M_{id} \oplus s_k \oplus S_{key}\right)$ to compare with $N_i' = N_i$. If the comparison is successful, then the rest of the operation will be proceeded. Otherwise, the operation will be aborted.

Step 3: If validation is successful, $User_i$ is asked to enter a new secret-key s_k^{new}.

Step 4: Compute $N_i^{new} = H\left(M_{id} \oplus s_k^{new} \oplus S_{key}\right)$ to replace N_i with N_i^{new} in the smart-card.

10.3.2 Cloud-Based CSIP Architecture

The architecture of the cloud-based side of CSIP is depicted in Figure 10.2. Figures show the client-based CSIP consists of about thirteen main modules described as follows:

- **CSIP Clients Communication Proxy:** The smart device might not be supporting the Hypertext Transfer Protocol (HTTP) protocol (such as Contiki devices and devices adopting CoAP protocol); hence, this module must do the mapping between the protocols, and synchronize transferring the blocks with the cloud-based side.
- **Data Processor:** It is the module responsible for collecting the data blocks and requests from the CSIP clients, then pass them to the history manager and reasoner modules. The data processor consists of a data collector and a data inspector. The data collector collects the data from different CSIP clients, while the inspector checks if the proper encryption and compression techniques are adopted, also it performs pre-processing steps on the collected data before passing it to the reasoner and history manager modules.
- **History Manager:** It is the module responsible for storing the inhabitants' data blocks and requests, so they can be accessed by the reasoning module for learning purposes.

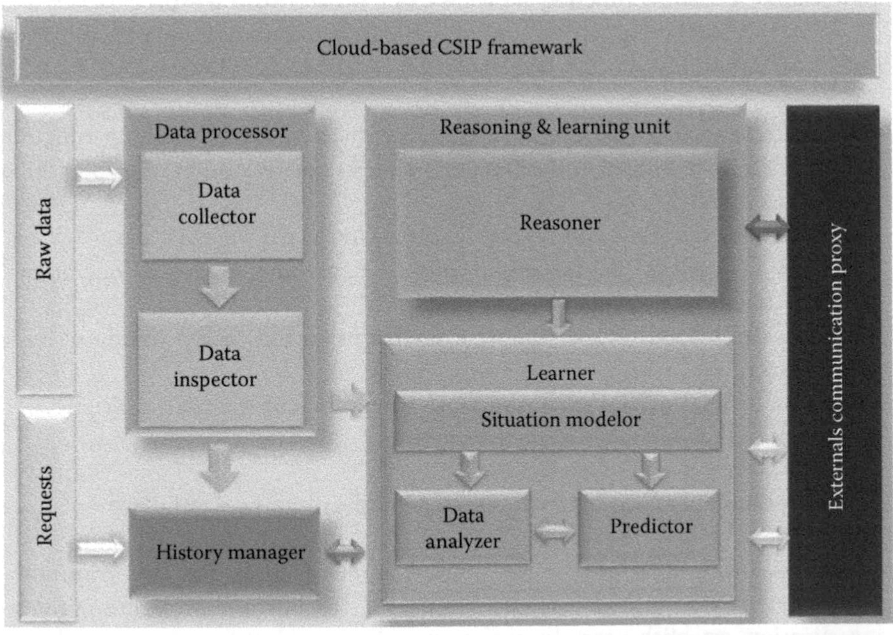

FIGURE 10.2 Cloud-based CSIP architecture.

- **Reasoning & Learning Unit:** It is the module responsible for detecting and predicting the inhabitants' usage and access patterns. This module should apply different on-line learning techniques to be able to model inhabitants' situations based on the coming inputs and previous history of inhabitants' interactions. The reasoner consists of a data analyzer, a predictor, and a situation modeler. The data analyzer applies different data mining techniques to extract the access and usage patterns. The predictor uses the given data to generate the expected inhabitant behavior in the detected context. The data analyzer and predictor send their results to the situation modeler to create a model for the current inhabitant situation. If the predictor model and the data analyzer results are not matching, the situation is considered a conflict and the reasoner has to make a decision to resolve such conflict.

10.4 Performance Evaluation

In this section, the proposed CSIP framework is evaluated and compared with related authentication schemes. To evaluate the various cryptographic operation, an extensive verification is done using MIRACLE C/C++ library with the system features of 32-bit Windows 7 Operating Systems and Microsoft Visual C++ 2008 software edition. To examine realistically, the execution time of symmetric key encryption/decryption $(AES-128)$, elliptic-curve point scalar multiplication over finite-field f_p and $SHA-1$ hashing function are set as $T_{ED} \approx 0.1303$ ms, $T_M \approx 0.7.3529$ ms and $T_{SH} \approx T_{MH} \approx 0.0004$ ms as referred in Ref. [26]. To better understand the evaluation criteria of communication cost, some notation is defined as follows:

- T_{SH} is defined as the execution time of one-way secure hashing function $H_2(\cdot)$. T_{MH} is defined as the execution time of one-way point to map hashing function $H_1(\cdot)$.
- T_P is defined as the computation time of bilinear pairing function.
- T_A is defined as the execution time of one-point additional operational function.
- T_{ED} is defined as the execution time of encryption and decryption algorithmic function.
- T_M is defined as the execution time of elliptic-curve scalar multiplication function.

This subsection will explain the requirements in detail that are as follows:

Mutual Authentication: In the mutual authentication, the two communication parties authenticate each other for the purpose of secure communication establishment. In the proposed scheme, the patient and doctor authenticate each other using $ssk_{a-gw} = H\left(S_N \oplus ID_{gw}\right)$. In the system login and authentication phase, the authentic-gateway access server authenticates the patient/doctor using the computation of $D = D_J\left[M_{id}^{*} \| ID_{gw}^{*}\right]$ and $D_k\left[CID_i\right] = E_k[H\left(M_{id} \| M \| S_N\right]$ to verify whether the patient/doctor satisfies the conditional expression of $V_i = E_{SK_{gw}}\left[M_{id} \| S_N \| M \| T_c'''\right]$ to access the data being transmitted. Though the adversary intercepts the login request-message of either patient or doctor and wishes to forge as a legitimate authentic-gateway server, the adversary cannot compute the parameters' like $\left\{H(\cdot), C, N_i, S_{key}\right\}$. Subsequently, the adversary cannot send a valid response-message to the authentic-gateway server. Hence, the proposed scheme asserts that it holds the security property of mutual authentication.

Secret Key Generation: In the proposed scheme, the patient and doctor can share a secure session-key via an authentic-gateway access after the successful execution of authentication phase. By then, the patient and doctor can exchange the real-time data securely using the establishment of secure-session key, which is used to encrypt the real-time data gathered by the smart medical sensor. The secure session-key is determined from the computation of $ssk_{a-gw} = H\left(S_N \oplus ID_{gw}\right)$ and validated from the computation of $V_i = E_{SK_{gw}}\left[M_{id} \| S_N \| M \| T_c'''\right]$. Since the parameter of T_c changes over a period of time; thus, the different set of a session—key will be used to authenticate the session. Hence, the proposed scheme asserts that it holds the security property of the session-key agreement.

Resilient to Privileged-Insider Attack: The proposed authentication scheme never sends the parameters, such as $\left\{H(\cdot), C, N_i, S_{key}\right\}$ to the authentic-gateway access server as the plaintext. The user, like patient/doctor sends the parameter as $H\left(S_N \oplus ID_{gw}\right)$, thus the authentic-gateway access cannot acquire the users' secret-key without the knowledge of $H\left(M_{id}\right)$ and $CID_i = E_k[H\left(M_{id} \| M \| S_N\right]$; since it is highly secured. Moreover, the expression $H\left(S_N \oplus ID_{gw}\right)$ is eventually verified using $V_i = E_{SK_{gw}}\left[M_{id} \| S_N \| M \| T_c'''\right]$ to authenticate the access of the session to the patient/medical expert. Thus, the adversary cannot compute a valid session-key without the proper deduction of S_N, ID_{gw}. Hence, the proposed scheme asserts that it is resilient to the attack of a privilege-insider.

Resilient to Replay Attack: Assume that the adversary uses a previous message to authentic-gateway $\{CID_i, C, T'\}$, medical sensor node $\left\{V_i, T_c'''\right\}$ and user $\left\{L, T_c^{*}\right\}$ to gain the access privilege. As the messages are validated using fresh timestamp, the adversary cannot be succeeded with the entries of previous message. Thus, the proposed scheme assert that it is resilient to the relay attack.

Resilient to User Masquerading Attack: As an instance, the adversary has forged a message of login $\{CID_i, C, T'\}$. So, he/she tries to login with an updated message $\left\{CID_i^{new}, C, T'\right\}$ to duplicate the context of the user identity CID_i^{new}. However, the context identity CID_i^{new} cannot decrypt the original message using $\left[M_{id}^{*} \| S_N^{*} \| ID_{gw}^{*} \| M^{*}\right]$. Thus, the proposed scheme claims that it can resist the user masquerading attack.

Resilient to Secret Gateway Guessing Attack: The proposed scheme has three different master keys, namely k, M and J. These keys are not transmitted as the plaintext; and thus the proposed scheme asserts that it is resilient to the secret gateway guessing attack.

By the above procedure, the medical expert and sensor node can authenticate one another to access the WMSN. Eventually, the medical expert/doctor can access the private info of the patient from the medical sensor node/smart device via authentic-gateway node. Since the secure session-key is tightly bound between the entities, the proposed CSIP mechanism can achieve the security goals, namely mutual authentication, session-key establishment and resilient to privileged-insider, replay, user masquerading, and secret gateway guessing attack for the WMSN as shown in Table 10.1. Unlike other competitive baselines in the literature, CSIP achieves all the aforementioned security goals. Table 10.2 shows the computational efficiencies of proposed authentication scheme. Since the proposed scheme has less computation cost, it mitigates the execution time of the phase to improve the performance of the WMSN in comparison with the other authentication schemes [20–22]. Moreover, we compare our proposed CSIP approach against the other authentication schemes in terms of number of required messages to deliver a specific amount of data (e.g., 512 bits), bandwidth utilization and overhead percentage as shown in Table 10.3. In this table, we assume three security levels per approach, which are L_1–L_3, respectively, and we assume that the corresponding length of the message authentication codes is 2/4/6 (in bits). Since our CSIP approach allows data with different levels of security requirement to be packed together while other

TABLE 10.1 Comparison of Security Properties

	[20]	*[21]*	*[22]*	*Proposed CSIP Scheme*
Mutual authentication	Not provided	Yes	Yes	Yes
Session-key agreement	Not provided	Partial	Yes	Yes
Resilient to privileged-insider attack	Not provided	Yes	Partial	Yes
Resilient to replay attack	Not provided	Yes	Yes	Yes
Resilient to user masquerading attack	Not provided	No	No	Yes
Resilient to secret gateway guessing attack	Not provided	No	No	Yes

TABLE 10.2 Computational Efficiency

Scheme	*System Registration*		*System Login and Authentication*		
	Medical Experts	*Authentic-Gateway*	*Medical Experts*	*Authentic-Gateway*	*Medical Sensor Node*
Proposed CSIP scheme	-	1H + 1S	4H + 2S	1H + 3S	1H + 2S
[20]	2H	6H + 1S	7H + 3T_E	3H + 1S + 1T_E	4H + 1T_E
[21]	1H	(n + m) H	2H + 1ECC	5H + 2ECC	3H + 1ECC
[22]	1H	(n) 2H	4H	8H + 1E	8H + 1E

H, denotes one-way hash function; S, denotes symmetric crypto-system; E, denotes exponential operation; ECC, denote elliptic curve encryption or decryption operation; and T_E, denote exponential operation.

approaches do not allow this, both the obtained number of messages and the bandwidth overhead are smaller in general for CSIP as compared to others. The reason is that compared with the bandwidth overhead brought by increasing the length of the message authentication code to guarantee the message security, the experienced bandwidth overhead by dividing data into different messages is much bigger. We can see from Table 10.3 also that along with the increasing of the security requirement, the obtained number of messages and the bandwidth overhead also increase. However, as CSIP is based on learning approach, even when the level of security requirement reaches L_3, the bandwidth overhead is limited to 3.1%. Thus, CSIP

TABLE 10.3 Comparison of Design Factors under Varying Security Levels

	[20]			*[21]*			*[22]*			*Proposed CSIP Scheme*		
	L_1	L_2	L_3	L_1	L_2	L_3	L_1	L_2	L_3	L_1	L_2	L_3
Number of messages	20	21	23	18	19	21	18	18	20	16	17	17
Bandwidth utilization (%)	20.19	21.5	22.3	19.67	20.12	22.6	21.93	22.6	24.1	23.67	24.12	25.6
Bandwidth overhead (%)	5.17	6.2	7.9	3.99	4.1	5.2	4.73	3.8	5.1	2.73	2.91	3.1

solves the tradeoff problem, where the bandwidth utilization is improved and the requirements in both security and real-time are met.

We remark also that CSIP divides the protected resources in the smart environment into a sum of assertion/security levels (n). All the following tests that have been used in assessing the CSIP method presumed the total of assertion levels n equal to 3. The verge values of the assertion levels are set as follows: $g_{t1} = 0$, $g_{t2} = 0.33$, and $g_{t3} = 0.66$. Where $g_{ti} \in [0,1]$ and represents the assertion level threshold value of the ith assertion level that can be calculated based on the following equation:

$$g_{ti} = (i-1)^{*}. \tag{10.1}$$

It imitates how assured the system must be about a user in order to state his/her identity before he/she is allowed to access the resource.

10.4.1 The Effect of Δg

Altering the parameter Δg marks how the system assurance is dispensed as shown in Figure 10.3. In this figure when Δg is set to values between 0.1 and 0.5, the scheme assurance increases gradually and needs at least six events for the global assertion value to reach the following threshold value. This situation would be useful in extremely protected locations. However, when Δg has to experience values such as 0.7–1.0, the scheme assurance increases much faster with less number of events to grasp the second level; this situation would be useful for more relaxed surroundings.

In Figure 10.4, the consequences of trying the identification for major events are examined. This figure shows big differences for the three examined Δg values

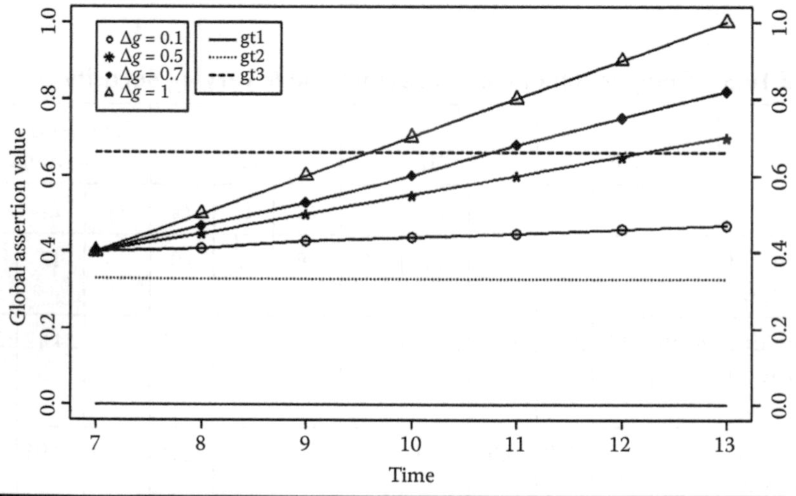

FIGURE 10.3 CSIP response for repeated events with altered Δg values.

(0.1, 0.5, and 1). In case of Δg value equal to 0.1, the assertion value increases slowly and necessitates almost three significant events to reach the next assertion level; this is considered to be undesirable as it indicates the CSIP scheme could not believe the identified event. For Δg equal to 0.5 and 1, the affirmation value rises faster and necessitates one major event to reach the equal effect, which is more preferable.

10.4.2 The Effect of Assertion Value on the CSIP Response

Picking diverse settings to denote the local assertion values for repeated events will yield alike responses for strict, moderate, and relaxed surroundings per Figure 10.5.

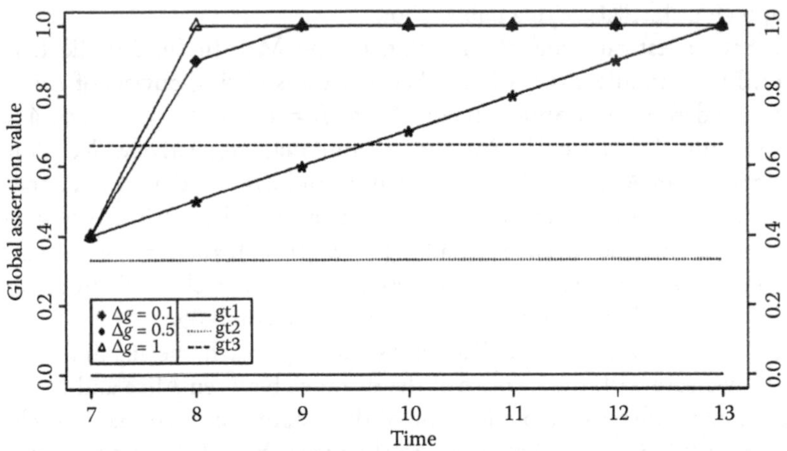

FIGURE 10.4 CSIP response for identifying major events with altered Δg settings.

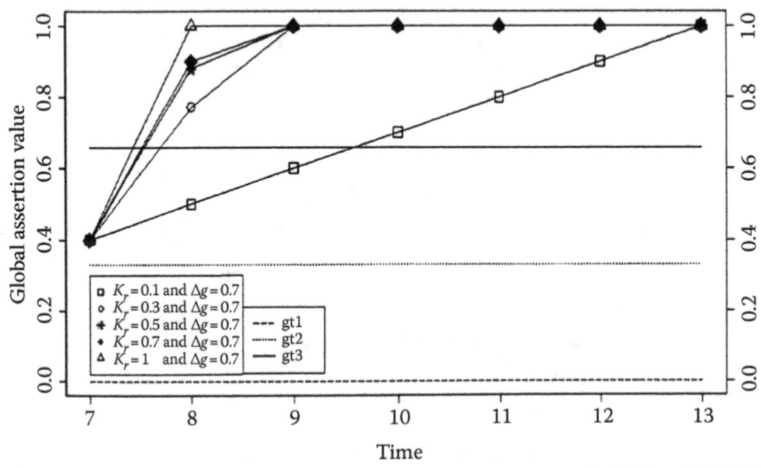

FIGURE 10.5 CSIP response for diverse repeated event values.

As depicted in the figure, selecting K_r values between 0.7 and 1.0 renders this type of event to have an identical effect to that of the major event type. Consequently, K_r is suggested to be assigned a value between 0.1 and 0.5.

To verify the security and suitability of the proposed Secret Key Generation (SKG) scheme, we compare the SKG Rate (SKGR) of the proposed scheme based on private pilot with the approaches based on random beam-forming described in Ref. [33], which are based on a public pilot, using Monte Carlo simulations. We assume Rayleigh fading channels and the entries obey Gaussian distribution with zero mean and unit variance. In addition, we consider the receiver noises as white Gaussian noises with zero mean and unit variance, and a middle person, Eric, can estimate channels H_{AE} between Alice and Eric, and H_{BE} between Bob and Eric, while applying the SKG with a public pilot.

We first investigate how the location of the Man-In-The-Middle (MITM) attacker; Eric, can affect the SKGR when the transmitting powers of Alice, Bob and Eric are described by SNR = 10 and $N_A = N_E = 4$, and $N_B = 1$, 2, and 4, where Alice, Bob, and Eric have N_A, N_B, and N_E antennas, respectively. We normalize the distance from Alice and Bob to Eric as 1, and assume that the channel gain between antenna i of the transmitter and antenna j of the receiver is $h_{ij} = \chi d_{ij}^{-l/2}$, where d_{ij} is the distance between i and j, $\chi \sim \mathcal{N}\left(0, \sigma^2\right)$, and $l = 2$ is the path-loss exponent. We assume that Eric moves along the wireless channels from Alice to Bob, the distance changes from 0.1 to 0.9 in the interval 0.05.

Figure 10.6 shows that the SKGRs rise at first, reach the maximum values, and then decrease with the increase of the distance between Alice and Eric, while there are some differences about location of the maximum; such that it reaches the maximum at 0.6, 0.5, and 0.45, respectively, when $N_B = 1$, 2, and 4. It indicates

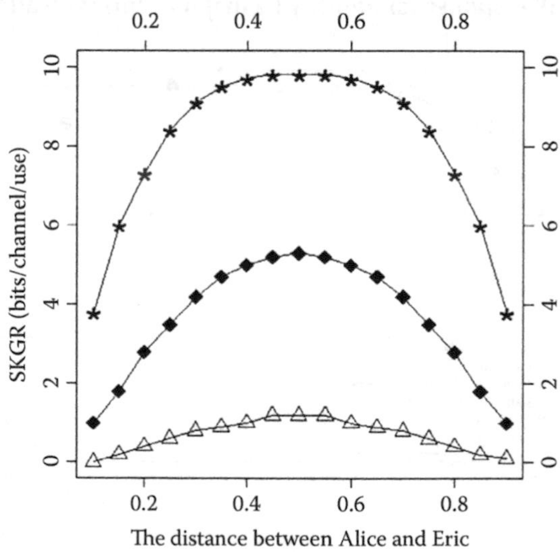

FIGURE 10.6 SKGR varies with the location of attacker.

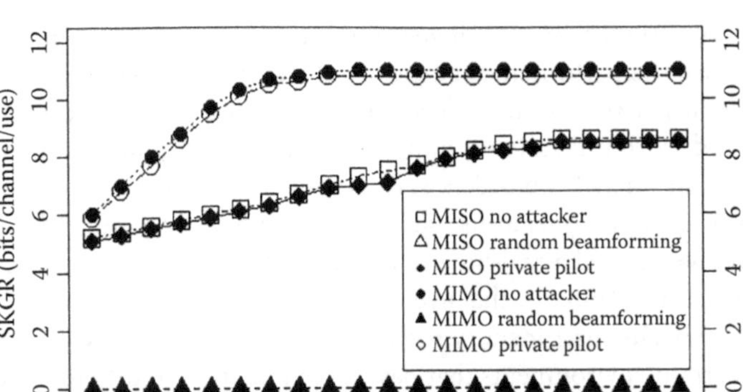

FIGURE 10.7 SKGR varies with SNR of transmitter.

Eric can intercept more information when it is closer to Bob than Alice, and Eric moves to the middle location with the antenna count increase at Bob. Since both multi-antenna and high SNR can increase the SKGR, this result is expected.

Second, we investigate the security of the Reconstitution Wireless Channels (RWC) in Static or Quasi-Static Environments (SQSE) by comparing the SKGR based on private pilot and reconstitution channels with that based on the random beam-forming described in Ref. [33] when $N_A = N_E = 4$, and $N_B = 1$, and $N_A = N_E = N_B = 4$, respectively, which are corresponding to the Multiple-Input Multiple-Output (MIMO) and Multiple-Input Single-Output (MISO) scenarios, respectively. We explore the SKGR for three cases under the MITM attacks: (1) no attacker; (2) based on Random Beam-Forming (RBF) described in Ref. [33]; and (3) based on the proposed private pilot. The former approaches use public pilot to estimate the wireless channels.

Figure 10.7 shows that the SKGRs based on the private pilot and reconstitution channels rise with the increasing of the SNR of the transmitter, which is close to the SKGR without attacker. However, the SKGR based on random beam forming with public pilot described in Ref. [33] are almost equal to zero under the MITM attacks, which shows that the MITM attack is a serious threat to SKG.

10.5 Conclusion

In this chapter, a secure-CSIP mutual authentication framework has been proposed for the health care system using WMSN. In addition, this chapter has exploited the two-factor (namely medical expert and sensor node) strategy for the mitigation of computational cost and fulfillment of WMSN security goals, such as session-key agreement and resilient to privileged-insider, replay, user masquerading, and secret gateway guessing attack. The proposed authentication meets the demand of platform adaptability, and thus suitable for all the real-time mission critical applications systems. Besides, the proposed scheme shows that it has less computational overhead to improve the performance efficiency of the systems. Moreover, it solves the natural tradeoff between security level and the added communication overhead, where the bandwidth utilization is improved and the requirements in both security and real-time are met. It considers several security threats including privileged-insider, replay, user masquerading, secret gateway guessing, and MITM attacks. Furthermore, we showed a real-life use case for CSIP adoption while investigating critical design factors. Finally, the investigated use case scenarios show that CSIP is successful in verifying user identity with high degree of reliability. It is particularly effective in certain environments that have a combination of different security level requirements and the user is authenticated automatically based on history/learning mechanisms.

References

1. B. Karschnia, Industrial internet of things (IIoT) benefits, examples | control engineering, Controleng.com, 2017. [Online]. Available: http://www.controleng.com/single-article/industrial-internet-of-things-iiot-benefits-examples/a2fdb5aced1d779991d91ec-3066cff40.html, [Accessed: 31-Aug-2017].
2. C. Qingping, H. Yan, C. Zhang, Z. Pang, and L. Xu, A reconfigurable smart sensor interface for industrial WSN in IoT environment, *IEEE Trans. Ind. Inform.*, vol. 10, no. 2, pp. 1417–1425, 2014.
3. V. Shnayder, B.-R. Chen, K. Lorincz, T.T.F. Fulford-Jones, and M. Welsh, Sensor networks for medical care, Harvard University, Technical Report TR-08-05, Apr. 2005.
4. D.S. Lee, Y.D. Lee, W.Y. Chung, and R. Myllyla, Vital sign monitoring system with life emergency event detection using wireless sensor network, *IEEE Conference on Sensors,* Daegu, pp. 518–521, 2006.
5. W. Chung, C. Yau, and K. Shin, A cell phone based health monitoring system with self-analysis processing using wireless sensor network technology, *Proceedings of the International Conference of the IEEE EMBS*, Lyon, France, pp. 3705–3708, Aug. 2007.
6. S. Koch and M. Hagglund, Health informatics and the delivery of care to older people, *Maturitas*, vol. 63, no. 3, pp. 195–199, 2009.
7. O. Omeni, O. Eljamaly, and A. Burdett, Energy efficient medium access protocol for wireless medical body area sensor networks, *4th IEEE/EMBS International Summer School and Symposium on Medical Devices and Biosensors*, Cambridge, UK, pp. 29–32, Aug. 2007.
8. G. Manes, G. Collodi, R. Fusco, L. Gelpi, and A. Manes, A wireless sensor network for precise volatile organic compound monitoring, *Int. J. Distrib. Sens. Networks*, vol. 8, no. 4, p. 820716, 2012.
9. Y.M. Yuang, M.Y. Hsieh, H.C. Chao, S.H. Hung, and J.H. Park, Pervasive, secure access to a hierarchical sensor-based healthcare monitoring architecture in wireless heterogeneous networks, *IEEE J. Sel. Areas Commun.*, vol. 27, pp. 400–411, 2009.

10. G. Zhao, Wireless sensor networks for industrial process monitoring and control: a survey, *Network Protoc. Algorithms*, vol. 3, pp.46–63, 2011.
11. G. Manes, G. Collodi, R. Fusco, L. Gelpi, and A. Manes, Continuous remote monitoring in hazardous sites using sensor technologies, *Int. J. Distrib. Sens. Networks*, vol. 8, no. 7, 2012. doi:10.1155/2012/317020.
12. K. Lu, Y. Qian, M. Guizani, and H.H. Chen, A framework for a distributed key management scheme in heterogeneous wireless sensor networks, *IEEE Trans. Wireless Commun.*, vol. 7, pp. 639–647, 2008.
13. X. Du, Y. Xiao, M. Guizani, and H.H. Chen, An effective key management scheme for heterogeneous sensor networks, *Ad Hoc Networks*, vol. 5, pp. 24–34, 2007.
14. P. Traynor, R. Kumar, H. Choi, G. Cao, S. Zhu, and T.L. Porta, Efficient hybrid security mechanisms for heterogeneous sensor networks, *IEEE Trans. Mobile Comput.*, vol. 6, pp.663–677, 2007.
15. M.L. Das, Two-factor user authentication in wireless sensor networks, *IEEE Trans. Wireless Commun.*, vol. 8, no. 3, pp. 1086–1090, 2009.
16. Y. Cheng and D.P. Agrawal, An improved key distribution mechanism for large-scale hierarchical wireless sensor networks, *Ad Hoc Networks*, vol. 5, pp. 35–48, 2007.
17. X.H. Le, M. Khalid, R. Sankar, and S. Lee, An efficient mutual authentication and access control scheme for wireless sensor networks in healthcare, *J. Networks*, vol. 6, pp. 355–364, 2011.
18. D.M. Lal, Two-factor user authentication in wireless sensor networks, *IEEE Trans. Wireless Commun.*, vol. 8, pp. 1086–1090, 2009.
19. K.M. Khurram and K. Alghathbar, Cryptanalysis and security improvement of 'two-factor user authentication in wireless sensor networks', *Sensors*, vol. 10, pp. 2450–2459, 2010.
20. D. He, N. Kumar, and N. Chilamkurti, A secure temporal-credential-based mutual authentication and key agreement scheme with pseudo identity for wireless sensor networks, *Inf. Sci.*, vol. 321, no. 10, pp. 263–277, 2015.
21. A. Hamed and M. Nikooghadam, Three-factor anonymous authentication and key agreement scheme for telecare medicine information systems, *J. Med Syst.*, vol. 38, no.12, pp.1–12, 2014.
22. H. Debiao, Y. Zhang, and J. Chen, Cryptanalysis and improvement of an anonymous authentication protocol for wireless access networks, *Wireless Pers. Commun.*, vol. 74, no. 2, pp. 229–243, 2014.
23. F. Al-Turjman, Impact of user's habits on smartphones' sensors: an overview, *IEEE HONET-ICT*, Kayrenia, pp. 70–74, 2016.
24. I. Elgedawy and F. Al-Turjman, Identity provisioning framework for smart environments, *IEEE HONET-ICT*, Kayrenia, pp. 12–16, 2016.
25. C. Liqun, Z. Cheng, and N.P. Smart, Identity-based key agreement protocols from pairings, *Int. J. Inf. Secur.*, vol. 6, no. 4, pp. 213–241, 2007.
26. M.C. Chuang and M.C. Chen, An anonymous multi-server authenticated key agreement scheme based on trust computing using smart cards and biometrics, *Expert Syst. Appl.*, vol. 41, no. 4, pp. 1411–1418, 2014.
27. R. Watro, D. Kong, S. Cuti, C. Gardiner, C. Lynn, and P. Kruus, TinyPK: securing sensor networks with public key technology, *Proceedings of the 2nd ACM Workshop on Security of Ad Hoc and Sensor Networks (SASN 2004)*, Washington, DC, pp. 59–64, 2004.

28. D. Wang, N. Wang, P. Wang, and S. Qing, Preserving privacy for free: efficient and provably secure two-factor authentication scheme with user anonymity, *Inf. Sci.*, vol. 321, pp. 162–178, 2015.
29. J. Yuan, C. Jiang, and Z. Jiang, A biometric-based user authentication for wireless sensor networks, *Wuhan Univ. J. Nat. Sci.*, vol. 15, no. 3, pp. 272–276, 2010.
30. E.-J. Yoon and K.-Y. Yoo, Robust biometrics-based multiserver authentication with key agreement scheme for smart cards on elliptic curve cryptosystem, *J. Supercomput.*, vol. 63, no. 1, pp. 235–255, 2013.
31. H.R. Tseng, R.H. Jan, and W. Yangand, An improved dynamic user authentication scheme for wireless sensor networks, *Proceedings of IEEE Global Communications Conference Exhibition & Industry Forum*, Washington, DC, pp. 986–990, 2007.
32. T.H. Lee, Simple dynamic user authentication protocols for wireless sensor networks, *Proceedings of the International Conference on Sensor Technologies and Applications*, Cap Esterel, pp. 657–660, 2008.
33. L. Cheng, W. Li, and D. Ma, Secret key generation via random beamforming in stationary environment, *International Conference on Wireless Communications & Signal Processing (WCSP)*, Nanjing, China, pp. 1–5, 2015.

Chapter 11

Green Femtocells in the IoT Era—An Overview*

Fadi Al-Turjman

Antalya Bilim University

Contents

* F. Al-Turjman, E. Ever, and H. Zahmatkesh, "Green femtocells in the IoT era: Traffic modelling and challenges – An overview", *IEEE Networks Magazine*, 2017. doi:10.1109/MNET.2017.1700062.

11.1 Green Femtocell

Mobile networks are used in smart grids in order to provide energy efficiency. However, it is a fact that the energy they consume cannot be regarded as negligible. In addition, carbon emission is also a result of energy usage and as mentioned in Ref. [1], Information and Communication Technologies (ICT) is responsible for between 2% and 2.5% of total carbon emission, which is surprisingly same as global carbon emission rate of aviation industry. It is expected that this rate will be doubled over the next decade. The 10% of ICT carbon emission is produced by mobile phone networks and between 55% and 60% of the energy is consumed by Radio Access Network (RAN) [1]. To illustrate, the power consumption of each macro base station (BS) is between 2.5 and 4 kW. As the number of these stations reach to tens of thousands, the amount of energy consumption and carbon emission becomes quite significant.

It is well known that macrocell BSs consume considerable amounts of energy and most of this energy is used by driving high power Radio Frequency (RF) signal. However, it is estimated that only between 5% and 10% of this energy is used to create useful signal. The reason behind that issue is that macrocells provide high area coverage and most of the area is empty space with no users that need radiated signals. Moreover, RF sections of BSs are run at high power, with low efficiency to obtain required linearity. In other words, a lot of energy is used "just in case" [2]. Femtocells can be solutions to prevent these energy losses. Femtocells provide local coverage while macrocells provide huge area coverage. In other words, femtocells deliver power where there is a need. Therefore, they require less RF power to provide high bandwidth since they are closer to the user. An average femtocell consumes total 2 W per hour while it is very high in microcells. For example, based on the power model presented in Ref. [3], the average power consumption of a macrocell BS per hour for weekdays and for the weekend are between 1016 and 1087 W [3]. The other advantage is that user equipment consumes less energy when it is connected to femtocell since it is closer to the BS. Therefore, femtocell is energy-efficient on top of well-known features such as providing more coverage, capacity, better QoS, and longer battery life.

Although the use of femtocells for IoT based infrastructures looks very promising as discussed in the existing studies, the integration of femtocells to the existing systems should be performed by considering relevant criteria such as the performance as well as energy efficiency and availability of the services to be provided. In that sense, traffic modelling of femtocells is an important factor in the integration of femtocells with current macrocellular networks. It can be used to simulate the behavior of a particular application or service provided by femtocells. The main objective of traffic modelling is to provide analytical models that describe the most important parameters of a given traffic type to present the characteristics of a real femtocell network in the IoT environments. In turn, these interactions can easily be employed in optimization of system parameters as well as deployment of new algorithms. In recent years, Mobile Network Objects (MNOs) and vendors have also been working to create smart applications that require introduction of femtocells in the market. Using traffic modelling, new optimization algorithms can be investigated to improve the existing infrastructures with femtocells in terms of the performance measures of interest. In the rest of this chapter, we first present some of the typical applications of femtocells in the IoT environments. Then, we review the elements of femtocells in terms of communication technologies, services, and hardware units, which are essentially incorporated to analytical models with high levels of accuracy. Accordingly, we present, summarize, and analyze various traffic-modelling approaches available in the literature for femtocells in the IoT.

11.2 Femtocell Applications and Elements in IoT

11.2.1 Applications

Femtocell technology is one of the popular infrastructures employed in Heterogeneous Network (HetNet) environments. Various applications can make use of femtocells to enhance coverage and capacity of the network as well as to reduce the traffic loads from the macrocell layer. In the following sections, we list and describe some typical ones of these applications.

Indoor Environments (Shopping malls, Airports): Femtocells are commonly utilized inside home or office environments to improve indoor coverage and provide higher data rates for mobile applications. By deployment of indoor femtocells, receivers and transmitters become closer to each other, which considerably reduce penetration- and packet-loss. Therefore, energy consumption can be effectively decreased. Femtocells can also be deployed in indoor public places such as shopping malls or airports to enhance the users' Internet experience [4].

Public Transportation Vehicles: Femtocells can be utilized in public transportation vehicles such as busses and trains to improve coverage and capacity, and provide better Internet experience for the users while on the move. Using femtocell technology inside a vehicle such as a bus or a train is a concept called Mobile Femtocell (MFemtocell). By the implementation of the MFemtocells, Spectral Efficiency (SE) of the entire network can be improved [5]. In addition, the battery life of the mobile users can last for a longer time because of the shorter-range communication with the serving MFemtocell. It should also be taken into account that MFemtocells are located inside vehicles and the antennas are placed outside of the vehicles. This provides better signal quality inside the vehicles.

Smart Cities: In smart cities, a wide range of services is expected to be available to users. These services include e-commerce, e-health, e-banking, e-government, Intelligent Transportation Systems (ITSs), etc. Smart cities are extremely dependent on the infrastructure of the network. Mobile networks have to support increased coverage and excellent QoS. In this regard, femtocells play an important role in smart city projects. In addition, wireless networking requirements of smart cities cannot be fulfilled with traditional macro-only networks because of the spectrum efficiency and regularity issues to indoor coverage.

Smart Homes: A smart home can be seen as an environment where various communication technologies are utilized to improve the quality of life of the residents. Smart homes can also be viewed as a sub-category of smart cities. In typical smart home scenarios, data are gathered by sensors and actuators and are sent to a processing unit in order to achieve desired goals. Various IoT devices such as cameras and electrical appliances can communicate with an FAP and be monitored by a smartphone or a computer remotely using Bluetooth or the Internet.

Another use of femtocell in smart homes is illustrated in Figure 11.1. Different devices on the network can be accessed from the mobile device, once the users enter the zone under the femtocell registration and coverage. For instance, the user can choose music from the home media server and play on the user's mobile phone.

Health Services: In recent years, electronic health (e-health) monitoring systems have attracted attention since they can provide instant information about physical and psychological fitness while being far from a health center [1]. Using body sensor networks, health information of a mobile user such as body temperature, blood pressure, and heart rate are captured and sent to the mobile device, which is under registration of a femtocell [1]. The information can be checked using a database inside the femtocell. If the user's health is abnormal, the data can be

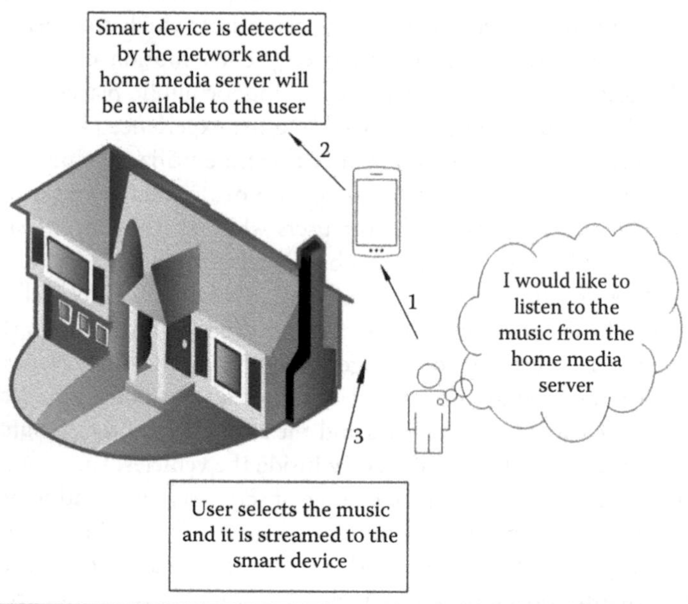
Smart device is detected by the network and home media server will be available to the user
2
1
3
I would like to listen to the music from the home media server
User selects the music and it is streamed to the smart device

FIGURE 11.1 An example of network music server in a smart home.

sent to the health center using a cloud architecture. By using femtocells in e-health monitoring systems, significant reduction in power consumption can be obtained compared to the use of macrocells, microcells, or picocells [1]. Figures 11.2 and 11.3 are provided to quantitatively show the potentials of using femtocells in e-health monitoring systems [1].

Figure 11.2. shows total power consumption in GW for accessing the large amount of health data in GB through the mobile station. It is indicated by Figure 11.2 that by using femtocells for accessing the large amount of health data through the mobile station, significant amount of power can be saved compared to the use of macrocells, microcells, and picocells. Figure 11.3 explicitly illustrates the amount of savings when femtocells are used instead of other technologies.

For all the aforementioned femtocell applications, an accurate traffic-modelling technique is required in order to investigate the performance of the network. Performance measures such as Mean Queue Length (MQL), throughput, and energy consumption are popularly employed in existing studies depending on the objectives of the IoT system with femtocell support.

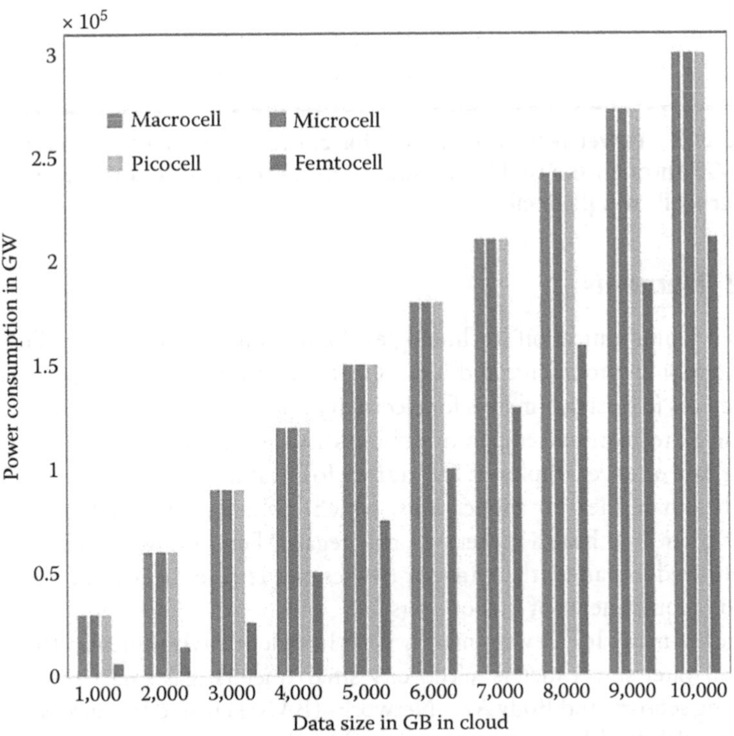

FIGURE 11.2 Total power consumption in GW for accessing the large amount of health data in GB through the mobile station.

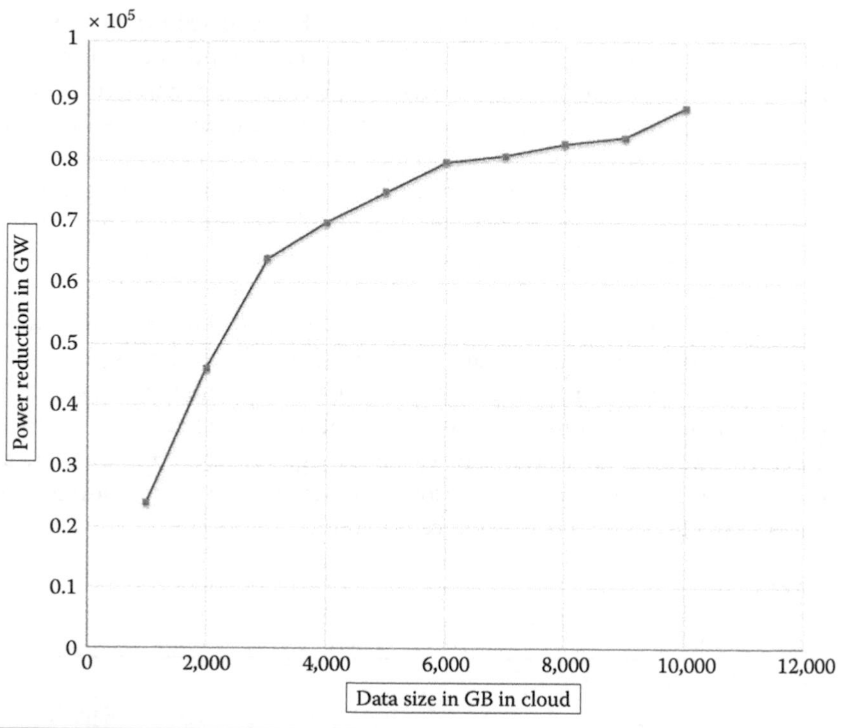

FIGURE 11.3 Power reduction in GW for accessing the large amount of health data in GB through the mobile station by using femtocells compared to macrocell, microcell, and picocell.

11.2.2 Elements

Femtocell Communication Technologies: Significant part of IoT traffic is generated in indoor environments and is designed over cellular technologies. Data from smart devices (e.g., smart meters for electricity and water), from home's monitoring sensors (e.g., to monitor temperature, humidity, light, and pollution level inside a building) are a few examples of IoT traffic. IoT traffic can also be generated from mobile health (m-health) applications, which gathers data of elderly people and transfers them into health centers. In this regard, Femtocell Access Points (FAPs) can be utilized to handle these indoor traffics and reduce the load of macro BSs to meet QoS requirements of indoor users.

There are many IoT devices mainly used in indoor environments. These devices consist of water smart meters, electricity smart meters, gas smart meters, home's monitoring sensors, and Body Area Networks (BANs) created by sensors to monitor elderly people's health parameters such as blood pressure, temperature, and heart rate for m-health applications. Bluetooth, Wi-Fi, ZigBee, Ultra-Wide Bandwidth (UWB), LTE-Advanced (LTE-A), and Light Fidelity (Li-Fi) are examples of communication technologies that these IoT devices use in order to communicate with the network. With the development and existence of 5G network and the expected increase in the number of IoT devices, these devices, using cellular technology, can communicate with an FAP.

For instance, in the case of BAN, the sensors use the technologies such as Bluetooth, Wi-Fi, or ZigBee to communicate with the smartphone of the patient and then the smartphone communicates with the FAP. With the LTE-A, this can happen using Device-to-Device communications (D2D). Other devices such as laptops and mobile phones can directly communicate with the FAP. This allows these IoT devices to benefit from 5G features, which guaranty high levels of Quality of Service (QoS) and provides wireless connectivity in indoor environments without any extra costs. This is because the communication between an FAP and IoT devices can be free of charge.

Femtocell Services and Hardware Units: In general, femtocell services can be classified into four categories. This classification includes Carrier services, Internet services, Femtozone services, and Connected home services as illustrated in Figure 11.4. Carrier services are the common and basic services offered by MNOs. Examples of these services are voice calls, Short Message Service (SMS), and push-to-talk. Internet services such as e-mail, web browsing, ftp, and mobile TV are accessible by any Internet access method such as mobile broadband, and are based on third-party platforms. Femtozone services are services where the femtocell provides or helps in the delivery of the service. This is done through service information from the femtocell or service execution at the femtocell. Examples of these services can be virtual home phone and home presence. Connected home

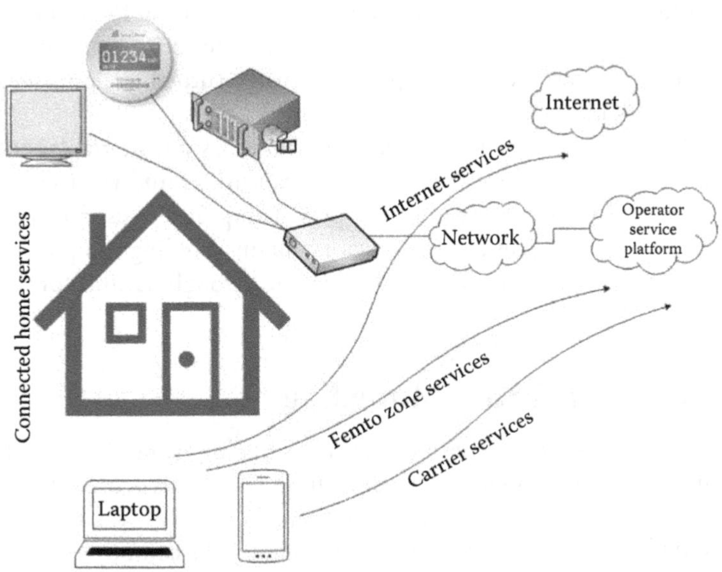

FIGURE 11.4 A classification of femtocell services.

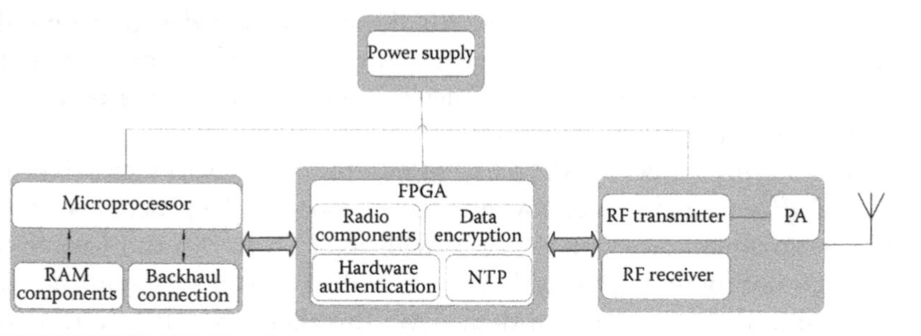

FIGURE 11.5 A typical femtocell hardware model.

services are delivered to a user's mobile device if he/she is in the home environment. Examples of this type of services are automated music synchronization, use of the mobile device as a media player to listen/watch home music/videos collection, use of the mobile device as a home automation remote control, etc.

A good understanding of the femtocell hardware units is crucial to design efficient algorithms and techniques such as sleep mode technique that can be utilized in low traffic periods to switch off unnecessary hardware components and consequently reduce power consumption. A typical hardware model for an FBS is shown in Figure 11.5 [6]. In this model, implementation and management of baseband processing and radio protocol stack as well as administration of the backhaul connection to the core network are done through a microprocessor, which is also connected to some Random Access Memory (RAM) and on chip memory components. In addition to the microprocessor, this model contains a Field Programmable Gate Array (FPGA) and a number of integrated circuitry. This component is responsible for various features such as data encryption, hardware authentication, and Network Time Protocol (NTP).

In the FPGA part, the radio component works as an interface between the microprocessor and RF transceiver. Both RF transmitter and receiver consume a certain amount of power. Hence, an RF power amplifier is used to pass a high power signal to antenna. This model does not contain a cooling component because the cooling process of the hardware units are done through natural convection [6].

11.3 Femtocell Traffic Modelling

Based on the aforementioned communication technologies, services, and hardware unit specifications, an accurate traffic-modelling strategy is required. Integration of femtocell technology with the current macrocellular networks helps MNOs to reduce the traffic loads of the macrocell layer. Femtocells can be used to aggregate traffic load and relay to the macrocells or other access networks. Unless it is planned carefully this can result in extensive amounts of energy consumption. Considerable amounts of energy can be wasted unless an appropriate traffic-modelling strategy is used for the femtocells, which are deployed to serve static/mobile users in IoT environments such as smart cities. Therefore, an accurate traffic-modelling strategy is required to predict the performance of the system especially in terms of energy consumption in an IoT environment. These traffic-modelling strategies are classified and discussed in the following sections. In recent decade, many research works have been carried out on the integration of femtocells with macrocellular networks to investigate the performance of the network [5]. It is possible classify these modelling approaches into static versus dynamic approaches as shown in Table 11.1.

11.3.1 Static Approaches

The static traffic models assume a fixed number of mobile terminals that communicate with a fixed number of base stations. In these models, the mobility of mobile terminals in terms of the arrivals and departures are not captured [7].

Moreover, in the static models, the call-level or packet-level dynamics in terms of packet arrival and call duration are not considered. When the position and numbers of the terminals considered are fixed, the deployment modelling and the positions of the terminals become essential for performance optimization. For these scenarios, traffic modelling is mainly employed for the optimization of the deployment strategies considering important traffic modelling related system parameters such as arrival rate per unit area, and mean value of independently distributed

TABLE 11.1 A Summary of Traffic Models

<table>
<tr><th colspan="3">Model</th><th>Comments</th></tr>
<tr><td colspan="3">Static</td><td>The mobility of mobile terminals in terms of the arrivals and departures are not captured.</td></tr>
<tr><td rowspan="3">Dynamic</td><td colspan="2">Traffic spatial fluctuation models</td><td>Different traffic models such as stochastic geometry can be used to obtain the spatial fluctuations in call traffic load.</td></tr>
<tr><td rowspan="2">Traffic temporal fluctuation models</td><td>Short-scale traffic models</td><td>The packet arrivals and departures of the mobile terminals are modeled in short terms.</td></tr>
<tr><td>Long-scale traffic models</td><td>Traffic fluctuations are captured over the days of the week.</td></tr>
</table>

file sizes [7]. The traffic models of these systems are significantly affected by the information rate that can be transmitted over a given bandwidth and since it is expected to have a trade-off between the SE and energy efficiency, it is also important to have analytical approximations to analyze these factors [8]. For example, in Ref. [8], the authors derive a generic closed form approximation of the energy/SE trade-off for the uplink of coordinated multi point system using static traffic model and demonstrate its efficiency for the deployment of small cells such as femtocells compared to non-cooperative systems. The energy/spectral trade-off for the uplink of coordinated multi point system in Ref. [8] is considering parameters such as the achievable SE, the ergodic per-cell SE of the uplink channel, the circuit power and amplifier efficiency of user terminals, the cooling parameter, the number of backhaul links and finally the base station signal processing power as well as additional backhauling induced power for supporting coordinated multi point system.

In Ref. [9], Generalized Stochastic Petri Nets (GSPN) technique is used to model and investigate performance of small-scale networks (e.g., femtocell networks) with channel breakdowns. A number of performance measures such as mean number of busy channels, failure probability of channels, and blocking probability are derived from the models introduced, while all of the users are assumed to be static.

11.3.2 Dynamic Approaches

Unlike static traffic models, the spatial and temporal fluctuations of the traffic load are captured in the dynamic models, which are discussed in the following subsections. In order to incorporate the effects of traffic models to scenarios where fluctuations are expected, stochastic geometry and Poisson processes are used together with system characteristics such as interference, capacity, rates of traffic arrivals, SE to specify the amount of the information rate that can be transmitted, range of the base stations and the data rates that can be supported, and various service classes/rates [10–13]. Using stochastic geometry for these studies is particularly suitable since their focus is on the spatially random deployment of femtocells, which can be modeled by a 3-D Poisson process where the employed traffic model can drastically change the optimum configuration.

11.3.2.1 Traffic Spatial Fluctuation Models

It is well known that traffic characteristics can be quite different even among closely placed base stations. Therefore, in literature various traffic models have been suggested to obtain the spatial fluctuations in call traffic load. For example, modelling of the spatial distribution of small cells such as femtocells using stochastic geometry has been quite popular in the literature [10,11]. In this model, base stations are placed according to homogeneous Poisson point process with intensity λ_n in the Euclidean plane and distribution of mobile terminals are according to various independent stationary point processes with intensity λ_m. With a stationary Poisson point process, the distance between a mobile terminal and its serving base station is distributed regardless of the exact location of the mobile terminal. In Ref. [11], the probability density function is derived by considering λ_n and the distance between the mobile terminals and its serving base stations. In the above model, the traffic spatial variability among different cells is captured.

In Ref. [12], the authors propose a mixed spatial traffic model based on the available K-tier HetNet of a macro cell and small cells, in which each tier has its own density, path loss exponent, Signal to Interference Ratio (SIR) coverage threshold, and transmission power. Both uniform and non-uniform User Equipments (UEs) are included simultaneously for a more realistic scenario. In this model, each UE connects to a base station that provides the highest downlink SIR. Also, in Ref. [12] UE locations are surrounded by different macro and small cells, and the downlink SIR is derived by considering the location of any interfering base station, UE downlink SIR with base station, the transmit power of the base stations, the magnitude squares of the channel gains, path loss exponent, and spatial Poisson point processes with various densities.

11.3.2.2 Traffic Temporal Fluctuation Models

Traffic temporal fluctuations can be viewed in two different time-scales [7]; long-scale traffic models and short-scale traffic models. In long-scale traffic model such as the one presented in Ref. [13], traffic fluctuations are captured over the days of the week, which is useful for MNOs in investigating energy efficient solutions. For example, traffic is much higher during daytime than that during night time [13]. In addition, traffic during weekends, holidays, and pick hours is much lower than that of a normal weekday [13]. In order to model such a behavior, an activity parameter $\psi(t)$ can be used to determine the percentage of active subscribers over time t [7]. Therefore, the traffic demand in bit per second per km^2 can be calculated by considering the number of users per km^2, the number of operators, and the fraction of subscribers with an average data rate for a terminal type such as tablet and smart phone. The activity parameter $\psi(t)$ can also be estimated from historical traffic statistics.

Another time-scale in traffic temporal fluctuation models is a short-scale traffic model that models the packet arrivals as well as departures of the mobile terminals [7]. In models following this approach, packet arrivals and departures are modeled using the concept of queuing theory and Markov chains [14,15]. It is quite common to use Markov chain analysis for various types of communication systems including femtocells since the arrival processes with an infinite population usually do not exhibit intercorrelation and service times are usually dependent on the type and size of incoming tasks. For example, in Ref. [14] arrivals of packets are modeled using a Poisson process with rate λ and departures of packet follow exponential distribution with rate μ. In Ref. [14], the state transition probability of having of having h_i mobile terminals within cell *i* at time slot *t*+1 given that g_i mobile terminals were available at time slot *t* is given by $\Pr\{M_i(t+1)=h_i \mid M_i(t)=g_i\}$ where h_i, g_i are members of *M*, and *M* is the state set of the number of packets in a given cell *i*. Similarly, in Ref. [15], a hybrid wireless cellular HetNet consisting of a macrocell and several femtocells are modeled using two dimensional Markov processes. In this study, femtocells are modeled as fault tolerant wireless communication systems where mobility of the mobile users, multiple channels for the femtocells as well as failure/repair behavior of the channels are considered for more realistic performance measures in IoT environments. Using spectral expansion solution approach, all the state probabilities ($P_{i,j}$) are calculated which can be used to calculate a number of important performance measures such as MQL and throughput (γ). MQL can be calculated by multiplying the sum of all the state probabilities ($P_{i,j}$) and the number of available packets in each state. Similarly, throughput can be derived by multiplying the sum of all the state probabilities ($P_{i,j}$), the number of available channels, and total service rate of completed packet departures in the cell [15].

According to the objectives of the system, these models can be used for traffic modelling and performance evaluation of the system under study. For example, stochastic geometry model presented in Ref. [10] provides accurate and tractable information for multi-tier cellular networks to simplify the modelling approaches. This technique is very powerful when utilized in the networks modeled with Poisson point processes with Rayleigh fading which results in general closed-form expressions. However, the tractability of the technique is reduced by generalizing the network models.

On the other hand, steady state analysis of traffic temporal fluctuation models considers the systems when $t \to \infty$ instead of comparing them for a limited time. In other words, regardless of the observation period it identifies the effects of various parameters which can be utilized to monitor the data transmission throughput, the obtained bit error probability, and the mobile terminal radio interface state in order to balance energy saving and QoS [7].

All these modelling approaches are achieved via at least one of the following techniques for performance evaluation of the communication systems: analytical modelling, simulation, or benchmarking. Analytical modelling simulates behaviors of a system using mathematical concepts and language. For example, studies such as [4,5,16,17] used analytical techniques to evaluate the performance of the system under study. In Ref. [5], closed-form expressions describing SE and energy efficiency are derived for MFemtocell network. Authors also investigate the SE for multi-user system-level MFemtocells using opportunistic scheduling schemes. In Ref. [4], exponential path loss and fast fading models are used to predict energy efficiency in static femtocells against the macrocells. Meanwhile, a cellular network of a single macrocell and several femtocells has been considered in Ref. [17]. The authors used an M/M/1 queuing model for their analytical study and the Matrix Geometric Method is used to analyze the femtocell performance in terms of average system delay and power savings. Similarly, authors in Ref. [16] analyze the performance characteristics of a finite capacity femtocell network in terms of a number of QoS parameters such as average packet delay, packet blocking probability, and buffer utilization for different sizes. The system is modeled as an M/M/1/K queue and the results presented show that the mentioned QoS parameters are highly dependent on traffic intensity as well as the buffer size.

In addition to the analytical model, the study in Ref. [4] used simulation technique to simulate the operations of a real-world system over time [18]. Simulation provides fairly precise results but this approach requires high computation times. Comparing to simulation, analytical modelling technique is computationally more efficient [18]. Benchmarking is another technique, which is performed by actual measurements and it is only possible when there already exists something similar to the system under study [18]. Other attempts such as those in Ref. [19] have used benchmarking for the performance evaluation of the femtocell networks. The main problem associated with employment of test beds is the difficulty in extrapolation of obtained results to various scenarios.

11.4 Open Research Issues

Despite of many research works carried out, and the rapid development that has been achieved recently, there are still several challenges and opens issues that need to be taken into account regarding the deployment and performance improvement of femtocells. One of them is Cognitive Radio (CR), which is based on Dynamic Spectrum Access (DSA). CR is a promising solution having positive effects on interference mitigation, energy consumption level, and network lifetime. The importance of CR comes from the fact that there is trade-off between power consumption and bandwidth. This means that more bandwidth is required in order to reduce power consumption. In other words, dynamic and optimal spectrum management is required.

Deployment of femtocells in the existing macrocellular networks needs extra spectrum resource or causes interference in the current cellular networks. Spectrum provides a wide range of services and delivers social and economic benefits, so the effective use of spectrum and scheduling is also a vital objective for many MNOs.

Another important challenge regarding IoT-femtocell based applications is mobility since the problem of frequent handover increases by such scenarios. Since coverage of the UE frequently changes, this causes a significant issue regarding optimization, as well as achieving enhancement in energy saving and bandwidth utilization. Usually, femtocells are installed by users in indoor environments. Therefore, no particular mobility management scheme is required. However, it is essential to apply mobility management techniques and have handover procedures for dense deployment of femtocells. In the case of dense deployment of femtocells, this is one of the main challenges since it would be very difficult for a femtocell to keep track of all the neighboring femtocells, because every FBS may have a large number of mobile neighbors that dynamically change the network topology. The dynamically changing network topology and high probability of frequent handovers also introduce challenges in modelling and evaluation of femtocells especially from the analytical point of view. There are also variety of issues on the energy efficiency of femtocell networks that need to be more investigated. For example, energy metrics and energy consumption models considering not only femtocells but also a combination of femtocells and macrocells in HetNets need to be examined in order to make more accurate estimations as well as more optimized deployment strategies in significantly large areas. A comprehensive comparison of energy efficiency of femtocells when they are deployed with different access types would also help operators to have more energy efficient networks. Investigation on energy efficient schemes is another important issue, which helps in the reduction of energy consumption in both, femtocell BS and user equipment. The effect of femtocell handover decision and interference problem on femtocell energy consumption should also be investigated. Moreover, it is also helpful to evaluate the efficiency of the femtocell in terms of energy consumption as a communication technology in IoT applications and compare it with other evolving technologies such as Li-Fi and 5G standards.

11.5 Conclusion

Nowadays, the emerging paradigm of the IoT is rapidly growing, intending to improve the quality of life by connecting several smart tools, applications, and technologies. In this regard, femtocells play a significant role towards the fast realization of the IoT. Femtocells have shown to be a great solution for MNOs to improve coverage and capacity, and provide high quality services to mobile users at low cost while maintaining high level of QoS. In this chapter, we provided an overview of the femtocell communication technologies, services, and hardware units as the most important elements of femtocells in the IoT era. We also listed and described some of the femtocell applications in IoT. Moreover, we presented a classification of modelling traffics in femtocells and discussed some open research issues.

References

1. A. Mukherjee and D. De, "Femtocell based green health monitoring strategy", *IEEE General Assembly and Scientific Symposium (URSI GASS)*, August 2014, pp. 1–4.
2. R. Baines (n.d.), "Femtocells – Reducing Power Consumption in Mobile Networks". Retrieved November 23, 2016, from http://www.low-powerdesign.com/article_baines_092811.htm.
3. M. Deruyck, W. Joseph, and L. Martens, "Power consumption model for macrocell and microcell base stations", *Transactions on Emerging Telecommunications Technologies*, 25(3), 2014, pp. 320–333.

4. F. Cao and Z. Fan, "The tradeoff between energy efficiency and system performance of femtocell deployment", *IEEE Seventh International Symposium on Wireless Communication Systems (ISWCS)*, September 2010, pp. 315–319.
5. F. Haider, C. X. Wang, B. Ai, H. Haas, and E. Hepsaydir, "Spectral/energy efficiency tradeoff of cellular systems with mobile femtocell deployment", *IEEE Transactions on Vehicular Technology*, 65(5), 2016, pp. 3389–3400.
6. I. Ashraf, F. Boccardi, and L. Ho, "Sleep mode techniques for small cell deployments", *IEEE Communications Magazine*, 49(8), 2011.
7. M. Ismail, W. Zhuang, E. Serpedin, and K. Qaraqe, "A survey on green mobile networking: From the perspectives of network operators and mobile users", *IEEE Communications Surveys & Tutorials*, 17(3), 2015, pp. 1535–1556.
8. O. Onireti, F. Héliot, and M. A. Imran, "On the energy efficiency-spectral efficiency trade-off in the uplink of CoMP system", *IEEE Transactions on Wireless Communications*, 11(2), 2012, pp. 556–561.
9. N. Gharbi, "Modeling and performance evaluation of small cell wireless networks with base station channels breakdowns", *Proceedings of the Eighth International Conference on Wireless and Mobile Communications (ICWMC)*, Venice, Italy, pp. 42–48, 2012.
10. H. ElSawy, E. Hossain, and M. Haenggi, "Stochastic geometry for modeling, analysis, and design of multi-tier and cognitive cellular wireless networks: A survey", *IEEE Communications Surveys & Tutorials*, 15(3), 2013, pp. 996–1019.
11. Y. S. Soh, T. Q. Quek, M. Kountouris, and H. Shin, "Energy efficient heterogeneous cellular networks", *IEEE Journal on Selected Areas in Communications*, 31(5), 2013, pp. 840–850.
12. C. Li, A. Yongacoglu, and C. D'Amours, "Mixed spatial traffic modeling of heterogeneous cellular networks", *IEEE International Conference on Ubiquitous Wireless Broadband (ICUWB)*, Montreal, QC, Canada, Oct. 2015, pp. 1–5.
13. E. Oh, B. Krishnamachari, X. Liu, and Z. Niu, "Toward dynamic energy-efficient operation of cellular network infrastructure", *IEEE Communications Magazine*, 49(6), 2011, pp. 56–61.
14. L. B. Le, D. Niyato, E. Hossain, D. I. Kim, and D. T. Hoang, "QoS-aware and energy-efficient resource management in OFDMA femtocells", *IEEE Transactions on Wireless Communications*, 12(1), 2013, pp. 180–194.
15. E. Ever, F. M. Al-Turjman, H. Zahmatkesh, and M. Riza, "Modelling green HetNets in dynamic ultra-large-scale applications: A case-study for femtocells in smart-cities", *Computer Networks*, 128, 2017, pp. 78–93.
16. W. Kumar, S. Aamir, and S. Qadeer, "Performance analysis of a finite capacity femtocell network", *Mehran University Research Journal of Engineering & Technology*, 33(1), 2014, pp.129–136.
17. W. Kumar, P. Kumar, and I. A. Halepoto, "Performance analysis of an energy efficient femtocell network using queuing theory", *Mehran University Research Journal of Engineering & Technology*, 32(3), pp. 535–542, 2013.
18. J. Banks, J. S. Carson, B. L. Nelson, and D. M. Nicol, *Discrete-Event System Simulation*, 4th ed., Prentice-Hall: Upper Saddle River, NJ, 2005.
19. P. Kulkarni, S. Gormus, W. H. Chin, and R. J. Haines, "Distributed resource allocation in small cellular networks-test-bed experiments and results", *IEEE Seventh International Wireless Communications and Mobile Computing Conference*, Istanbul, Turkey, July 2011, pp. 1262–1267.

Chapter 12

Conclusions and Future Directions

Fadi Al-Turjman
Antalya Bilim University

Contents

The rapid increase in counts of communicating devices such as smartphones, Personal Digital assistants (PDAs), and notebooks, causes the demand for mobile data traffic to grow significantly. In recent years, mobile operators have been trying to find solutions to increase the network capacity in order to satisfy mobile users' requests and meet the requirements in terms of various Quality of Service (QoS) measures for the case of high mobile data traffic. With ever-increasing demand from the mobile users and implementations in the area of Internet of Things (IoT), femtocells are proved to be a promising solution for network operators to enhance coverage and capacity, and they provide high data rate services in a less expensive manner. This book describes possible femtocell deployment strategies, data delivery approaches, and traffic modelling methods in the IoT environment and highlights potentials and challenges for IoT-femtocell based applications.

12.1 Summary of the Book

We started the work in this book with a comprehensive overview about the use of femtocells in the IoT era, while discussing its traffic modelling and deployment alternatives in Chapter 2. The vision of next generation IoT is to provide real-time multimedia applications using small cells (e.g., Femtocells) for long-term usage. However, QoS assurances for both best effort data and green applications introduced new design and performance challenges, which have been overview in this chapter. Femtocells will help mobile operators to provide a basis for the next generation of services, which are a combination of voice, video, and data services to mobile users.

In Chapter 3, we describe prominent performance metrics in order to understand how the energy efficiency is evaluated. Then, we elucidate how energy can be modeled in terms of femtocell and provide some models from the literature. Since femtocells are used in heterogeneous networks to manage energy efficiency, we also express some energy efficiency schemes for deployment. The factors that affect the energy usage of a femtocell Base Station (BS) are discussed, and then the power consumption of user equipment under femtocell coverage is mentioned.

In Chapter 4, we present an energy-based analysis for mobile femtocells in Ultra-Large Scale (ULS) applications such as the smart grid. The potential reduction of the consumed energy and service interruption due to mobility and multihop communication effects are considered as well as various performance metrics such as throughput, availability, and delay. In that sense, the smart grid is modeled as a green wireless communication system for traffic offloading while considering mobility of the Femtocell Base Station (FBS) as well as the FBS cut-offs for energy-saving aspects. A typical scenario, called *e-Mobility*, is considered as a case study where a set of mobile femtocells are utilized inside a single macrocell to achieve optimized utilities' usage. Numerical results achieved from the proposed smart-grid model have been verified and validated via extensive simulation results while considering typical operational conditions/parameters.

In Chapter 5, we propose an agile vehicular-cloud & IoT framework. The framework is employing the standardized Long-Term Evolution (LTE) while improving the QoS in modern mobile applications. It minimizes the communication delay in sensitive real-time applications while maintaining the highest network throughput via an Enriched-real time Polling System (E-rtPS). The objective of E-rtPS is to improvise the service connection between the vehicles, which are equipped with real-time devices. The proposed approach integrates the Real Time-Batch Markovian Arrival Process (RT-BMAP) with IEEE 802.16d/e to analyze various QoS measures, namely throughput and average packet delay via real-time systems.

The probing results show that the proposed approach is superior in comparison to other traditional services of IEEE 802.16d/e based systems.

In Chapter 6, we present the main motivations in carrying these smart devices, and the correlation between the user-surrounding context and the application usage. We focus on context-awareness in smart systems and space discovery paradigms: online versus offline, the femtocell usage, and energy aspects to be considered, and about the ongoing social IoT applications. Moreover, we highlight the most up-to-date open research issues in this area.

To guarantee the connectivity among these objects and people, fault tolerance routing has to be significantly considered. In Chapter 7, we propose a bio-inspired Particle Multi-Swarm Optimization (PMSO) routing algorithm to construct, recover, and select k-disjoint paths that tolerates the failure while satisfying QoS parameters. Multi-swarm strategy enables determining the optimal directions in selecting the multipath routing while exchanging messages from all positions in the network. The validity of the proposed algorithm is assessed and results demonstrate high-quality solutions compared with the Canonical Particle Swarm Optimization (CPSO), and Fully Particle Multi-Swarm Optimization (FPMSO).

In Chapter 8, we present a hybrid pervasive sensing framework for data gathering in IoT-enabled smart-cities' paradigm. This framework satisfies service-oriented applications in smart cities where data is provided via data Access Points (APs) of various resources. Moreover, public vehicles are used in this work as Data Couriers (DCs) that read these APs data packets and relay it back to a BS in the city. Accordingly, we propose a hybrid fitness function for a genetic-based DCs selection approach. Our function considers resource limitations in terms of count, storage capacity and energy consumption as well as the targeted application characteristics. Extensive simulations are performed and the effectiveness of the proposed approach has been confirmed against other heuristic approaches with respect to total travelled distances and overall data-delivery cost.

In Chapter 9, we propose a Cognitive Caching approach for the Future Fog (CCFF) that takes into consideration the value of the exchanged data in Information Centric Sensor Networks (ICSNs). Our approach depends on four functional parameters in ICSNs. These four main parameters are as follows: age of the data, popularity of on-demand requests, delay to receive the requested information, and data fidelity. These parameters are considered together to assign a value to the cached data while retaining the most valuable one in the cache for prolonged time periods. This CCFF approach provides significant availability for most valuable and difficult to retrieve data in the ICSNs. Extensive simulations and case studies have been examined in this research in order to compare to other dominant cache management frameworks in the literature under varying circumstances such as data popularity, cache size, data publisher load, and node connectivity degree. Formal fidelity and trust analysis has been applied as well to emphasize the effectiveness of CCFF in Fog paradigms, where edge devices can retrieve unsecured data from the authorized nodes in the cloud.

In Chapter 10, we propose a Context-sensitive Seamless Identity Provisioning (CSIP) framework for the IIoT. CSIP proposes a secure mutual authentication approach using Hash and Global Assertion Value to prove that the proposed mechanism can achieve the major security goals of the mobile sensor network in a short time period. From the obtained results, we concluded the capability of the proposed CSIP approach in the IoT era.

With ever-increasing demand from the mobile users and implementations in the area of IoT, femtocells are proved to be a promising solution for network operators to enhance coverage and capacity, and they provide high data rate services in a less expensive manner. Chapter 11 describes possible femtocell applications and traffic modelling approaches in the IoT environment and highlights potentials and challenges for IoT-femtocell based applications.

12.2 Future Directions

With the femtocell-based research done in this book, mobile service providers will be able to provide better infrastructure support for smarter IoT paradigms across the world. Not only this, several future research directions and open issues can also be derived from the work done in this book so far. In the following, we outline some of these future directions and open research issues.

12.2.1 Exploiting the Femtocell Placement Problem More and More

One of the key aspects in any smart IoT paradigm is the positioning of the key enabling technology (e.g., a femtocell BS) and its availability across the network. Such femtocells would rise an issue with the mobility-enabled node presence. Thus, exploring the role of mobile femtocells in smart environments, and the study of their replacement techniques that suit the mobile IoT paradigm is still a direction to explore.

12.2.2 More Developed and Commercialized Cognitive Femtocells

In terms of enhancing the data delivery while supporting diverse application platforms, this work can be reconsidered for further enhancement in domain-specific technology for better connectivity and coverage capacity in the IoT era. The existence of such technology can contribute toward the development of an enterprise architecture that can be applied to different application domains using the same underlying Internet infrastructure.

12.2.3 Femtocell and Smart Environments Integration with the NGNs

More functions could be incorporated at the femtocell BS to integrate it with Next Generation Networks (NGNs) such as the 5G toward more capable and smarter environments. The expansion of the femtocell functions would then be able to take requests directly from different network users such as cellphones, access points and BSs, and thus making the IoT paradigm more accessible to end-users in smart paradigms.

12.2.4 Mobility-Enabled Femtocell BSs

In general, the effect of the BS mobility on the QoS, and its impact on the IoT adaptability and longevity shall be investigated more for safer smart applications such as those found in ITS and healthcare. While the discussed mobile BS architecture can help in overcoming the limitations of the cross-layer design, we strongly recommend further investigations in the physical layer components and configurations of the femtocell.

12.2.5 Cognition in the Femtocell

The idea of cognition can be investigated more while being applied to femtocell BS of the current cellular networks' infrastructure to realize the cognitive network concept in general. eServices and user requests need not to be retrieved from a specific host/service provider, nor it has to travel end-to-end in the network. Instead, it can be cached at the network's edge (Fog). Femtocell BSs could be used to understand the user request patterns and manage the cache content intelligently.

12.2.6 Security and Privacy Open Issues

Security and privacy are key issues to be addressed while spreading the adoption of femtocell in IoT smart applications. Unluckily, security solutions have not been incorporated in a planned way in the early IoT version. Where IoT enabling technologies such as cellular and sensor networks were simply integrated across with the existing Internet infrastructure, without considering key design factors such as the trust and privacy issues while gathering and delivering the data traffic. These security considerations span not only along the data management, application, and service levels, but also at the communication and networking level, and are significant aspects to be investigated in any future work.

Index

I

K

L

M

N

O

P

Q

R

S

T